Next Generation Wireless Network Security and Privacy

Kamaljit I. Lakhtaria
Gujarat University, India

A volume in the Advances in Information Security, Privacy, and Ethics (AISPE) Book Series

Published in the United States of America by
Information Science Reference (an imprint of IGI Global)
701 E. Chocolate Avenue
Hershey PA 17033
Tel: 717-533-8845
Fax: 717-533-8661
E-mail: cust@igi-global.com
Web site: http://www.igi-global.com

Library of Congress Cataloging-in-Publication Data

Next generation wireless network security and privacy / Kamaljit I. Lakhtaria, editor.
pages cm
Includes bibliographical references and index.
ISBN 978-1-4666-8687-8 (hardcover) -- ISBN 978-1-4666-8688-5 (ebook) 1. Computer networks--Security measures. 2. Wireless communication systems--Security measures. I. Lakhtaria, Kamaljit I., 1982-
TK5105.59.N347 2015
005.8--dc23

2015015525

This book is published in the IGI Global book series Advances in Information Security, Privacy, and Ethics (AISPE) (ISSN: 1948-9730; eISSN: 1948-9749)

British Cataloguing in Publication Data
A Cataloguing in Publication record for this book is available from the British Library.

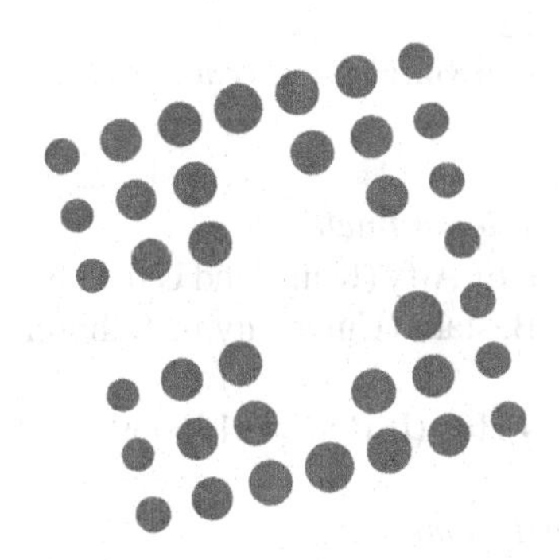

Advances in Information Security, Privacy, and Ethics (AISPE) Book Series

ISSN: 1948-9730
EISSN: 1948-9749

MISSION

As digital technologies become more pervasive in everyday life and the Internet is utilized in ever increasing ways by both private and public entities, concern over digital threats becomes more prevalent.

The **Advances in Information Security, Privacy, & Ethics (AISPE) Book Series** provides cutting-edge research on the protection and misuse of information and technology across various industries and settings. Comprised of scholarly research on topics such as identity management, cryptography, system security, authentication, and data protection, this book series is ideal for reference by IT professionals, academicians, and upper-level students.

COVERAGE

- Network Security Services
- Data Storage of Minors
- Technoethics
- CIA Triad of Information Security
- Information Security Standards
- Device Fingerprinting
- Tracking Cookies
- IT Risk
- Cookies
- Global Privacy Concerns

IGI Global is currently accepting manuscripts for publication within this series. To submit a proposal for a volume in this series, please contact our Acquisition Editors at Acquisitions@igi-global.com or visit: http://www.igi-global.com/publish/.

The Advances in Information Security, Privacy, and Ethics (AISPE) Book Series (ISSN 1948-9730) is published by IGI Global, 701 E. Chocolate Avenue, Hershey, PA 17033-1240, USA, www.igi-global.com. This series is composed of titles available for purchase individually; each title is edited to be contextually exclusive from any other title within the series. For pricing and ordering information please visit http://www.igi-global.com/book-series/advances-information-security-privacy-ethics/37157. Postmaster: Send all address changes to above address.

Titles in this Series

For a list of additional titles in this series, please visit: www.igi-global.com

Improving Information Security Practices through Computational Intelligence
Wasan Shaker Awad (Ahlia University, Bahrain) El Sayed M. El-Alfy (King Fahd University of Petroleum and Minerals, Saudi Arabia) and Yousif Al-Bastaki (University of Bahrain, Bahrain)
Information Science Reference • copyright 2016 • 328pp • H/C (ISBN: 9781466694262) • US $210.00 (our price)

Handbook of Research on Security Considerations in Cloud Computing
Kashif Munir (King Fahd University of Petroleum & Minerals, Saudi Arabia) Mubarak S. Al-Mutairi (King Fahd University of Petroleum & Minerals, Saudi Arabia) and Lawan A. Mohammed (King Fahd University of Petroleum & Minerals, Saudi Arabia)
Information Science Reference • copyright 2015 • 409pp • H/C (ISBN: 9781466683877) • US $325.00 (our price)

Emerging Security Solutions Using Public and Private Key Cryptography Mathematical Concepts
Addepalli VN Krishna (Stanley College of Engineering and Technology for Women, India)
Information Science Reference • copyright 2015 • 302pp • H/C (ISBN: 9781466684843) • US $225.00 (our price)

Handbook of Research on Emerging Developments in Data Privacy
Manish Gupta (State University of New York at Buffalo, USA)
Information Science Reference • copyright 2015 • 507pp • H/C (ISBN: 9781466673816) • US $325.00 (our price)

Handbook of Research on Securing Cloud-Based Databases with Biometric Applications
Ganesh Chandra Deka (Ministry of Labour and Employment, India) and Sambit Bakshi (National Institute of Technology Rourkela, India)
Information Science Reference • copyright 2015 • 434pp • H/C (ISBN: 9781466665590) • US $335.00 (our price)

Handbook of Research on Threat Detection and Countermeasures in Network Security
Alaa Hussein Al-Hamami (Amman Arab University, Jordan) and Ghossoon M. Waleed al-Saadoon (Applied Sciences University, Bahrain)
Information Science Reference • copyright 2015 • 450pp • H/C (ISBN: 9781466665835) • US $325.00 (our price)

Information Security in Diverse Computing Environments

This books is dedicated to my father, Ishwarlal J. Lakhtaria, my mother, Prabhaben I. Lakhtaria, and my wife, Panna K. Lakhtaria.

Editorial Advisory Board

List of Reviewers

Table of Contents

Detailed Table of Contents

A study and identification of vulnerabilities during the set-up procedure of the Universal Mobile Telecommunication System (UMTS) and how some of them can be exploited. For accomplishment a good understanding of the security messages exchange, a part of UMTS architecture is developed firstly. After the explanation of the security mode set-up procedure debilities, the chapter identify attacks that take advantage of the fact that some messages during their exchange in the process are not protected. The attacks indicated in the chapter are mostly of Denial of Service (DoS) kind, and mainly are performed with a rogue BTS.

Wireless Body Sensor Network is a collection of physiological sensors connected to small embedded machines and transceivers to form a monitoring scheme for patients and elderly people. Intrusion and foolproof routing has become mandatory as the Wireless Body Sensor Network has extended its working range. Trust in Wireless Body Sensor Network is greatly determined by the Encryption key size and Energy of the Node. The Sensor Nodes in Wireless Body Sensor Network is powered by small battery banks which are to be removed and recharged often in some cases. Attack to the implanted node in Wireless Body Sensor Network could harm the patient. Finite State Machine helps in realizing the Trust architecture of the Wireless Body Sensor Network. Markov model helps in predicting the state transition from one state to other. This chapter proposes a Trustworthy architecture for creating a trusted and confidential communication for Wireless Body Sensor Network.

Spread spectrum modulation (SSM) finds important place in wireless communication primarily due to its application in Code Division Multiple Access (CDMA) and its effectiveness in channels fill with noise like signals. One of the critical issues in such modulation is the generation of spreading sequence. This chapter presents a design of chaotic spreading sequence for application in a Direct Sequence Spread Spectrum (DS SS) system configured for a faded wireless channel. Enhancing the security of data transmission is a prime issue which can better be addressed with a chaotic sequence. Generation and application of chaotic sequence is done and a comparison with Gold sequence is presented which clearly indicates achieving better performance with simplicity of design. Again a multiplierless logistic map sequence is generated for lower power requirements than the existing one. The primary blocks of the system are implemented using Verilog and the performances noted. Experimental results show that the proposed system is an efficient sequence generator suitable for wideband systems demonstrating lower BER levels, computational time and power requirements compared to traditional LFSR based approaches.

In the absence of central authority, dynamic topology, limited bandwidth and the different types of vulnerabilities secured data transmission is more challenging in Mobile Ad hoc Network (MANET). A node in MANET acts as a host as well as a router. Routing is problematic due to node mobility and limited battery power of node. Security mechanisms are required to support secured data communication. It also requires mechanism to protect against malicious attacks. In recent times multipath routing mechanisms are preferred to overcome the limitation of the single path routing. Security in routing dealt with authentication, availability, secure linking, secure data transmission, and secure packet forwarding. In this chapter, we discuss different security requirements, challenges, and attacks in MANETs. We also discuss a few secured single path and multipath routing schemes.

A typical WSN contains spatially distributed sensors that can cooperatively monitor the environment conditions, like second, temperature, pressure, motion, vibration, pollution and so forth. WSN applications have been used in several important areas, such as health care, military, critical infrastructure monitoring, environment monitoring, and manufacturing. At the same time. WSN Have some issues like memory, energy, computation, communication, and scalability, efficient management. So, there is a need for a powerful and scalable high-performance computing and massive storage infrastructure for real-time processing and storing the WSN data as well as analysis (online and offline) of the processed information to extract events of interest. In this scenario, cloud computing is becoming a promising technology to provide a flexible stack of massive computing, storage, and software services in ascalable and virtualized manner at low cost. Therefore, sensor-cloud (i.e. an integrated version of WSN & cloud computing) infrastructure is becoming popular nowadays that can provide an open flexible, and reconfigurable platform for several monitoring and controlling applications.

Wireless sensor networks (WSN) have become one of the most attractive research areas in many scientific fields for the last years. WSN consists of several sensor nodes that collect data in inaccessible areas and send them to the base station (BS) or sink. At the same time sensor networks have some special characteristics compared

At present a majority of computer and telecommunication systems requires data security when data is transmitted the over next generation network. Data that is transient over an unsecured Next Generation wireless network is always susceptible to being intercepted by anyone within the range of the wireless signal. Hence providing secure communication to keep the user's information and devices safe when connected wirelessly has become one of the major concerns. Quantum cryptography algorithm provides a solution towards absolute communication security over the next generation network by encoding information as polarized photons, which can be sent through the air security issues and services using cryptographic algorithm explained in this chapter.

Wireless Networks are vulnerable in nature, mainly due to the behavior of node communicating through it. As a result, attacks with malicious intent have been and will be devised to exploit these vulnerabilities and to cripple MANET operation. In this chapter, we analyze the security problems in MANET. On the prevention side, various key and trust management schemes have been developed to prevent external attacks from outsiders. Both prevention and detection method will work together to address the security concern in MANET.

Organizations focuses IDPSes for respective purposes, e.g. identifying problems with security strategies, manually presented threats and deterring individuals from violating security policies. IDPSes have become a necessary technique to the security infrastructure of approximate each association. IDPSes typical record information interrelated to practical events, security administrators of essential observed events and construct write up. Many IDPSes can also respond to a detected threat by attempting to thwart it succeeding. These use several response techniques, which involve the IDPS restricting the attack, changing the security environment or the attack's content. Sensor node should diverge in size from a shoebox down to the small size, although functioning "motes" of genuine microscopic dimensions have to be formed. The cost of sensor nodes is variable, from a few to thousands of dollars, depend on the complexity of the sensor nodes. Size and cost constraints on sensor nodes represent in corresponding constraints on resources such as energy, memory, computational velocity and communications bandwidth. The arrangement of the WSNs alters itself from a star network to efficient multi-hop wireless mesh network. The proliferation technique between the hops of the network can be routing or flooding.

Foreword

Next Generation Network (NGN) nowadays plays an important role in the sector of telecommunication and has huge importance for future research. NGN communication is making the world life easier by giving round the clock connectivity with high data transfer. The NGN communication comprises number of promising architectures as Internet of things and Personal wireless networks (including body area wireless sensor network, wireless sensor and actuator network, Wireless Mesh networks, Next generation Internet, LTE advanced, and Machine-to-Machine networks). NGN communications must be secure and preserve the privacy of its users and require innovative methods to address the continuous and emerging security threats.

Many organizations have over time experienced network sprawl, particularly in terms of a large growth of where things are connecting from and to, the types of connections and the influx of cloud and mobility. As networks continue to grow, many things are becoming more condensed in the data center. More companies are using virtualization and an internal cloud, which has resulted in consolidated and more powerful technology. These things are leading to discussions and evaluations of new ways to design and secure network architecture.

With more than fourteen self-contained chapters, this volume provides a complete survey of the state-of-the-art research that encompasses all areas of Security and Privacy for Next Generation Wireless Networks. Written by distinguished researchers in the field, these chapters focus on the theoretical and experimental study of advanced research topics involving security, privacy, security architecture, security standard.

This book is a great reference tool for graduate students, researchers, and mathematicians interested in studying Next Generation Wireless Network Security and Privacy.

This is the best book to give you a quick overview of all important aspects of Security and Privacy for NGN as it covers Technical Study and Real Time Performances in the area of Network security and privacy management.

I feel proud of Dr. Kamaljit and his authors, for working on such project and gather all at most emerging information on NGN technology in a single Book.

I recommend this book for four reasons; First, this book's structure and chapter flow is organized based on its the technology. Second, this book includes such basic concept for each aspect and raises this concept to advance most level of experiment and usage for Next Generation Wireless Networks technology. Third, this book shows many real-world implementations and four, the information presented in this book is definitely demanded by researchers and scientists.

Jaime Lloret Mauri
Polytechnic University of Valencia, Spain

Preface

The world is currently witnessing rapid advancements in wireless communications and networking technologies. These technologies are gradually making the dream of ubiquitous high-speed network access a reality. The Next Generation Wireless Networks include communication networks such as Next Generation Networks (NGN), 3G, 4G, Long Term Evolution (LTE), Wireless LANs, PANs, WiMax, ad hoc networks, Mobile ad hoc networks (MANETs), and Wireless Sensor Networks (WSNs). Unlike the wired networks, the unique characteristics of wireless networks pose a number of nontrivial challenges to security design, such as open peer-to-peer network architecture, shared wireless medium, stringent resource constraints, highly dynamic network topology, and absence of a trusted infrastructure.

Ubiquitous roaming impacts a radio access system by requiring that it supports handover between neighboring cells and different networks. Also, mobile networks are more exposed to interferences than wired networks. Such ubiquitous network access allows vandals and criminals to exploit vulnerabilities in networked systems on a widespread basis. This situation makes security and privacy designs for emerging wireless networks both critical and challenging. Security mechanisms in wireless networks are indispensable to secure access control, authentication, authorization, confidentiality, data security, user confidentiality, quality of service, and quality of experience for users.

As a result, today's research is directed towards the preservation of wireless network trust, security, and privacy all at the same time. This book looks to discuss recent research focused on Next Generation Wireless Network security including topics in network security, cryptography, user authentication, various types of attacks and their counter measures, secure routing algorithm, intrusion detection, security algorithms, performance analysis of various tools and proposed techniques, security issues, privacy management, smart phone application security, and key management.

This book consist of contributions from experts working on real time problem solving towards security and privacy of Next Generation Networks. Chapters focus on recent developments and solutions towards the problem of managing network security and privacy.

This book will be beneficial to academicians, research supervisors, researchers, advanced level students, and technology developers, and corporate organizations will find this reference source useful for their research, advanced concept study, and product development. Chapters of this book will provide researchers and technology developers with pertinent topics and assist in their own research efforts in this field.

The Editor,

Kamaljit I. Lakhtaria
Gujarat University, India

Acknowledgment

The concept of this book project was born in my mind when I was guiding my Ph.D. student in the area of Network Security. Rapid growth in wireless technology and the realization of pervasive computing provide users 360 degree Internet accessibility. But with this threat on user's data, organizational data increases as well.

This book is based on research conducted for providing Security and Privacy to Next Generation Networks.

I am thankful to Kayla Wolfe, Caitlyn Martin, Jan Travers, and Christine Smith from IGI Global for their support and guidance for this edited book project.

My best supporters were the Editorial Advisory Board Members and Reviewers. Without their constant support, this project would never have been possible. I heartily thank them all.

I would like to thank my mother, father, and my beloved wife, Panna, for their love and inspiration to me.

Editor,
Kamaljit I. Lakhtaria
Gujarat University, India

Chapter 1
Debilities of the UMTS Security Mode Set-Up Procedure and Attacks against UMTS/HSPA Device

Diego Fernández Alonso
University of Vigo, Spain

Ana Vázquez Alejos
University of Vigo, Spain

Manuel García Sánchez
University of Vigo, Spain

ABSTRACT

A study and identification of vulnerabilities during the set-up procedure of the Universal Mobile Telecommunication System (UMTS) and how some of them can be exploited. For accomplishment a good understanding of the security messages exchange, a part of UMTS architecture is developed firstly. After the explanation of the security mode set-up procedure debilities, the chapter identify attacks that take advantage of the fact that some messages during their exchange in the process are not protected. The attacks indicated in the chapter are mostly of Denial of Service (DoS) kind, and mainly are performed with a rogue BTS.

DOI: 10.4018/978-1-4666-8687-8.ch001

INTRODUCTION

The UMTS is the acronym of Universal Mobile Telecommunication System developed by the 3rd Generation Partnership Project (3GPP). The UMTS standard is based on WCDMA (Wideband Code Division Multiple Access) for the radio interface (Uu). This interface uses CDMA (Code Division Multiple Access) as access method and can operate in two ways, FDD (Frequency Division Duplex) and TDD (Time Division Duplex).

Nowadays HSPA (High Speed Packet Access) is used for the Up Link (HSUPA) and the Down Link (HSDPA). HSPA is an upgrade of WCDMA and was introduced by parts. The HSDPA (High Speed Down Link Packet data Access) was introduced in 3GPP Release 5 and the HSUPA (High Speed Up Link Packet data Access) was introduced in 3GPP Release 6. HSPA is the combination of both.

HSPA represents the 3rd generation of mobile communication technology and provides mutual authentication, confidentiality and integrity.

One of the part of UMTS architecture more vulnerable is the network access, since is where there are the mainly threats. The mainly vulnerabilities over UMTS are in relation with the security mode set-up and transmission of keys and data in text clear in its process. The Authentication and Key Agreement (AKA) and encryption and integrity algorithms are very robust, the vulnerabilities exist due the way the system have for start the security establishment with those algorithms.

To take advantage of the UMTS vulnerabilities is not easy due it is necessary to have knowledge about specific resources, like the software to configure a Base Transceiver Station (BTS). Although there are vulnerabilities through the complete architecture, for an intruder the easier interface to perform the attacks is the radio interface or Uu (in UMTS).

UMTS ARCHITECTURE

The architecture of Universal Mobile Telecommunications System (UMTS) includes three different domains, UMTS Terrestrial Radio Access Network (UTRAN), Core Network (CN) and User Equipment (UE).

The UTRAN provides the air interface access method for the User Equipment through the Base Station (BS) or Node-B. Core Network provides switching, routing and transit for the user traffic, and it contains the databases and network management functions. And finally User Equipment is the terminal that allows the mobile communication of the user through the air interface.

The UMTS Terminal

In 3rd Generation the device used by the user for the communication is called 3G User Equipment (UE), and it is compound for the UMTS device or Mobile Equipment (ME), and the Universal Subscriber Identity Module (USIM). The UE also is called Mobile Station (MS), terminology used primarily in GSM.

The USIM is a smart card with user's configurations and data. The module that stores, among other information, the International Mobile Subscriber Identity (IMSI). The IMSI is a 15-digit number that uniquely identifies a particular mobile station, and it is sensitive to receive attacks. The terminal is also characterized by the International Mobile Equipment Identity (IMEI), 15-digit number assigned by the manufacturer. In GSM the equivalent card for the USIM is the Subscriber Identity Module (SIM).

UTRAN Architecture

The UMTS Terrestrial Radio Access Network (UTRAN) consists of Radio Network Subsystems (RNSs) and each RNS contains different numbers of Base Stations (BSs or Node-Bs) and one RNC.

Radio interface Uu, situated between UE and UTRAN, is based on the Wideband Code Division Multiple Access (WCDMA).

The downlink (DL) is in the direction Node-B to UE and the uplink (UL) is in the direction UE to Node-B.

RNSs are separated from each other by the UMTS interface which is situated between RNC (*Iur*) interfaces that form connections between two RNCs. The *Iur*, which has been specified as an open interface, carries both signaling and traffic information.

Figure 1. Architecture of UMTS with PLMN and PDN connections

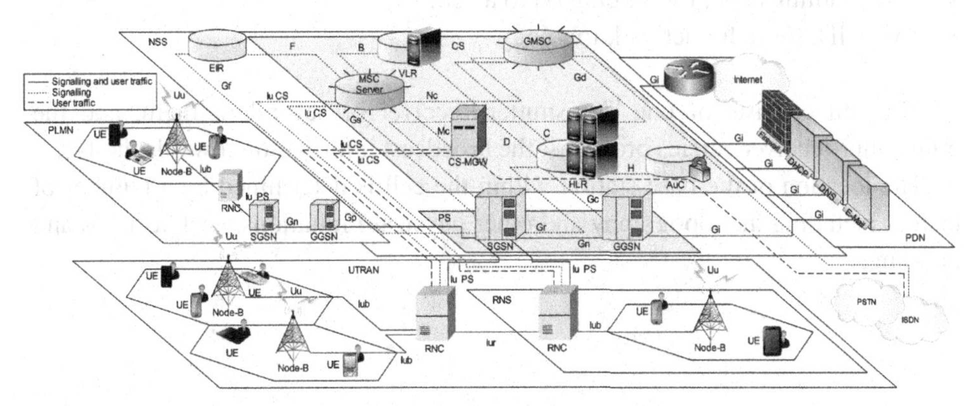

The UMTS Core Network (CN) consists on a number of elements interconnected by a big number of interfaces, and the configuration supports Circuit Switched (CS) and Packet Switched (PS) traffic. There are interfaces that only support user traffic, only signalling, or both.

The architecture can be completed with the connection with other networks.

In the Table 1 the elements of the architecture are described in detail.

The structure and the interfaces depend on each release, varying slightly some elements and the way how they are connected, therefore the UMTS architecture can be different from one reference to another.

Node-B or Base Station (BS)

The Base Station (BS) or Node-B is the element of the architecture located between the Uu interface (radio) and the Iub interface, that connects with Radio Network Controller. The BS implements WCDMA radio access physical channels and transfers information from transport channels to the physical.

The Node-B performs (Kaaranen, Ahtiainen, Laitinen, Naghian, & Niemi, 2005, pp. 101-103):

- Radio signal in reception and transmission.
- Signal filtering and amplifying.
- Signal modulation and demodulation.
- Interfacing to the Radio Access Network (RAN).

The Node-B provides coverage to an area known as "cell". A cell is the smallest radio network entity having its own identification number (cell ID) that is public visible to the UE. Every cell has one scrambling code, and the UE recognizes a cell by two values:

- Scrambling code, for logging on to a cell.
- Cell ID, for radio network topology.

The cell consists of one Transmitter-Receiver or several of them, and the transmitter-receiver carries broadcast the broadcast information towards the UE.

The location of the base station within the cell is determined by a number of factors including the topography and other physical limitations, such as trees and buildings.

Table 1. Architecture components

AuC (Authentication Center)	It is the entity responsible for keeping record information, allowing the access of user to the system. It interacts with HLR.
B	Interface responsible for MSC and VLR interconnection used to manage the database of subscriber using of area controlled by MSC that is associated with VLR. It offers location services, update existing UE and data associated to Supplementary Services (SS).
C	Interface that makes the interconnection between MSC and HLR, used when GMSC needs to interrogate HLR associated with the subscriber to obtain information necessary for routing calls or sending of SMS.
CS-MGW (Circuit Switched - Media Gateway Function)	PSTN/PLMN transport termination point for a defined network and interfaces UTRAN with the core. It may terminate bearer channels from a switched circuit network and media streams from a packet network. Over Iu, it may support media conversion, bearer control and payload processing.
D	Interface that connects HLR and VLR to exchange data related to UE location or to subscriber management. The main service provided is the ability of UE to originate and receive calls within the service area. It is used in most signaling procedures.
E	Interface that connects two MSC (or MSC with . It was used in the handover procedures among different MSCs for h. (This interface does not appear in the graphics)
EIR (Equipment Identity Register)	It is the database with UE information.
F	Interface used to connect MSC and EIR, and is used to check IMEI status of UE.
Gc	Interface that connects GGSN to HLR, being optional and used by GGSN to request location information to HLR. The main interface is Gr.
Gd	Interface that connects SGSN to GMSC, providing access to SMS (SMS-GMSC).
Gf	It is used between SGSN and EIR to exchange data, in order that the EIR can verify the status of the IMEI retrieved from the Mobile Station.
GGSN (Gateway GPRS Support Node)	It is between UTMS and PDN network (or other PLMN).
Gi	Interface between SGSN and the *Packet Data Networks* (PDNs), Internet, e-mail, DNS and DHCP.
GMSC (Getaway Mobile Switching Centre)	It is the device that makes the interface with PSTN, ISDN or other MSCs.
Gn	Interface between GGSN located in the same PLMN (or between two SGSN), which aims at transmitting data traffic (packets)
Gp	Interface for interconnections between SGSN and GGSN or different PLMNs. Also, it provides the security function.
Gr	Interface used by SGSN to access data residing in HLR. This is the main interface, while Gc is optional.
Gs	Optional interface used by SGSN to send location information to all MSC/VLR. SGSN can receive the search request (paging) from MSC via this interface.

continued on following page.

Table 1. Continued.

HLR (Home Location Register)	it is the database in which UE is recorded (main database).
ISDN	Integrated Services Digital Network.
Iu (Iu CS and Iu PS)	It is the interface between UTRAN and CN (NSS), being specified to support the Radio Network Subsystem (RNS) interconnection formed by RNC and *Node-B* with access points to a single PLMN.
Iub	Interface that connects Node-B to RNC. Here, all information exchanged between UE and NSS occurs.
Iur	Interface that allows the exchange of signalling information between two RNCs within a same UTRAN. It supports RNC interconnection of different manufactures, offers support services provided by RNS and UTRAN and separates the functionalities between RF and transport networks
MSC (Mobile Switching Center)	It is the mechanism that transports circuit-switched information.
Nc	Reference point over wich the Network-Network based call control is performed. Different options for signalling transport on Nc shall be possible including IP.
Node-B or BS (Base Station)	Equipment equivalent to the BTS in GSM. It concentrates the air traffic.
NSS	Network Subsystem.
PDN	Packet Data Network.
PLMN	Public Land Mobile Network.
PSTN	Public Switched Telephone Network.
RNC (Radio Network Controller)	Network smart mechanism that coordinate the Node-B.
RNS (Radio Network Subsystem)	Base Stations and one RNC.
SGSN (Serving GPRS Support Node)	It is one of the elements responsible for the network operation switched to packets in UMTS.
UE (User Equipment)	Mobile terminal or cell phone.
UMTS (UMTS Terrestrial Radio Access Network)	It is based on Wideband Code Division Multiple Access (WCDMA).
Uu	Link of the air interface between UE and Node-B. It carry voice and data traffic. It is in this interface that WCDMA works.
VLR (Visitor Location Register)	It is the temporary database necessary for when UE is out of the record area. Through MSC, it interacts with HLR.

Radio Network Controller (RNC)

The Radio Network Controller (RNC) is the element of the architecture that switches and controls some elements of the UTRAN. The RNC is located between three interfaces: Iub, Iu and Iur. The Iub interface connects Node-B and RNC, the Iu (PS) connects the RNC and SGSN, and the Iur connects RNCs. The RNC sees the BS as two entities, common transport and a collection of Node-B communication contexts (Kaaranen et al., 2005, pp. 110-111).

RNC can be classified into two parts:

- UTRAN Radio Resource Management (RRM), RRM is a collection of algorithms used to guarantee the stability of the radio path and the QoS of radio connection by the efficient sharing and managing of radio resources.
- UTRAN control functions. These functions include all the functions that are related to set-up, maintenance and release of RBs, including the support functions for RRM algorithms.

UMTS Core Network Architecture

The UMTS Core Network (CN) domain is an entity directly interfacing one or more access networks and consists of equipment entities called ''domains'' and ''subsystems'' connected by several links. The UMTS CN contains the following entities (Kaaranen et al, 2005, pp. 146-152):

- Circuit Switched (CS) domain.
- Packet Switched (PS) domain.
- IP Multimedia Subsystem (IMS).

The mainly elements of the Core Network are:

- Home Location Register (HLR);
- Authentication Centre (AuC);
- Visitor Location Register (VLR);
- Equipment Identity Register (EIR);
- Mobile Switching Centre (MSC);
- Gateway MSC (GMSC);
- Serving GPRS Support Node (SGSN);
- Gateway GPRS Support Node (GGSN).

There are three distinguishable networks in the Core:

- Home Network (HN);
- Serving Network (SN);
- Visited Network.

The Home Network (HN) is the network of the Core that contains the static and security information about the users subscriptions and contains the HLR and the AuC. The Serving Network (SN) is in charge of the communication with the UE and provides the CS domain and the PS domain. Home and the serving network are controlled by different mobile network operators.

In case of the mobile is roaming the Visited Network contains the MSC, VLR and SGSN and the Home Network the HLR and AuC. The GGSN is usually located in the home network, but may be in the visited one if the home network permits access to the visited network's services (Cox, 2008).

Home Location Register (HLR)

This element consists of a database that contains the administrative information about each subscriber along with their last known location. The UMTS network is able to route calls to the relevant RNC/Node B (Poole, n.d.). When a user switches on, the UE is registered with the network, and for routing appropriately the calls, the Node B is determined for the communication.

Authentication Centre (AuC)

The AuC is a protected database that contains the secret key also contained in the user's USIM card. This database is associated to HLR and provides the validation and authentication of network users.

Visitor Location Register (VLR)

The Visitor Location Register (VLR) is the location database, associated to a MSC, where dynamic information about transient users in the MSC place is stored.

Home Subscriber Server (HSS)

The entity called as Home Subscriber Server (HSS) contains functionalities and the Home Location Register (HLR) with the Authentication Centre (AuC) for the CS and PS Domains. The HSS contains the information of subscription and location for handling the calls and sessions.

Mobility Management (MM) functionality supports user mobility through the CS domain, PS domain and IMS.

Equipment Identity Register (EIR)

The EIR Equipment Identity Register (EIR) is a functionality common to all domains and subsystems. The EIR stores information about end-user equipment and the status of this equipment. This register manages three lists:

- **The White List:** This list contains information about approved, normal terminal equipment.
- **The Black List:** It stores information about stolen equipment.
- **The Grey List:** This list contains serial number information about suspect equipment.

The EIR maintains these lists and provides information about user equipment to the CN Domain on request. If the EIR indicates that the terminal equipment is blacklisted, the CN domain refuses to deliver traffic to and from that terminal. In case the terminal equipment is on the grey list, the traffic will be delivered but some trace activity reporting may occur.

CS Domain

The Circuit Switched (CS) domain uses switched technology and transports calls. This domain of the UMTS architecture has the interfaces for the communication with Public Switched Telephone Networks (PSTNs) and with circuit switched domains running by other network operators. The circuit switched elements of the UMTS core network architecture include the Mobile Switching Centre (MSC), and the Gateway MSC (GMSC).

Mobile Switching Centre (MSC)

The Mobile Switching Centre (MSC) is essentially the same as that within GSM, and it manages the circuit switched calls. In the 3GPP Release 99, the set of MSC and VLR is the combination of the CS Media Gateway (CS-MGW) entity and the MSC Server.

Gateway MSC (GMSC)

The Gateway MSC (GMSC) is the interface to the external networks, and in the 3GPP Release 99, the GMSC is the combination of the CS Media Gateway (CS-MGW) entity and the GMSC Server.

The division of the MSC in GMSC Server and CS-MGW makes that the MSC Server can control many CS-MGSWs, separating the control and user plane from each other. This division provides scalability and geographical optimization.

PS Domain

The packet switched (PS) domain uses packet switching and transports data streams. This domain communicates with external Packet Data Networks (PDNs) like the Internet, and with packet switched domains controlled by other network operators.

The two main elements of the PS domain are types of mobile network-specific servers: Serving GPRS Support Node (SGSN) and Gateway GPRS Support Node (GGSN), and they are used for routing.

Serving GPRS Support Node (SGSN)

Serving GPRS Support Node (SGSN) contains the location registration function, which maintains data needed for originating and terminating packet data transfer. These data are subscription information containing the International Mobile Subscriber Identity (IMSI), various temporary identities, location information, Packet Data Protocol (PDP) addresses (but not necessarily IP addresses), subscripted QoS and so on.

In order to transfer data, the SGSN must know with which GGSN the active PDP context of a certain end-user exists so the SGSN stores the GGSN address for each active PDP context. The SGSN provides mobility management, session management, interaction with other areas of the network and Billing.

Gateway GPRS Support Node (GGSN)

The Gateway GPRS Support Node (GGSN) also holds some data about the subscriber. These data may contain the IMSI number, PDP addresses, location information and information about the SGSN that the subscriber has registered. Packet traffic requires additional elements/functionalities for addressing, security and charging.

For security reasons operators now use dynamic address allocation for end-users, being able be allocated in many ways, but the standard way to do this is to use Dynamic Host Configuration Protocol (DHCP) functionality/server. Depending on the operator's configuration, the DHCP allocates either IPv4 or IPv6 addresses for the end-user's terminal equipment.

The GGSN has a Firewall (FW) facility integrated for providing security. Every connection to and from the PS domain is done through the FW in order to guarantee security for end-user traffic. There are many networks that contain a PS domain and roaming between these networks is a most vital issue as far as business is concerned. The PS domain contains a separate functionality in order to enable roaming and to make an interconnection between two PS domains belonging to separate networks. This functionality is called the Border Gateway (BG). GPRS Roaming Exchange (GRX) is a concept designed and implemented for General Packet Radio Service (GPRS) roaming purposes.

IP Multimedia Subsystem

The IM Subsystem comprises all CN elements for provision of IP multimedia services comprising audio, video, text, chat, etc. and a combination of them delivered over the PS domain. The IMS enables applications in mobile devices to establish peer-to-peer connections and introduces multimedia session control using Session Initiation Protocol (SIP) in the PS domain, this allows users to establish connections with various Application Servers (Ass) and to use the IP based services between the terminals (Kaaranen et al, 2005, pp. 180-181).

The entities related to IMS are CSCF, MGCF, MRF. Points that determine the IMS architecture are:

- IP connectivity.
- Access independence.
- Layered design.
- Quality of Service (QoS).
- IP policy control.
- Secure communication.
- Charging.
- Possibility to roam.
- Interworking with other networks.
- Service development and service control for IP-based applications.

A fundamental requirement is that a terminal has to have IP connectivity to access it. Peer-to-peer applications require end-to-end reachability and this connectivity is easiest attainable with IP version 6 (IPv6) because IPv6 does not have address shortage. However, early IMS implementations and deployments may use IP version 4 (IPv4).

Public Land Mobile Network

The Public Land Mobile Network (PLMN) is a mobile network composed of an access network and a core network. Each PLMN in the world is uniquely identified by a PLMN identifier composed of two fields (Sanchez & Thioune, 2007, pp. 188-189):

- **The Mobile Country Code (MCC):** Indicating the country where the network is located. Attribution of MCCs is under the unique responsibility of ITU. Due to this centralization, the uniqueness of the MCC attributed to a network is guaranteed.
- **The Mobile Network Code (MNC):** Which is used to differentiate networks within a same country. MNCs attribution is under the responsibility of the regulation authority of the network country.

Equivalent PLMNs

PLMNs are equivalent to each other with regard to PLMN selection/reselection, cell selection/reselection and handover procedures An operator owning both GSM and UMTS network in the same country could use a different PLMN code for each of them. If the two networks are located in different countries, it will be compulsory to use different PLMN codes as well. The UE can select a cell belonging to an equivalent network regardless of the radio access technology.

Types of PLMNs

A given mobile will have to differentiate:

- **The Home PLMN (HPLMN):** For which the user has subscribed a network service provision. The operator of the HPLMN is the owner of the USIM in the user equipment.
- **The Visited PLMN (VPLMN):** Which is a PLMN the UE has accepted in a roaming situation and which is different from the HPLMN.
- **The Registered PLMN (RPLMN):** Which is the last PLMN on which the UE has registered successfully. It could be either the HPLMN or a VPLMN.

PLMN Selection and Roaming

When an UE is switched on, it attempts to make contact with a PLMN. The particular PLMN to be contacted may be selected either automatically or manually and follows rules defined in 3GPP specification (ETSI 3rd Generation Partnership Project,

2013). The UE normally operates on its Home PLMN (HPLMN) or Equivalent Home PLMN (EHPLMN). However a Visited PLMN (VPLMN) may be selected, for example if the MS loses coverage. There are two modes for PLMN selection:

- **Automatic Mode:** This mode utilizes a list of PLMNs in priority order. The highest priority PLMN which is available and allowable is selected.
- **Manual Mode:** Here the MS indicates to the user which PLMNs are available. Only when the user makes a manual selection does the MS try to obtain normal service on the VPLMN.

To prevent repeated attempts to have roaming service on a not allowed area, when the MS is informed that an area is forbidden, the Location Area (LA) or Tracking Area (TA) is added to a list of "forbidden LAs for roaming" or "forbidden TAs for roaming" respectively which is stored in the MS. These lists, if existing, are deleted when the MS is switched off or when the SIM is removed. The structure of the routing area identifier supports area restriction on LA basis.

EXCHANGED OF SECURITY MESSAGES IN UMTS

The security messages are studied in function of the two mainly process of security:

- Mutual Authentication.
- Ciphering / Integrity.

The specifications about security procedures and algorithms started to be described in the 3rd Generation Partnership Project (1999) of the UMTS 3GPP Specification.

UMTS provides:

- Confidentiality.
- Integrity.
- Mutual authentication.
- Access control.
- Non repudiation.

The authentication works between the UMTS subscriber and the Serving Network (SN). Confidentiality is achieved through the use of encryption algorithms for data traffic, using a mechanism for Temporary Mobile Subscriber Identity (TMSI), causing the IMSI cannot be spied on. The insertion of the data integrity mechanism for signaling enables protection against attacks from false Base Stations.

Data messages and signals are exchanged between the UE and the set SGSN/VLR via radio frequency links and wired links. Depending on the kind mechanism, the messages are exchanged in different layers:

- Integrity mechanism exchanges signaling messages at Radio Resource Control (RRC) level between UE and RNC.
- Encryption mechanism exchanges signaling messages in layers Radio Link Control (RLC) and Medium Access Control (MAC) between UE and RNC.

The transport of Cipher Key (CK) and Integrity Key (IK) uses the signaling control. The information sent to UMTS network has two targets:

- Data protection through the encryption and integrity towards UE ↔ RNC.
- Mutual authentication towards USIM ↔ SGSN/VLR and HLR/AuC ↔ USIM.

The Authentication process contents:

- Authentication Key K.
- Authentication functions f1 and f2.
- Key generation function f3, f4 and f5.
- Sequence number management SQn.

The Ciphering/Integration protection contents:

- Ciphering Key (CK).
- Integrity Key (IK).
- Ciphering function f8.
- Integrity function f9.

The HLR stores the environment data where UE is recorded for security policy, and AuC is part or the network where the sequence numbers (SQNs) in authentication messages are generated, in addition to the random numbers (RANDs) to be co-related with the secret key K, which is shared between UE and the Home Network (HN).

Security Mode Set-Up Procedure

The set-up procedure defines the security establishment process with their messages, and indicates the integrity-protected messages at both uplink and downlink.

Encryption and decryption take place in the terminal and in the RNC on the network side, and the CK has to be transferred from the Core Network (CN) to the Radio Access Network (RAN). This is done in a specific Radio Access Network Application Protocol (RANAP) message, called the "security mode command". After the RNC has obtained the CK, it can switch encryption on by sending a Radio Resource Control (RRC) security mode command to the terminal.

For encryption at downlink the first encrypted message is the one sent after the Radio Access Network Application Protocol (RANAP) security mode complete (Niemi & Nyberg, 2003). At uplink the first encrypted message is the first message sent after the security mode complete message.

When integrity protection (and possibly encryption) is turned on, messages and security-relevant information parameters are transferred between the UE, RNC and SGSN/VLR during the set-up procedure.

The USIM shall support UMTS Authentication and Key Agreement (AKA) and may support backwards compatibility with the GSM system (3rd Generation Partnership Project, 2014).

The first connection request message is sent on Random Access Channel which is a common channel and can be intercepted easily posing a vulnerability point.

In Figure 2 and Figure 3 the steps of the security set-up procedure are described graphically between the involved network elements.

In the Table 2 the steps of the set-up procedure are shown with their descriptions (3rd Generation Partnership Project, 2014).

Figure 2. Security set-up

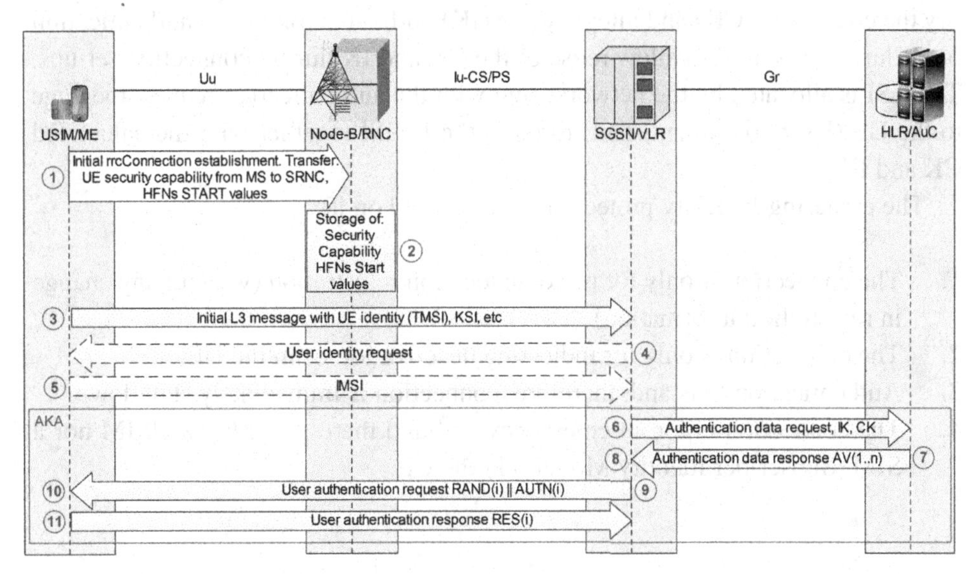

Figure 3. Security set-up (2)

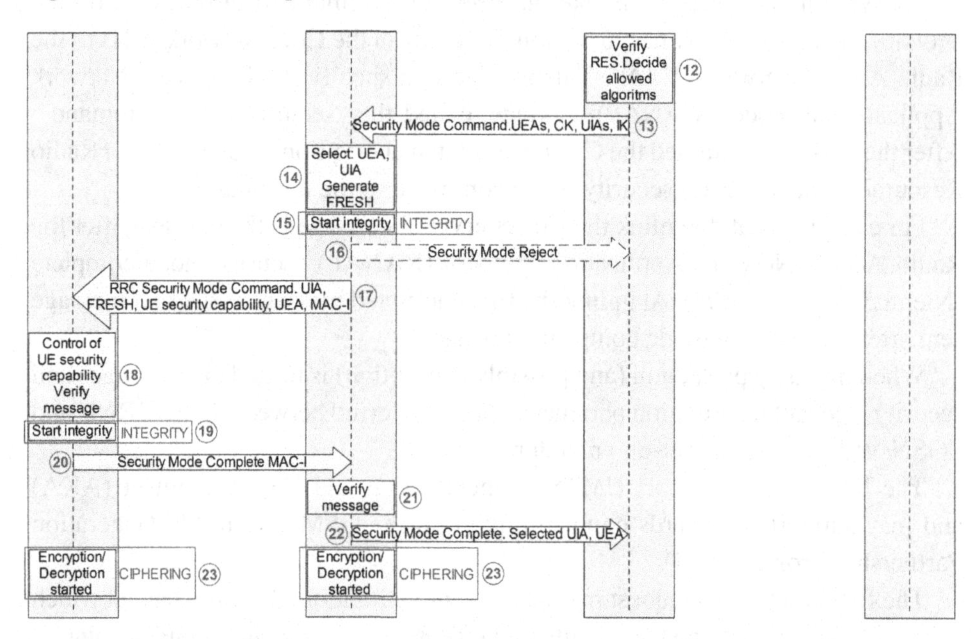

The "UE security capability" indicates UIAs and UEAs supported by MS, and the network must have this information before the integrity protection can start, for this reason "UE security capability" must be sent to the network in an unprotected message.

The Key Set Identifier (KSI) number is associated with the cipher and integrity keys derived during authentication. The KSI makes possible for the network to identify the cipher key (CK) and integrity key (IK) without invoking the authentication procedure and is used to allow reuse of the CK and IK during connection set-ups. The KSI is allocated by the network, sent with the authentication request message to the UE (User equipment) and stored in the USIM together with the calculated CK and IK.

The ciphering Integrity protection is not turned on if:

1. The connection is only for periodic location registration (without any change in registration information).
2. The connection is only for indicating deactivation from the UE.
3. Authentication fails and, therefore, connection is immediately shut down.
4. The connection is for an emergency call and there is neither a USIM nor a SIM (Subscriber Identity Module) in the UE.

Table 2. Messages and relevant information transferred during the set-up-procedure

Step	Description
1	The User Equipment (UE) security capability information includes: • Ciphering capabilities (UMTS Encryption Algorithms (UEAs)) of the MS. • Integrity capabilities (UMTS Integrity Algorithms (UIAs)) of the MS.
2	If the GSM Classmarks 2 and 3 are transmitted during the RNC Connection establishment, the NC must storage the GSM cyphering capability of the UE.
3	The Initial LA (Message of the Layer 3 or Network Layer) message includes: • Temporary Mobile Subscriber Identity (TMSI). • Key Set Identifier (KSI), allocated by the Circuit Switched (CS) service domain or Packet Switched (PS) service domain at the last authentication for this CN domain. • Location update request. • Communication Management (CM) service request. • Routing area update request. • Attach request. • Paging response.
4	If TMSI cannot be solved an user identity interrogation may be performed.
5	If the user identity request is performed, ME sends its International Mobile Subscriber Identity (IMSI).
6	Authentication of the user and generation of new security keys can be performed: • Integrity Key (IK). • Cipher Key (CK).
7	n-Authentication vectors are generated and sent from HLR/AuC to SGSN/VLR. Each authentication vector consist on: • A random number (RAND). • An expected response (ARES). • A Cipher Key (CK). • An Integrity Key (IK). • An Authentication Token (AUTN). Each authentication vector is for one Authentication and Key Agreement (AKA) between SIGNS/VELAR and USE.
8	The authentication vectors are stored.
9	SGSN/VLR initiates the AKA. It selects the next authentication vector from the ordered array and sends the parameters: • RAND. • AUTN.
10	The USIM checks whether AUTN can be accepted, and it computes CK and IK.
11	If the AUTN can be accepted USIM produces a response (RES).
12	The SGSN/VLR compares the received RES with ARES. If they match the SGSN/VLR considers the AKA exchange successfully completed. The SGSN/VLR determine which UIAs and UEAs are allowed to be used in order of preference. CK and IK are selected.
13	The SGSN/VLR initiates integrity and ciphering. The Security Mode Command message sent contains: • List of allowed UIAs in order of preference. • The IK to be used. If ciphering shall be started, the message contains: • Ordered list of allowed UEAs in order of preference. • The CK to be used. If a new authentication and security key generation has been performed, this shall be indicated in the message.
14	The Serving RNC (SRNC) decides which algorithms to use by selecting the highest preference algorithm from the list of allowed algorithms, and generates a random value FRESH (random number used in the Integrity Algorithm f9 input.).
15	The SRNC initiates the downlink integrity protection

continued on following page

Table 2. Continued.

Step	Description
16	If the requirements received in the Security Mode Command cannot be fulfilled, the SRNC sends a Security Mode Reject message to the requesting SGSN/VLR.
17	The SRNC generates the RRC message Security Mode Command, that includes: • ME security capability. • The GSM ciphering capability, if received during RRC Connection Establishment. • UIA (UMTS Integrity Algorithm). • FRESH. • UEA if ciphering shall be started. • CN type indicator information. This indicator is used because as the MS can have two ciphering and integrity key sets, the network must indicate which key set to use. • Message Authentication Code for Integrity (MAC-I). • Additional information may also be included.
18	The MS controls that "UE security capability" received is equal to the sent in the initial message. The GSM ciphering capability is also controlled if it was included in the RRC Connection Establishment. The MS computes XMAC-I using: • UIA. • COUNT-I, generated from the stored START. • FRESH. The MS verifies the integrity of the message comparing the received MAC-I with the generated XMAC-I.
19	The uplink integrity protection is initiated.
20	If all controls are successful, the MS compiles the RRC message Security mode complete and generates the MAC-I for the message. If any control is not successful, the procedures ends.
21	The SRNC computes XMAC-I on the message and verifies the data integrity of the message by comparing the received MAC-I with the generated XMAC-I.
22	The transfer of the RANAP message Security Mode Complete response, including the selected algorithms, from SRNC to the VLR/SGSN ends the procedure
23	Finally the encryption/decryption can start between the UE and the Node-b/RNC.

In case of a faulty UEA1 implementation, the SGSN/VLR may configure the security mode command to establish security without ciphering (UEA0) for certain UEs. The UEs that support UEA1 shall have security established with ciphering (UEA1).

Authentication and Key Agreement (AKA)

Authentication and Key Agreement (AKA) achieves mutual authentication by the user and the network showing knowledge of a secret key K, which is shared between USIM and AuC in the user's HE, and only available to them. The method achieves maximum compatibility with GSM security architecture and facilitate migration from GSM to UMTS.

There are three goals for the UMTS AKA (AL-Fayoumi & AL-Saraireh, 2011, p. 36):

- Mutual authentication between the user and the network.

- Establishment of a cipher key and an integrity key upon successful authentication.
- Freshness assurance to the user of the established cipher (or, synonymously, encryption) and integrity keys.

AKA in UMTS uses five functions, f1, f2, f3, f4 and f5 carried out in Home Network (HN). The Sequence Number (SQN) has 48 bits in length and it is a random number generated by AuC. This number can be cancelled in the AKA algorithm using the Anonymity Key (AK), SQN $\oplus$ AK.

In the beginning of exchange of information is generated a recent sequence number SQN and a challenge RAND (128 bit). Functions f1 and f2 correspond to messaging authentication functions, functions f3, f4 and f5 are key generations functions, and f1* and f5* are functions for messaging authentication and AK generator respectively for synchronization. The secret is shared between HE and AuC.

For each user, HE/AuC generates a sequence number, SQN_{HE}. The values are computed by HE/AuC using the specific key of each K user and a specific operator called Authentication Management Field (AMF) (Wagner, M., Wagner, J., & Zucchini, 2010, pp. 88-90).

If MAC or RES are not in agreement with the expected, the process is restarted. If SQN is not recent, the synchronization with AuC is provided. If SQN in AuC is correct, the process is restarted, otherwise MAC is verified by SN. With MAC irregular, the process is restarted, however, being correct, SQN is reset in AuC, and then the process is restarted.

In AuC environment, the secret K key and SQN_{AuC} are generated, being encapsulated in AKA algorithms generating AVs that are sent to the network components (VLR for circuit-switched network and SGNS for packet-switched network) RNC selects the RAND and AUTN for ME, using algorithms f8 (encryption) and f9 (integrity). UE provides confidentiality and integrity in the communication. USIM uses AKA algorithm to generate RES and send to VLR/SGSN, besides generating SQN_{USIM}.

Cipher Key and Integrity Key Setting

The Authentication procedure sets:

- Authentication.
- Key Setting.

The process is initiated by the network operator and the key setting can occur as soon as the identity of the mobile subscriber (TMSI or IMSI) is known by the VLR/SGSN. The CK and IK are stored in the VLR/SGSN and transferred to the

RNC when needed. The CK and IK are stored on the USIM and updated at the next authentication (3rd Generation Partnership Project, 2014). The Mobile Equipment (ME) shall store CK and IK security context data on the USIM.

Ciphering and Integrity Mode Negotiation

In the Initial RRC Connection establishment message, where the UE try to establish a connection with the network, the UE indicates the basic information set about its capabilities for letting know to the network wich cipher and integrity algorithms the UE supports. This information is indicated in the UE/USIM Classmark. This UE/USIM Classmark information is stored in the RNC due the RNC does not have the integrity key IK for performing the integrity protection when receives the Classmark. The data integrity is performed. When the IK is generated in the security mode setup procedure, the data integrity of the Classmark is carried out because the Classmark information must be integrity protected.

The network shall compare its integrity and ciphering capabilities, preferences and special requirements of the UE subscription, with the appropriate indicated by the UE. The network acts, depending upon the case, following some rules. This rules are defined in the Releases (Table 3) where this mode is described (3rd Generation Partnership Project, 2014).

The preferences and special requirements for the ciphering and integrity mode setting shall be common for CS and CN domains

Table 3. Mode negotiation rules

	Comparison of Integrity Capabilities, Preferences, and Requirements
1	If the UE and the network have no versions of the UMTS Integrity Algorithms (UIA) algorithm in common, then the connection shall be released.
2	If the UE and the network have at least one version of the UIA algorithm in common, then the network shall select one of the mutually acceptable versions of the UIA algorithm for use on that connection.
	Comparison of Ciphering Capabilities, Preferences, and Requirements
1	If the UE and the network have no versions of the UMTS Encryption Algorithms (UEA) algorithm in common and the network is not prepared to use a not ciphered connection, then the connection shall be released.
2	If the UE and the network have no versions of the UEA algorithm in common and the user (respectively the user's HE) and the network are willing to use an not ciphered connection, then an not ciphered connection shall be used.
3	If the UE and the network have at least one version of the UEA algorithm in common, then the network shall select one of the mutually acceptable versions of the UEA algorithm for use on that connection.

Cipher Key and Integrity Key Lifetime

The UE have a mechanism to limit the amount of data protected by a ciphering/integrity key set for avoid that a particular key set is used for an unlimited period of time, exceeding the limit set by the operator. The mechanism prevents the malicious re-use of compromised keys for performing attacks.

The operator sets a value called "THRESHOLD" stored in the USIM and is used by the mechanism for carrying out checks with the values $START_{CS}$ and $START_{PS}$ (3rd Generation Partnership Project, 2014). The verifications are performed when an RRC connection is released, when the UE has powered and when an RRC connection is established. If $START_{CS}$ and $START_{PS}$, are greater than or equal to THRESHOLD, the ME sets the START value in the ME to zero, deletes the cipher key and the integrity key on the USIM and the ME and sets the KSI to invalid.

VULNERABILITIES AND EXPLOITS IN THE SECURITY MODE SET-UP PROCEDURE

The security architecture of UMTS offers some protection against threats including false base station attacks, man-in-the-middle attacks and replay attacks, even the UMTS system protection is difficult. The system also successfully ensures user data confidentiality and signalling data integrity. However, the architecture presents vulnerabilities in relation with the security mode set-up procedure because one of the used interface is radio. In that interface the intruders can carry out most of the 3G attacks. Other attacks can be executed against other parts of the system, on wireless and wired interfaces.

The complete security threats list is defined by the ETSI 3rd Generation Partnership Project (2001). In that list the threats on the radio interface, where the intruders have more capability for accessing, is shown in the Table 4.

In the security mode set-up procedure the "RRC connection establishment" message is not protected. This initial RRC connection message includes the "UE security capabilities" because the network must have that information before the beginning of the integrity protection. The UE security capabilities include Ciphering capabilities (UMTS Encryption Algorithms (UEAs)) and Integrity capabilities (UMTS Integrity Algorithms (UIAs)) of the MS.

The AKA provides two authentications simultaneously, ME and the network. But not every message is protected, the messages: International Mobile Subscriber Identity (IMSI), the quintet Authentication Token (AUTN), Random Number (RAND), Expanded Response (XRES), Cipher Key (CK) and Integrity Key (IK), the pair RAND and AUTN, and the User Response (RES), are neither encrypted

Table 4. Threats on the radio interface

Unauthorized Access to Data	
Eavesdropping user traffic	To eavesdrop user traffic on the radio interface.
Eavesdropping signalling or control data	To eavesdrop signalling data or control data to access security management data or other information.
Masquerading as a communication participant	To masquerade a network element to intercept user traffic, signalling data or control data.
Passive traffic analysis	To observe the time, rate, length, sources or destination of messages to obtain access to information.
Active traffic analysis	To initiate communications sessions and then obtain access to information through observation of the time, rate, length, sources or destinations of associated messages.
Threats to Integrity	
Manipulation of user traffic	To modify, insert, replay or delete user traffic, for accidental or deliberate manipulation.
Manipulation of signalling or control data	To modify, insert, replay or delete signalling data or control data, for accidental or deliberate manipulation.
Denial of Service Attacks	
Physical intervention.	To prevent user traffic, signalling data and control data from being transmitted by physical means.
Protocol intervention	To prevent user traffic, signalling data and control data from being transmitted by inducing specific protocol failures.
Do's by masquerading as a communication participant	To deny service to a user by preventing user traffic, signalling data or control data from being transmitted by masquerading as a network element.
Unauthorised Access to Services	
Masquerading as another user	To masquerade as a user towards the network. The intruder masquerade as a base station towards the user, then hijacks his connection after the authentication.

nor integrity protected because they are transmitted before key agreement (Khan, Ahmed, Raza Cheema, 2008, p. 351). But the authentication data response where AUTN, XRES, CK, IK, and RAND are included, not is transmitted over the Uu interface in which the mainly threats are located.

One important security breach about the IMSI is due to the process of the identification by a permanent identity. In this process when the VLR/SGSN sends "the user identity request" the UE/USIM send a response with the IMSI in clear text inside the "user identity response" message.

The messages are transmitted through air on Uu interface between ME and Node-B and hence are susceptible to interception, insertion and modification, and so may cause user specific Denial of Service (DoS).

The signalling messages that are neither encrypted nor integrity protected (3rd Generation Partnership Project, 2014):

- Handover to UTRAN complete.
- Paging type 1.
- Push capacity request.
- Physical shared channel allocation.
- RRC connection request.
- RRC connection setup.
- RRC connection setup complete.
- RRC connection reject.
- RRC connection release (CCCH only).
- System information (Broadcast information).
- System information change indication.
- Transport format combination control (TM DCCH only).

Any change in the way will result in non-authentication of network and/or user. The rest of messages after the RRC connection establishment and execution of the security mode set-up procedure shall be integrity protected.

For the performing or the attacks the intruders must have computational capabilities (3rd Generation Partnership Project, 2000) and one or more of the following:

- Eavesdropping.
- Impersonation of a user.
- Impersonation of the network.
- Man-in-the-middle.
- Compromising authentication vectors in the network.

The attacks can be classified according to the security threats. This classification is explained in Table 5 for unauthorized access to data, Table 6 for denial of service and Table 7 for unauthorized access to service.

In the following paragraphs the attacks in relation with the security mode set-up and are explained more in detail, and although there are more attacks classified and defined, only the attacks over the radio interface (Uu) are developed.

Table 5. Unauthorized access to data

Attack	Weakness	Requirement	3G Security Countermeasures
Passive identity catching	The network may request the user to send its identity in clear text	Modified MS	Yes. Identity confidentiality mechanism with temporary identities
Active identity catching	The network may request the MS to send its permanent user identity in clear text	Modified BS	Yes. Identity confidentiality mechanism using an encryption key shared by a group of users
Impersonation of the network by suppressing encryption between the target user and the intruder	MS cannot authenticate messages received over the radio interface	Modified BS	Yes. Cipher mode command with message authentication and replay inhibition
Impersonation of the network by suppressing encryption between the target user and the true network	The network cannot authenticate messages received over the radio interface	Modified BS/MS	Yes. Mobile station classmark with authentication and replay inhibition
Impersonation of the network by forcing the use of compromised cipher key	The user has no control upon the cipher key	Modified BS and a compromised authentication vector	Yes. Sequence number in the challenge. Architecture not protected against force use of compromised authentication vector not used to authenticate in USIM
Eavesdropping on user data by suppressing encryption between target user and intruder	The MS cannot authenticate messages received over the radio interface	Modified BS/MS	Yes. Cipher mode command with message authentication and replay inhibition
Eavesdropping on user data by suppressing encryption between target user true network	The network cannot authenticate messages received over the radio interface	Modified BS/MS	Yes. Message authentication and replay inhibition of the mobile´s ciphering capabilities.
Eavesdropping on user data by forcing the use of compromised cipher key	The user has no control the cipher key	Modified BS/ MS and a compromised authentication vector	Yes. A sequence number in the challenge. Architecture not protected against force use of compromised authentication vector not used to authenticate in USIM

Table 6. Denial of service attacks

Attack	Weakness	Requirement	3G Security Countermeasures
User de-registration request spoofing	Network cannot authenticate the received messages over the radio interface	Modified MS	Yes. Integrity protection of signalling messages, data authentication and replay inhibition of the de-registration request
Location update request spoofing	Network cannot authenticate the received messages over the radio interface	Modified MS	Yes. Integrity protection of signalling messages, data authentication and replay inhibition of the location update request
Camping on a false base station	A false base station can manipulate signalling	Modified BS	No. But this attack only persists while the attacker is active
Camping on a false base station/mobile station	A false base station can act as repeater and relay request between the network and a target user	Modified BS/ MS	No. But integrity protection of critical messages may help. This attack only persists while the attacker is active and

Table 7. Unauthorised access to services

Attack	Weakness	Requirement	3G Security Countermeasures
Impersonation of the through the use of by the network of a compromised authentication vector	Architecture vulnerable to attacks using compromised authenticate vectors.	Modified MS and a compromised authenticate vector	Yes. Sequence number.
Impersonation of the user through the use by the network of an eavesdropped authentication response	An authentication vector may be used several times	Modified MS	Yes. Sequence number
Hijacking outgoing calls in networks with encryption disabled	The network can work with an un-enciphered connection	Modified BS/ MS	Partly. Integrity protection of signalling messages, data authentication and replay inhibition of the connection set-up request
Hijacking outgoing calls in networks with encryption enabled	Possibility of suppress the encryption	Modified BS/ MS	Yes. Integrity protection of signalling messages, data authentication and replay inhibition of the MS station classmark and connection set-up request
Hijacking incoming calls in networks with encryption disabled	The network can work with an un-enciphered connection	Modified BS/ MS	Partly. Integrity protection of signalling messages, data authentication and replay inhibition of the connection accept message
Hijacking incoming calls in networks with encryption enabled	Possibility of suppress the encryption	Modified BS/ MS	Yes. Integrity protection of signalling messages, data authentication and replay inhibition of the MS station classmark and connection accept message

Identity Catching

UMTS system offers little protection against identity catching or also called Disclosure of International Mobile Subscriber Identity (IMSI). Integrity protection of critical signaling messages avoids the DoS attacks, however, unprotected messages before security mode command may be utilized for launching DoS attacks. The protection of IMSI is an important issue in UMTS and must its use must be avoid or used rarely, but, although IMSI is replaced by Temporary Mobile Subscriber Identity (TMSI) after the initial connection request, IMSI is sent clear during the RRC Connection Request, and also on the occasions like VLR database crash, VLR's inability to identify the TMSI (Khan, Ahmed, Raza Cheema, 2008, p. 353).

The weakness that makes possible the catching of the IMSI is the fact that the network can make an identity request when the TMSI cannot be resolved during the RRC Connection Request. When the ME receives the request sends its IMSI in clear and the attacker catch it. As the goal is only catching the IMSI the attacker can disconnect himself.

There are two types depending on the mode for obtain the IMI, passive catching and active catching.

Passive Identity Catching

The attacker using a modified MS exploits the weakness that the network may sometimes request the user to send its identity, and when the identity is sent is done in clear text. In this attack there is not interference over the Uu link, using a jammer or a transmitting BTS with more power than the real one. The attacks only waits for a new registration or a database crash as in this state the user is requested to send its identity in clear text. The Figure 4 shows the passive IMSI catching attack process.

According with the 3rd Generation Partnership Project (2000) the identity confidentiality mechanism counteracts this attack.

Active Identity Catching

The attacker uses a modified BTS for making the identity request, because the UE will have to answer with its permanent user identity in clear text. The intruder in order to do this should attach the target user on its false BTS. The Figure 5 shows in this case the active IMSI catching attack process

3G security provides protection against this type of attack, specifically the identity confidentiality mechanism counteracts it by using an encryption key shared by a group of users to protect the user identity in the event of new registrations or temporary identity database failure in the serving network.

Figure 4. Passive IMSI catching

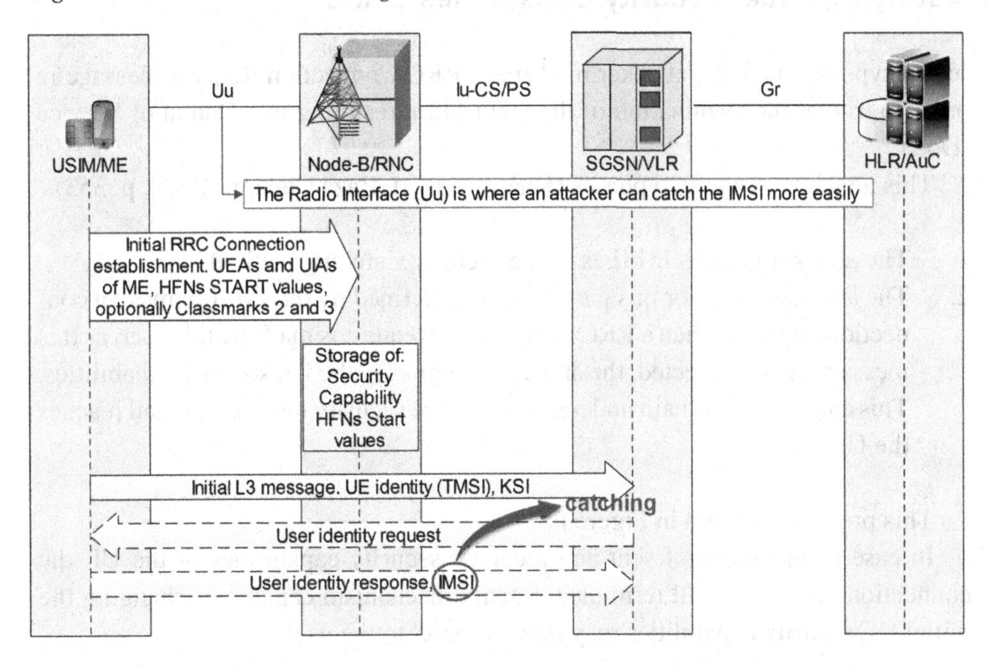

Figure 5. Active IMSI catching

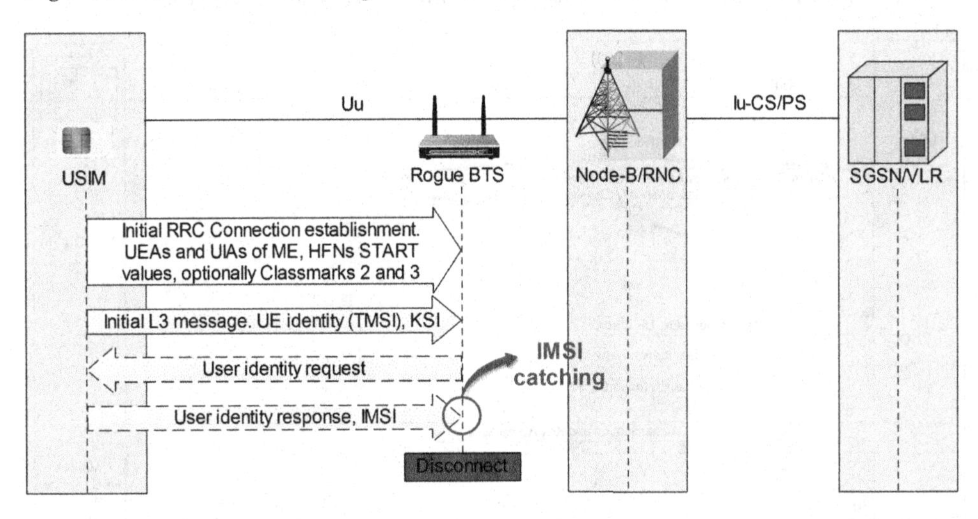

Modifying Initial Security Capabilities of ME

In this type of attack the attacker modifies a RRC Connection Request message in order to trigger the termination of the connection resulting in a Denial of Service (DoS).

This attack works in two phases (Khan, Ahmed, Raza Cheema, 2008, p. 353):

1. The attacker obtains IMSI as in the Identity Catching method.
2. The attacker waits for this particular user defined by the IMSI to make a connection request. When a RRC Connection Request is made by this user, as this message is not protected, the attacker can modify the UE security capabilities. This change will remain undetected until the security mode command reaches the UE.

This process is shown in Figure 6.

In case of mismatch of sent and received security capabilities of the UE the connection procedure will terminate. Such a mechanism capable of changing the initial UE security capabilities may result in DoS to the user.

Figure 6. Modification of RRC connection request and connection reject

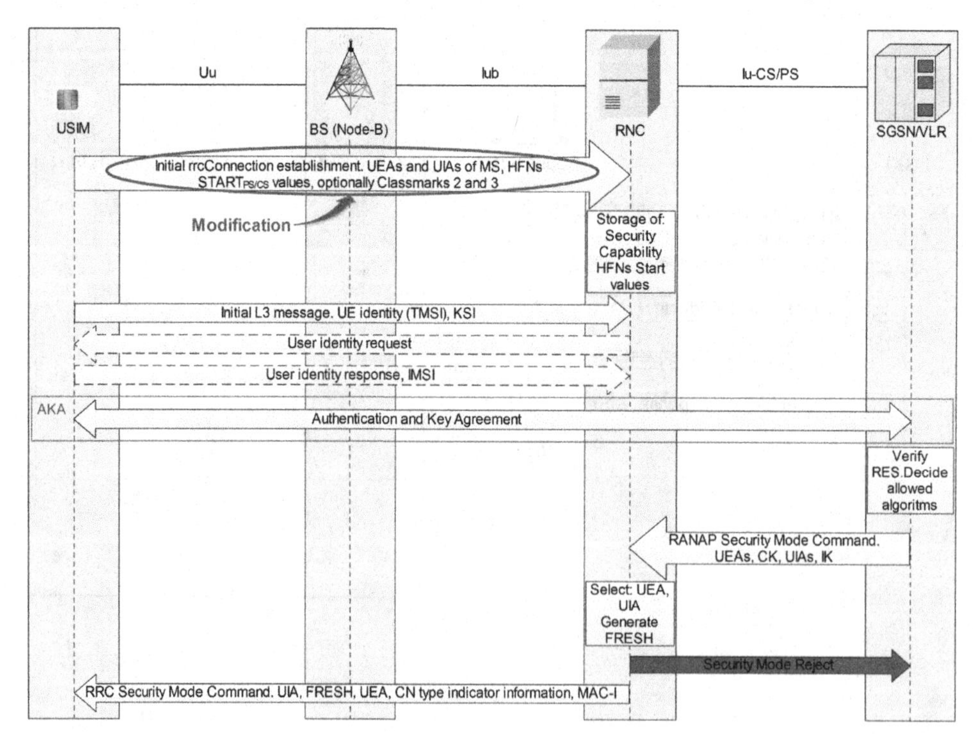

Bidding Down

Using the Identity Catching method an intruder can modify the UE security capabilities and force to the victim device to use an authentication mechanism weaker. In fact the network allows to some users to support security without ciphering, using UEA0.

This attack can be associated to the Man-in-the-Middle attack using a false base station for can eavesdropping the communication between the victim and network when the false BTS is connected to a real PLMN.

Modifying Authentication Parameters

The modification of any of the authentication parameters including AUTN, RAND, or RES may also result in DoS. Even the objective is not perform an impersonation, some of the AKA parameters can be modified:

- AUTN (SQN⊕AK, AMF, MAC-A).
- XRES.
- RAND.
- RES.

In this case when the verification of parameters is done, the result may cause a resynchronization process or a denial of service. The AUTN, XRES, CK, IK and RAND, are transmitted between the SGSN/VLR and HLR/AuC, so for an intruder is easier intercept the RAND, AUTN and RES between the USIM and the Node-b/RNC over the Uu interface.

Using RRC Connection Reject Message

This attack works in 3 phases (Khan, Ahmed, Raza Cheema, 2008, p. 353) and follows the same diagram than Figure 5:

1. The attacker obtains IMSI as in the Identity Catching method.
2. The attacker waits for RRC Connection Request by this user.
3. The attacker will respond with RRC Connection Reject message. This message is transmitted by the network when the RRC connection cannot be accepted and is not integrity protected. The UE compare the "Initial UE identity" value of the Information Element (IE) in the received RRC Connection Reject message with the value of the variable "INITIAL_UE_IDENTITY". When the checking of the values is performed:
    a. If they are different the UE shall ignore the rest of the message.

b. If the values are identical, the UE shall terminate the connection causing DoS to the user.

Flooding the HLR/AuC

This is a dangerous attack in which the services of a particular mobile operator can be blocked because for the exhausting of resources and bandwidth. This attack works in two phases (Khan, Ahmed, Raza Cheema, 2008, p. 354):

1. The attacker builds a database of IMSIs corresponding to the victim operator. Using the procedure of Identity Catching the attacker can obtain the IMSIs of the user victims.
2. The attacker generates RR Connection Request messages corresponding to each IMSI. For each request, (except already connected one) VLR/SGSN sends the IMSIs to the HLR/AuC and checks the validity of IMSIs, passing the validity test due to they were obtained in phase 1. At this point the HLR have to compute the AVs for each IMSI with the RAND, MAC, XRES, CK, IK and AK, and this process is costly. The AVs are sent to VLR/SGSN, as it is shown in step 7 of Figure 2, and the VLR/SGSN selects one AV and sends the RAND and AUTN parameters for authentication to UE.

The goal of the attacker is not break the authentication, it is to exhaust the computing resources of HLR/AuC by flooding more and more valid RRC Connection Requests and making it to compute AVs. This may also cause bandwidth exhaustion between VLR/SGSN and HLR/AuC. The exhaustion of resources will result in DoS to new users who are attempting to connect.

Camping on a False Base Station

In this DoS attack an attacker with a false BS equipment moves close to its target victims and if the signal is stronger than the legitimate BS, the ME will lose the connection with its network and will connect to the false BS (3rd Generation Partnership Project, 2000) When the UE and the false BTS are connected, the attacker can capture the packets that are transmitted from and towards the UE. UMTS security architecture does not counteract this attack.

The attack has the following difficulties:

- The attack persists only when the attacker is active.
- It affects only a small number of users.

- It cannot be directed to inflict specific targets (users) only, without affecting others as well.
- It is easy to detect the attack and locate the malicious element in the network.

For these indicated reasons can be used other kind of DoS attacks more effectives, but this use of the false base station is the first principle for other forms of attack, for example manipulating the signaling on the radio interface (Uu) in order to masquerade the BTS as network element. Doing this the attacker can develop attacks as "User identity catching", "Suppression of encryption between target and intruder", "Compromise of authentication data" and "Hijacking of services".

Man-in-the-Middle Attack

For doing this type of attack the attacker would have to put itself between the target user and a real network for impersonating the valid network to the user. When the attack is successfully done the intruder could do the next actions over the messages between the victim user and its valid network:

- Eavesdrop.
- Modify.
- Delete.
- Re-order.
- Replay.
- Spoof signalling.
- Spoof user data.

However, the UMTS architecture has two security mechanisms for protection of the mobile station from this attack, as was described in Meyer, & Wezel (2004):

- **The Authentication Token AUTN:** It ensures the timelines and origin of the authentication challenge and protects against replay of authentication data.
- **Integrity Protection of the Security Mode Command Message:** It prevents an attacker from relaying correct authentication information when it fool into not using encryption for communication.

Although in UTRAN the Security Mode Command message that informs the mobile station which algorithm to use, is integrity protected, in the GSM case like in hybrid GSM/UMTS networks, integrity protection is not supported, as a consequence, Cipher Mode Command message is not integrity protected, allowing an attacker forge this message and fool the victim mobile station into using either no encryption or a weak encryption algorithm. Thus an attacker can impersonate a GSM BTS and using the UMTS authentication procedure of the hybrid GSM/UMTS scenario. This attack does not work against mobile equipment that is capable of the UTRAN interface only.

The attacker has to know the IMSI of his victim, therefore as was described in the "Identity Catching" attack, the attacker with a false base station could initiate an authentication procedure and using a user identify request obtain the IMSI value. Doing this, the attacker also learns security capabilities of the mobile station.

As the mobile station have to connect with the base station that provides the best reception, it is relatively easy carry out this attack. To have more signal than the valid base station, the false BTS must locate close enough to UE victim, or with a jammer interfere with the user frequency. If it is successful the victim device will disconnect to the real base station and will connect with the false base station.

This attack is specified in the 3GPP Technical Specifications, but the detailed process is not provided completely, however there are well documented works and articles with this process, for example according with Meyer, & Wezel (2004), the attack would work in two phases:

1. The attacker obtains a valid authentication token AUTN from a real network. For this, the attacker pass himself off the victim mobile station and executes the protocol of the Table 8.

The procedure of obtaining vail AUTN is described in Figure 7.

None of the messages in Steps 1 to 7 are protected. The attacker obtains an authentication token for using in the second phase.

Table 8. Steps for the first phase in the Man-in-the-Middle attack

Step	Description
1	Sending of security capabilities of the victim UE.
2	Sending of TMSI of the victime UE. If the TMSI is unknown, the atacker sends a false TMSI.
3	Sending of IMSI in case the network cannot resolve the TMSI.
4	Requesting of authentication information for the victim device from its home network.
5	Response of authentication information to the visited network from the home network.
6	Sending of RAND and AUTN to the attacker.
7	Disconnection of the attacher to the visited network.

Figure 7. Obtaining valid AUTN

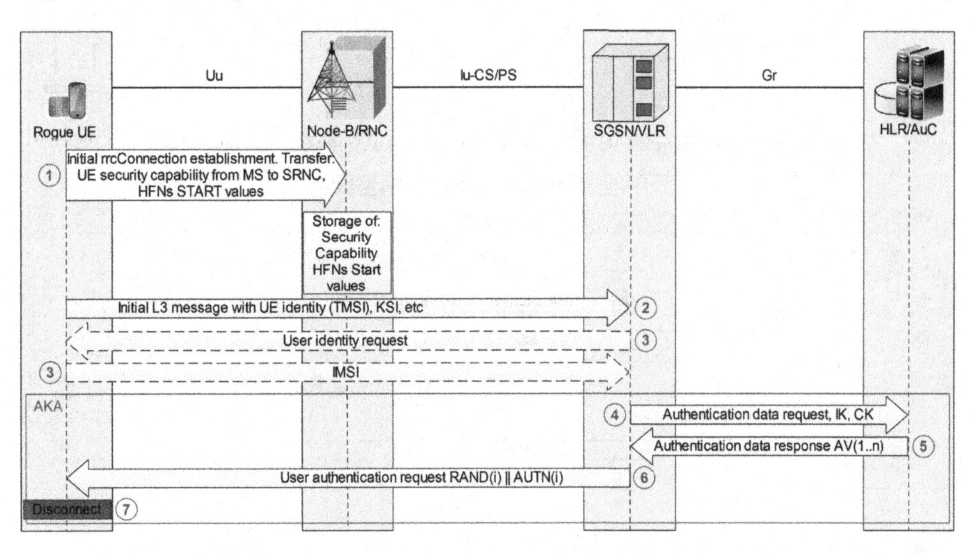

2. The attacker impersonates a valid GSM base station to the victim mobile station. The step are shown in Table 9 for the second phase in Man-in-the-Middle attack and the procedure for impersonation of valid GSM BTS in Figure 8.

The attacker has to establish a regular connection to a real network to forward traffic it receives from the mobile station.

The decision of requesting no encryption or weak encryption depends on the security capabilities of the mobile unit which are sent to the network during connection setup.

Table 9. Steps for the second phase in the Man-in-the-Middle attack

Step	Description
1	Establishment of connection between the victim UE and the BTS attacker. The UE sends its security capabilities.
2	Sending of the TMSI or IMSI of victim UE.
3	Sending of RAND and AUTN obtained from the real network in the first phase.
4	Verification successful of the AUTN.
5	Sending of authentication response.
6	The attacker decides to use “no encryption” or weak encryption (it can be a broken version of the GSM encryption algorithms).
7	Sending of GSM cipher mode command with the chosen encryption algorithm.

Figure 8. Impersonation of valid GSM base station to the victim

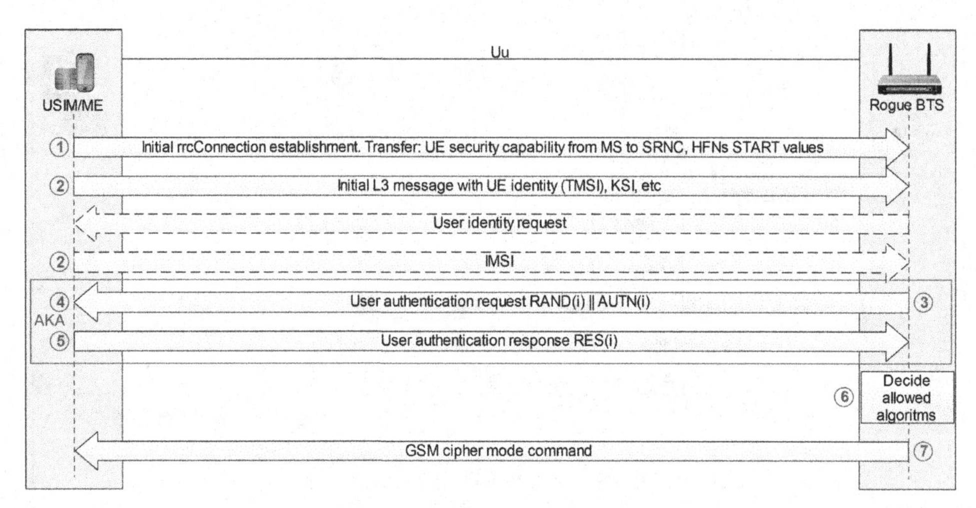

ATTACKS AGAINST UMTS/HSPA DEVICE WITH GSM/GPRS/EDGE CAPACITY

The attacks developed in the chapter are:

- The active user identity catching.
- Denial of Service (DoS).
- Suppression of encryption between target and intruder.
- Man-in-the-Middle.

These attacks use a modified base station against a 3G UMTS device with 2G/2.5G capacity. The implementation of these attacks aims to prove how take advantage of some UMTS system debilities, as well as study of real difficulties of carry out.

Theoretical Approach

For the performance of these attacks it is necessary a modified base station GSM/GPRS to impersonate a BTS of a real cell. The attacks are carry out against an UMTS device that is capable of connecting to the GSM/GPRS/EDGE network and supports GEA/0 encryption (GPRS Encryption Algorithm). In the first instance would be expected that the attack would be performed with a false UMTS station on 3G network, but the reality is that for such attack would have succeed the cryptography of the authentication process should break.

These attacks in the current approach exploit a logic weakness that many mobile devices have due the option to automatically connect to 3G or 2G network depending on the availability of both. On many devices the default option is configured to accept both networks.

The UE target should lose the connectivity with its real network and connect with the rogue BTS. In order to do the outage the attacker can use two ways:

- Using a "jammer".
- Transmitting the signal BTS with more power that the real one.

Doing so, the GSM weakness of not using bidirectional authentication but unidirectional is taken in advantage. The device have to authenticate itself to the network through its IMSI to show that its SIM card is valid as a subscriber, but the base station does not have to authenticate itself. Another weakness is that the base station in the process may decide to use "no encryption", and the device has to accept it, because it is obliged to accept the type of encryption specified for the base station.

In the first case with the "jammer" or frequencies inhibitor, the UMTS frequencies are interfered, performing a denial of service attack (Denial of Service Attack (DoS)). Unable to connect with the 3G network, the 3G device will connect the station to the false GSM/GPRS BTS. In the second case the BTS may be configured to transmit with the power enough in order to the target disengage of its PLMN and connect to false BTS. The BTS would have to approach enough to victim for having good signal.

Once the connection to the network is established it is possible capture its traffic. The weakness of the network, in this case, and of the UMTS/HSPA device, is that the standard requires to the third generation be compatible with earlier technologies 2G and 2.5G GSM/GPRS/EDGE, and in doing so inherits some security weakness.

These attacks do not work against devices that only support connections in UMTS/HSPA.

Practical Approach

The attacks are raised in a scenario where the goal is an UE (or MS) for several reasons:

- Capture the user IMSI.
- Make the target device lose the connection with his PLMN.
- Establish a connection with the false BTS.
- Capture the traffic of the UE.

The ability of configuring the system is important to perform the attack and successfully make the connection between the UE victim and the false BTS.

The 3G device can connect in normal operation to a PLMN through the Base Station (BS) or node-B that provides coverage in the cell in which it is operating.

The two options for radio access are:

- WCDMA, through the UMTS Terrestrial Radio Access Network (UTRAN).
- GSM/EDGE, through GSM/EDGE Radio Access Network (GERAN).

During the attack, the access to the UTRAN is interfered and the device is forced to connect via the GERAN.

Um is the air interface in GSM and GPRS, and the A-bis is the interface that connects the Base Transceiver Station (BTS) and Base Station Controller (BSC). BSC is connected to the Transcoder Rate Adapter Unit (TRAU).

Of particular importance is the Gb interface, interface that enable the connection of a BSC belonging to a network GERAN (GSM/EDGE Radio Access Network) with the SGSN (Serving GPRS Support Node) of the GPRS network core, or with the SGSN of the UMTS network core. In Figure 9 a complete architecture is represented with the elements of the network.

The victim mobile terminal, or MS, will be previously connected to a PLMN determined, but using a jammer, or jammer of frequencies, there will be implemented a denial of service attack. After the MS attack, the MS will lose its 3G connectivity and will attempt to connect through the GSM/GPRS GERAN. How-

Figure 9. UMTS and GERAN network architecture of the Man in the Middle attack

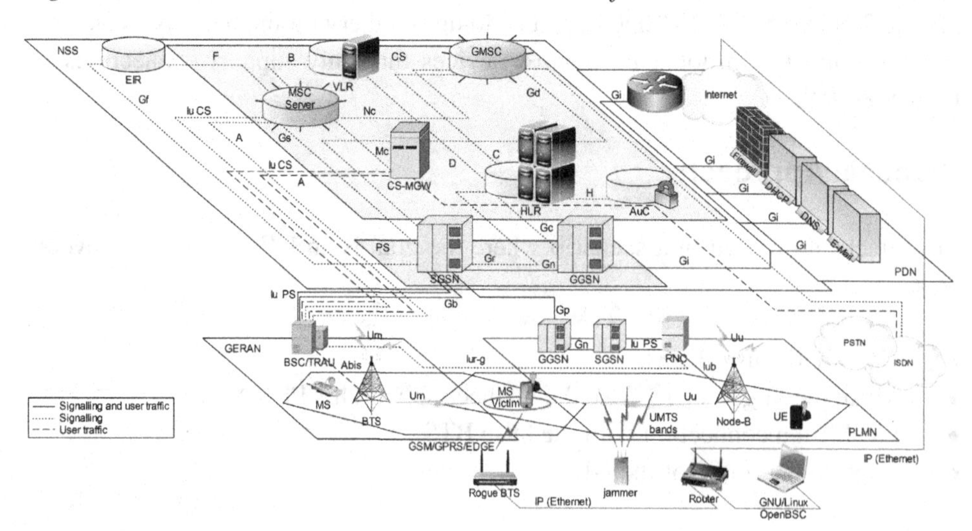

ever, being the false BTS located at a suitable distance and transmitting with more power than the actual BTS, the MS will connect to the attacker BTS.

Material Resources

Mobile Station

The victim Mobile Station (MS) or User Equipment (UE) should be a mobile device with 3G GSM/GPRS capacity and GEA0 encryption algorithm.

Base Transceiver Station

The attacks will force to the MS connect through the GSM/GPRS connectivity and use the GEA/0 encryption, therefore, the used BTS must be capable of GSM/GPRS(/EDGE) and *A-bis* interface IP. This is the main element of this kind of attacks.

As there is no GPRS protocol stack developed in open source for levels 1 and 2 of the Um interface, it should be used a commercial BTS with support for GPRS. Thereby, the BTS can be controlled through the *A-bis* interface from an open source solution as OpenBSC.

Jammer

The jammer is a device that transmits signal in the same frequency band in which the jammer attempts to interfere another device. The jammer must have enough signal power to interfere successfully in the desired band. In the case of mobile communications the interference is successful when the mobile phone cannot communicate in the area where the jammer is situated.

To achieve the MITM attack, the bands of UMTS/HSPA at the location of the victim MS must be interfered, and keep the GSM/GPRS/EDGE bands unchanged.

The jammer could be not necessary if the BTS transmits with enough power, or being close, as to make that the MS lose its connectivity.

BTS Software

Depending on the BTS, it would be necessary to management the BTS control and computer software. There are options of open source software, like the OpenBSC project of Osmocom. For using this kind of software, the BTS must be checked for guarantee the compatibility because OpenBSC gives service to a limited number or BTSs.

OpenBSC is the project of Osmocom to create free software for GSM/3GPP protocol stack and elements. It implements the minimal parts to build a small and self-contained GSM network. It Includes functionalities for the GSM network components, Base Station Controller (BSC), Mobile Switching Center (MSC), Home Location Register (HLR), Authentication Center (AuC), Visitor Location Register (VLR) and Equipment Identity Register (EIR) (Osmocom, 2014).

From its website, it is possible accessing to the configuration software of diverse BTSs, configuration libraries and additional information. Depending on the chosen BTS, must be accessed to the appropriate software.

Controlling properly the BTS software is essential for carry on the connection between any MS and the BTS. Some BTSs use an embedded Linux and on which is installed the OpenBSC. According with the Osmocom project a compatible BTS can run in three different modes:

- **OpenBSC Network-in-the-Box (NITB) Mode:** This mode uses the executable program osmo-nitb.
- **OpenBSC with a ISDN or SIP PBX:** This mode uses the executable program osmo-nitb.
- **OpenBSC in BSC-Only-Mode:** This mode uses the executable program osmo-bsc.

The second mode allows to route calls outside the OpenBSC network, although the basic configuration of NITB over the BTS is enough for the basic attacks as identity catching, bidding down or camping on a false base station

The osmo-nitb program runs in the first and second mode, and can be running inside the embedded Linux of the BTS or inside a Linux installed in a computer. Operating with the osmo-nitb in a computer involve that the computer will take the control of the BTS, and this computer must be working in the same BTS network for being able to connect each other. If the BTS is working with the NITB mode in the computer another program called osmobts-trx will run together with the program osmo-nitb.

If it is necessary run a bridge between the SIP (*Session Initiation Protocol*) and GSM interfaces the LCR (Linux Call Routing) paquet have to be installed in the same computer than NITB to install the software for the open source bridge ISDN (DSS1)/SIP/GSM (MNCC protocol) for running the LCR as this bridge.

To work with the GPRS service, the BTS must be configured for that, and if the BTS is working with the mode NITB over a computer, the following software must be installed:

- osmoSGSN.
- openGGSN.

Figure 10. BTS architecture connection
(Osmocom, 2014).

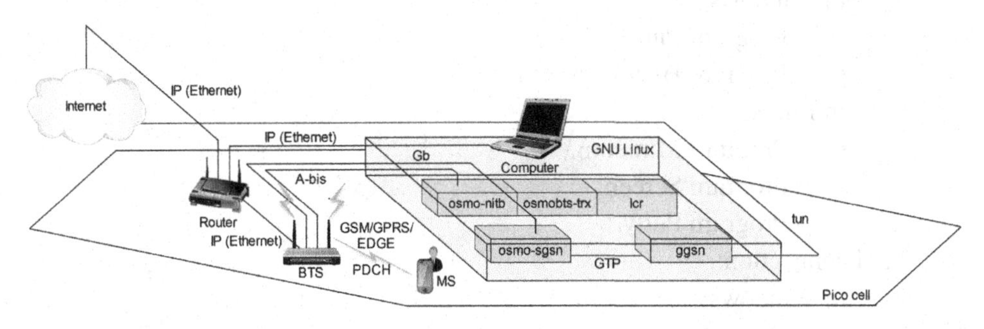

OsmoSGSN is part of OpenBSC git (software version control) repository, and establishes the implementation of free software of the Serving GPRS Support Node (SGSN), GPRS Mobility Management (GMM) and Session Management (SM) (Osmocom, 2011).

The SGSN is connected to the BSS (in this case the chosen BTS) via the Gb interface, and also connects to the Gateway GPRS Support Node (GGSN) via the GTP protocol.

The OpenGGSN is an implementation of the GPRS Gateway Support Node (GGSN), element of the GPRS core network used by mobile operators as an interface between the Internet and the rest of the mobile network infrastructure. In Figure 10 the BTS modes are shown graphically and how connect to the BTS and a Router.

Computer

In case of using a computer, this one can be a personal computer or laptop with the appropriate software.

The connection of computer and BTS is done for two reasons:

- Configure the BTS parameters configuration from the computer.
- Run de BTS NITB mode from the computer.

Although the BTS is running with its NITB mode, the computer will be necessary for able to configure the parameters of the mode, due to the BTS probably will not have a graphic display. In this first case of connection an Ethernet IP connection could be not necessary if the the BTS allows a USB connection, and a Windows operative system is enough to make the configuration. Depending if the connection is through USB cable or SSH (Secure Shell) over Ethernet, the computer software can be different.

- USB cable connection:
  - In Windows:
    - Program Putty.
    - Program Hyperterm.
  - In Linux:
    - Program Minicom.
    - Program Screen.
    - Program Putty.
- SSH connection:
  - In Windows:
    - Program Putty.
    - Program Hyperterm.
  - In Linux:
    - Program SSH.
    - Program Putty.

In the second case a Linux must be installed into the computer for run the programs of OpenBSC. The Linux can operate as native operative system or be installed over a virtual machine. Once all the programs and software have been installed, they must be started in order:

1. osmo-nitb.
2. osmobts-trx.
3. lcr (optionally).

Development of Man-In-The-Middle Attack

The attacker can develop the MITM attack in a few steps, trying to intercept the mobile data communication of a victim user. The process has been developed in depth in works like Picó & Pérez (2011):

1. The frequency inhibitor (jammer) is connected and then interferes with the frequency bands used in that location. The victim MS will lose 3G connectivity and will drop in the 2G connection. This point is optional due to the possibility of interrupt de connection only with the BTS transmitting with more power.
2. The attacker base station is located at a suitable distance and listens on the frequency band. The frequency to broadcast should be adequate, as the indicated by the bearer of a cell in a Public Land Mobile Network (PLMN) Real. If the frequency is already being used by a PLMN in the area, just the rogue base station emits more power to that location.

3. Once the frequency of the false BTS is found out, this one is configured to emit at that frequency. It is also necessary configure the BTS to impersonate the real PLMN. A PLMN is uniquely identified by two fields:
   a. The Mobile Country Code (MCC) indicating the country in which the PLMN is located.
   b. The Mobile Network Code (MNC), which is used to identify different networks in the same country. Its attribution is the responsibility of the appropriate regulatory authority in each country.

The BTS must be setting to accept mobile connection from attacked mobile, therefore the International Mobile Subscriber Identity (IMSI) must be known or the International Mobile Equipment Identity (IMEI), or being in a BTS state of open acceptance for any MS.One way to know the IMSI is having done a previous attack called "Attack IMSI disclosure" or having had access to the mobile terminal.Another way to discover the IMSI/IMEI is recording connection attempts of the Mobil Station (MS) in different locations when it is known that the target is in place. After several implementations, the IMSI that appears in all locations should correspond with the target MS.

4. The laptop connects to internet, using a 2G/3G connection, to a real PLMN, to a DSL connection, or to a cable connection.
5. In Linux, the OsmoSGSN and OpenGGSN programs are configured, and the routing tables. Thus the traffic of the victim may be forwarded to Internet, and the response traffic redirected toward the victim.
6. If it is desired only to establish an IP connection to the attacker and do not want to redirect traffic to internet, this step would not be necessary, as if is enough run the systems with the NITB system alone in the BTS to for example getting the IMSI.

The BTS is connected in the coverage area of the victim. If everything worked correctly, the MS victim would leave the connection to his PLMN and would connect to the BTS attacker. Because the MS should connect to the BTS with more signal, the attacker BTS must exceed the power level of the actual PLMN. When the victim is in the connection process to the BTS attacker, the BTS will indicate the GEA/0 encryption algorithm to MS, which means that the information is not encrypted.

7. When the MS is properly connected, the false BTS will have the control of communications, being able to read, modify, or redirect the IP packets sent or received during the communication between the MS victim and the GPRS/EDGE connection.

May be there will be necessary to install specific programs on the computer as Ettercap, Wireshark, Nmap and Tcpxtract. These programs are used in the Linux operating system for Man-In-The-Middle attacks in Local Area Networks (LANs).

CONCLUSION

UMTS architecture performs security features and advantages and provides improvements as mutual authentication, freshness and liveliness assurance of AKA, sufficient and suitable Integrity Key (IK) and Cypher Key (CK) sizes (128 bits) and data integrity of signaling messages in radio interface.

Part of the communications is over the radio interface (Uu) and is a shared medium between the users and the users and network, so it is necessary to use cryptographic methods for ensuring authentication, integrity and confidentiality. Due to this, one of the most important problems the UMTS system is the message exchange in the set-up-procedure, because the first messages are transmitted not encrypted. For that reason, they are target for different kind of attacks, mainly against the parameters and security capabilities of an UE.

Even the UMTS is not completely secure, taking advantages of the insecurities is not always easy. The core of the security methods, Milenage algorithm for Authentication and Key Agreement, and Kasumi for encryption and integrity, are not broken. The attacks attempts to break the security from the non-encrypted messages, but an attacker does not decipher the encrypted messages.

Carrying on the attacks is not easy neither. For performing the attacks it is necessary technical and special acknowledge, and the special and modified devices necessary for attempting the attacks may be very expensive. For example, a false base station, necessary for the Man-in-the-Middle attack, may cost thousands of dollars or euros. Preparing an attack is not available to everyone.

REFERENCES

Al-Fayoumi, M., & Al-Saraireh, J. (2011). An Enhancement of Authentication Protocol and Key Agreement (AKA) For 3G Mobile Networks. *International Journal of Security*, *5*(1), 35–51.

Cox, C. (2008). System architecture. In *Essentials of UMTS* (p. 37). New York, NY: Cambridge University Press. doi:10.1017/CBO9780511536731

ETSI 3rd Generation Partnership Project. (2001). Security threats. In *3G security; Security threats and requirements* (Technical Specification TS 21.133 version 4.1.0 Release 4). Retrieved from http://www.etsi.org/deliver/etsi_ts/121100_121199/121133/04.01.00_60/ts_121133v040100p.pdf

ETSI 3rd Generation Partnership Project. (2013). *Non-Access-Stratum (NAS) functions related to Mobile Station (MS) in idle mode* (Technical Specification TS 23.122 version 11.4.0 Release 11). Retrieved from http://www.etsi.org/deliver/etsi_ts/123100_123199/123122/11.04.00_60/

Kaaranen, H., Ahtiainen, A., Laitinen, L., Naghian, S., & Niemi, V. (2005). Base Station (BS, Node B). In *UMTS Networks Architecture, Mobility and Services* (2nd ed.; pp. 101–103). Chichester, UK: John Wiley & Sons. doi:10.1002/047001105X

Kaaranen, H., Ahtiainen, A., Laitinen, L., Naghian, S., & Niemi, V. (2005). Radio Network Controller (RNC). In *UMTS Networks Architecture, Mobility and Services* (2nd ed.; pp. 110–111). Chichester, UK: John Wiley & Sons. doi:10.1002/047001105X

Kaaranen, H., Ahtiainen, A., Laitinen, L., Naghian, S., & Niemi, V. (2005). UMTS Core Network Architecture. In *UMTS Networks Architecture, Mobility and Services* (2nd ed.; pp. 146–152). Chichester, UK: John Wiley & Sons. doi:10.1002/047001105X.ch6

Kaaranen, H., Ahtiainen, A., Laitinen, L., Naghian, S., & Niemi, V. (2005). Base Station (BS, Node B). In *UMTS Networks Architecture, Mobility and Services* (2nd ed.; pp. 180–181). Chichester, UK: John Wiley & Sons. doi:10.1002/047001105X

Khan, M., Ahmed, A., & Raza Cheema, A. (2008). Vulnerabilities of UMTS Access Domain Security. In *Proceedings of Architecture, Ninth ACIS International Conference on Software Engineering, Artificial Intelligence, Networking, and Parallel/Distributed Computing*, (pp. 350-355). doi:10.1109/SNPD.2008.78

Meyer, U., & Wezel, S. (2004, October 1). *A Man-in-the-Middle Attack on UMTS*. ACM.

Niemi, V., & Nyberg, K. (2003). *Set-up of UTRAN security mechanisms*. Chichester, UK: John Wiley & Sons.

Osmocom. (2011). *OsmoSGSN*. Retrieved from http://openbsc.osmocom.org/trac/wiki/osmo-sgsn

Osmocom. (2014). *OpenBSC GPRS/EDGE Setup page*. Retrieved from http://openbsc.osmocom.org/trac/wiki/OpenBSC_GPRS

Osmocom. (2014). *Welcome to Osmocom OpenBSC*. Retrieved from http://openbsc.osmocom.org/trac/

Picó, J., & Pérez, D. (2011). *A practical attack against GPRS/EDGE/UMTS/HSPA mobile data*. Black Hat. Retrieved from https://media.blackhat.com/bh-dc-11/Perez-Pico/BlackHat_DC_2011_Perez-Pico_Mobile_Attacks-wp.pdf

Poole, I. (n.d). UMTS Core Network. In *UMTS/WCDMA Network Architecture*. Retrieved from http://www.radio-electronics.com/info/cellulartelecomms/umts/umts-wcdma-network-architecture.php

3. rd Generation Partnership Project. (1999). *3G Security; Security architecture* (Technical Specification TS 33.102 version 2.0.0 Release 99). Retrieved from www.3gpp.org/ftp/Specs/archive/33_series/33.102/33102-200.zip

3. rd Generation Partnership Project. (2000). Counteracting envisaged 3G attacks. In *A Guide to 3rd Generation Security* (Technical Specification TR 33.900 version 1.2.0 Release 99). Retrieved from www.3gpp.org/ftp/Specs/archive/33_series/33.900/33900-120.zip

3. rd Generation Partnership Project. (2000). Denial of Service. In *A Guide to 3rd Generation Security* (Technical Specification TR 33.900 version 1.2.0 Release 99). Retrieved from www.3gpp.org/ftp/Specs/archive/33_series/33.900/33900-120.zip

3. rd Generation Partnership Project. (2014). Access link data integrity. In *3G Security; Security architecture* (Technical Specification TS 33.102 version 12.0.0 Release 12). Retrieved from www.3gpp.org/ftp/Specs/archive/33_series/33.102/33102-c00.zip

3. rd Generation Partnership Project. (2014). Cipher key and integrity key lifetime. In *3G Security; Security architecture* (Technical Specification TS 33.102 version 12.0.0 Release 12). Retrieved from www.3gpp.org/ftp/Specs/archive/33_series/33.102/33102-c00.zip

3. rd Generation Partnership Project. (2014). Cipher key and integrity key setting. In *3G Security; Security architecture* (Technical Specification TS 33.102 version 12.0.0 Release 12). Retrieved from www.3gpp.org/ftp/Specs/archive/33_series/33.102/33102-c00.zip

3. rd Generation Partnership Project. (2014). Ciphering and integrity mode negotiation. In *3G Security; Security architecture* (Technical Specification TS 33.102 version 12.0.0 Release 12). Retrieved from www.3gpp.org/ftp/Specs/archive/33_series/33.102/33102-c00.zip

3. rd Generation Partnership Project. (2014). Security mode set-up procedure. In *3G Security; Security architecture* (Technical Specification TS 33.102 version 12.0.0 Release 12). Retrieved from www.3gpp.org/ftp/Specs/archive/33_series/33.102/33102-c00.zip

3. rd Generation Partnership Project. (2014). The Core Network (CN) entities. In *Network architecture*. (Technical Specification 23.002 version 12.4.0 Release 12). Retrieved from www.3gpp.org/ftp/Specs/archive/23_series/23.002/23002-c40.zip

3. rd Generation Partnership Project. (2014). USIM. In *3G Security; Security architecture* (Technical Specification TS 33.102 version 12.0.0 Release 12). Retrieved from www.3gpp.org/ftp/Specs/archive/33_series/33.102/33102-c00.zip

Sanchez, J., & Thioune, M. (2007). PLMN selection. In *UMTS* (pp. 188–189). London, UK: ISTE Ltd.

Wagner, M., Wagner, J., & Zucchini, W. (2010). *Authentication - AKA. In 3G Performance and Security, Evolution towards UMTS Network and Security Mechanism* (pp. 88–90). Saarbrücken, Germany: VDM Verlag Dr Müller Aktiengesellschaft & Co. KG.

KEY TERMS AND DEFINITIONS

Bandwidth: Frequency range available for a signal, and corresponds where is located most of the signal power.

Channel: Physical medium where the information and signal are transmitted between transmitters and receivers.

Coverage: Geographical area where a mobile device has connectivity with its operator and the services.

Database: Set of stored information for their posterior use.

Gateway: Device used as interface for interconnection and communication of networks that use different protocols and architectures.

Roaming: Capacity of a mobile device to change from a cell to another without lose connectivity with the mobile operator.

Token: Set of characters, identifiers, numbers, signs or bits.

Chapter 2
Trustworthy Architecture for Wireless Body Sensor Network

G. R. Kanagachidambaresan
Dhanalakshmi Srinivasan College of Engineering, India

ABSTRACT

Wireless Body Sensor Network is a collection of physiological sensors connected to small embedded machines and transceivers to form a monitoring scheme for patients and elderly people. Intrusion and foolproof routing has become mandatory as the Wireless Body Sensor Network has extended its working range. Trust in Wireless Body Sensor Network is greatly determined by the Encryption key size and Energy of the Node. The Sensor Nodes in Wireless Body Sensor Network is powered by small battery banks which are to be removed and recharged often in some cases. Attack to the implanted node in Wireless Body Sensor Network could harm the patient. Finite State Machine helps in realizing the Trust architecture of the Wireless Body Sensor Network. Markov model helps in predicting the state transition from one state to other. This chapter proposes a Trustworthy architecture for creating a trusted and confidential communication for Wireless Body Sensor Network.

1. INTRODUCTION

Design of Trusted network has become mandatory for Wireless Body sensor Network since its role is prodigious in Health monitoring system. Wireless Networks hold the key for unlocking 24 × 7 monitoring of patients in and out of hospital environment. Physiological signals of the patients are monitored across the clock

DOI: 10.4018/978-1-4666-8687-8.ch002

using sensors sticked with the body of the subject (Kanagachidambaresan, SarmaDhulipala, Vanusha, & Udhaya, 2011; Akyildiz, Sankarasubramaniam, & Cayirici, 2002; Kanagachidambaresan, Chitra, 2014; Otal, Alonso, & Verikoukis, 2009; Kanagachidambaresan, SarmaDhulipala, & Udhaya, 2011). Wireless Body Sensor Network is mainly used for two major e-health application scenarios one for monitoring and collecting health data of the subject and delivering this data to the remote medical centre. Second major application is automatic treatment by the cooperation of various biosensor nodes with the help of actuators. Wireless Body Sensor Network helps the subject from Asthma to Cancer monitoring and has very large application. A sensor node could be placed to monitor nitric oxide emitted by cancer cell to monitor the progress of cancer in the human body. A Wireless Body Sensor Network could help the asthma patients by sensing the allergic agents in the air and reporting the patient himself and doctor continuously avoiding him from breathing trouble. Fatal conditions due to belated medical facility can be majorly avoided using Wireless Body Sensor Network. The main motto of the Wireless Body Sensor Network is to enhance the patients mobility without making them immobile. Wireless Body Sensor Network helps in monitoring patients continuously without disturbing their day to day life (SarmaDhulipala, Kanagachidambaresan, Chandrasekaran, 2012; Riaz, et al., 2009; Zhang, Das, & Liu, 2006; Momani & Alhmouz, 2008; Boukerche, Li, & Khatib, 2006). Wireless Body Sensor Network facilitates the patients to be monitored out of hospital environment, making the network facile to attackers. For example McAfee experts demonstrated an attack to the insulin pump causing a fatal dosage of insulin in Black Hat conference in 2012. Health Insurance Portability and Accountability Act (HIPAA) mandates the e-health data to be secured and routed through trusted nodes. BSN mandates a valuable trust to the system before being practiced in real time. Trust for these miniaturized embedded systems should also convince with the limited available resource. The Nodes in the Wireless Body Sensor Network are wearable, sticked and implanted in nature. The first implantable heart pacemaker was designed by 1958, In spite to the advancement to the technology of manufacturing of Implantable Sensor Nodes faces a series of challenges varying with person to person and environment to environment (Marsh, 1994; Hoffman, Lawson, & Blum, 2006, Ng, Sim, & Tan, 2006; Pirzada & McDonald, 2004). Rechargeable batteries in the implanted nodes are charged by the radio frequency, ultrasonic, infrared light, low-frequency magnetic field and so on. Recent technology introduces the energy harvesting mechanism with body motions and bio-heat generation (Sun, Yu, Han, & Liu, 2006; Shaikh, et al., 2006; Momani, Challa, & Abour, 2007; Liu, Joy, & Thompson, 2004; Gradison & Sloman, 2000; Shi & Perrig, 2004). Future design of implanted nodes concentrates battery less node design directly harvesting energy and serving the need of the Wireless Body Sensor Nodes. The Trust of the nodes in these cases mainly depends on the availability to harvest energy. Apart from sensing mechanism the implanted nodes in Wireless Body Sensor Network concentrates in drug release, mechanical adjustment in prosthetic devices which consumes a major part of energy from the battery. The Telemetric link supports the bidirectional usage of data transfer from and to the implanted sensor devices. The sensed data is recorded inside the small memory unit of the implanted node. The telemetric link between the implanted node and the sink enables Wireless Programming facility enabling over the air programming facility. The wireless programming facility ease the hackers to reprogram to implanted node and malfunction its usage (Walters, Liang, Shi, & Chaudhary, 2006; Buchegger & Boudec, 2005; Heinzelmean, Chandrakasan, & Balakrishnan, 2002; Olariu, et al., 2005; Misra & Xue, 2006; Cherukuri, Venkatasubramanian, & Gupta, 2003).

Wireless Body Sensor Network constitutes have its working by two major types of sensor, Wearable, Implanted. Apart from sensing mechanism, Wireless Body Sensor Network also constitutes to have actuators which help the subjects under extreme situations. Wearable sensors are sticked connected to the external body surface of the subject. The Electrical pulses generated by the neurons of the body is sensed by the electrode. The physiological signals are sensed by the sticky electrode and transmitted to the CMU (Ruzzelli, Juradak, Ohare, & Van Der Stok, 2007; Kim, et al., 2009; Maskooski, Soh, Gunawan, &Low, 2011; Sayrafian, et al., 2009; Liang, Balasingham, & Byun, 2008; Djenouri &Balasingham, 2009). Figure 1 illustrates the various Wearable sensors attached with the body surface.

Figure 1. Wearable body sensors, a) ECG b) EEG sensors

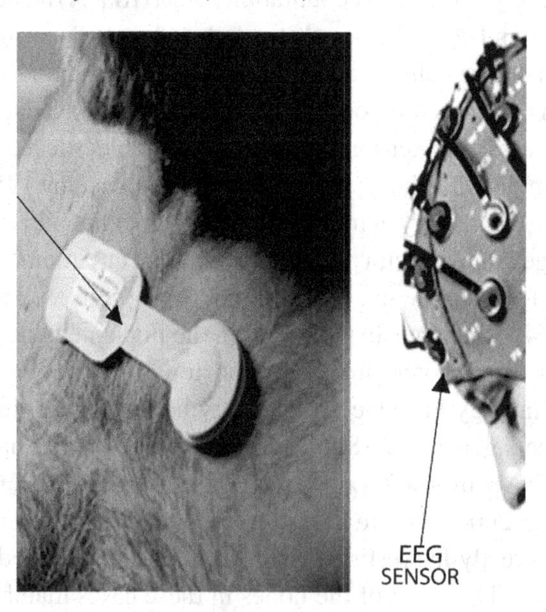

Figure 2. Pacemaker implanted to human body

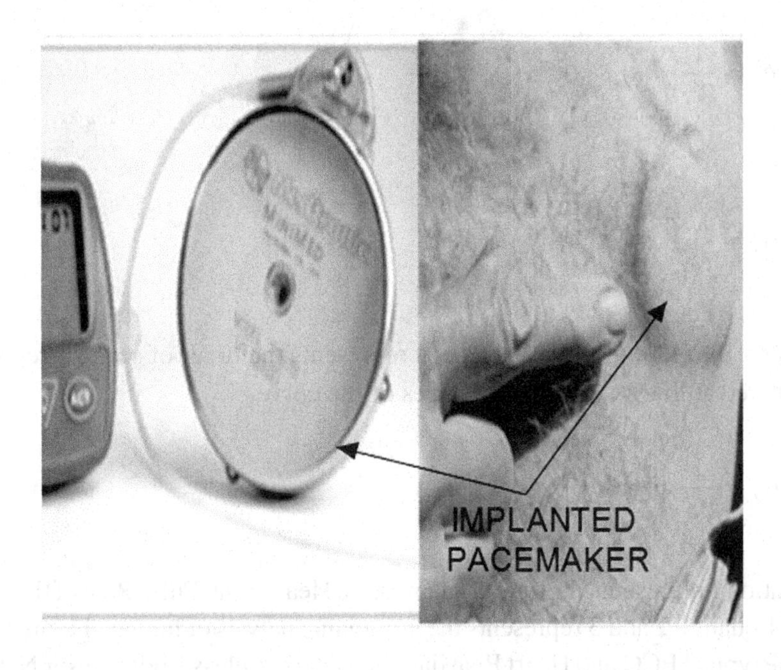

Electro Gel reduces the body resistivity making them easy to conduct the bio signals without incision. Wearable sensors are connected to processing unit, capable of processing bio signal consuming very less energy. Figure 2 represents the pacemaker and pacemaker implanted inside the human body.

Pacemaker is equipped with non-rechargeable battery which is to be replaced within a particular duration (Razzawue, Hong, & Lee, 2011; Tang, Tummala, Gupta, & Schwiebert, 2005; Bag & Bassiouni, 2006; Takahashi, Xiao, & Hu, 2007; Bazaka & Mohan, 2013; Bag & Bassiouni, 2007).

1.1. Health Data

The data from the Wearable and Implanted nodes is termed as Health Data. The physiological value includes Blood pressure (Systolic and Diastolic), Electrocardiograph (ECG), blood oxygen level (SpO_2), activity recognition (Kanagachidambaresan, SharmaDhulipala, Vanusha, & Udhaya, 2011; Kanagachidambaresan & Chitra, 2014; Kanagachidambaresan, SarmaDhulipala, & Udhaya, 2011; Ullah, et al., 2012; Hirata & Shiozawa, 2003; Havenith, 2001; Natarajan, de Silva, Kok, & Motani, 2009).

Time Domain Analysis of HB and PR.

$$HB_{\psi}(s,a) = \frac{1}{2\alpha}\left[S(s).\psi(s)\right] \quad (1)$$

$$HB_{\psi}(t,\alpha) = \frac{1}{a}\int_{-\infty}^{\infty} s(t)\psi\left(t - t_{\alpha}\right)dt \quad (2)$$

where $\psi(t)$ is width of order of scale, a represents the order of scale, t_{α} represents the Centered at line s(t) is Time Services of the curve.

$$PR_{\psi}(s,\alpha) = \frac{1}{2\alpha}[S(s)\psi(s)] \quad (3)$$

Equation 1 represents the time domain based Heart Beat, Pulse Rate of the typical human. Equation 2 and 3 represents the s-domain analysis of human. Figure 3 illustrates the typical ECG and Heart Beat data from the Wireless Body Sensor Network.

Figure 3. ECG signal and heart beat signal

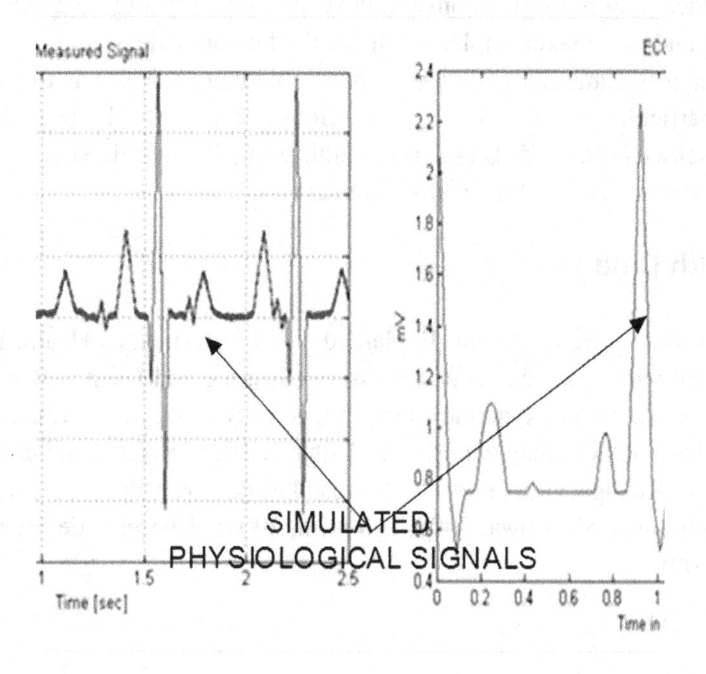

ECG and Pulse Rate signal of the patient data from the sensor is processed using Data Processing Unit (DPU).

Table 1 represents the physiological parameter specification used in Wireless Body Sensor Network.

The challenges in Wireless Body Sensor Network is not like conventional Wireless Networks. The critical challenges faced in Wireless Body Sensor Network is imposed by the architecture, data rate, density of the node, Heat dissipation of the node, size of the node, latency and mobility issues of the patient. The type of data from the sensor node also influences the Security Trust (Braem, Latre, Moerman, Blondia, & Demeester, 2006; Latre, et al., 2007; Quwaider & Biswas, 2009; Clupepper, Dung, & Moh, 2003; Gosalia, Weiland, Humayun, & Lazzi, 2004; Moneda, Ioannidou, & Chrissoulidis, 2003; Gao, et al., 2005).

Table 2 describes the type of sensing system and its monitoring nature of the Wireless Body Sensor Network.

Table 1. Physiological parameters specification

Physiological Parameter Specifications	
Electrocardiogram (ECG)	Frequency: 0.5Hz – 100 Hz Amplitude: 0.25 – 1mV
Electromyogram (EMG)	Frequency: 10Hz - 3KHz Amplitude: 50μV – 1mV
Electroencephalogram (EEG)	Frequency: 0.5Hz - 100Hz Amplitude: 1μV – 100μV
Blood Pressure (BP)	Systolic: 60 - 200mmHg Diastolic: 50 – 110mmHg
Body Temperature	32°C – 40°C
Galvanic Skin Response (GSR)	0 – 100 KΩ
Respiratory Rate (RR)	2 – 50 breaths/min Frequency 0.1 – 10Hz
Oxygen Saturation in Blood (SaO2)	0-100%
Heart Rate (HR)	40 – 220 Beats per minute

Table 2. Sensing system specification

Sensing System	Continuous/Intermittent
Organ monitoring	Continuous
Cancerous Cell monitoring	Continuous
Sugar level	Intermittent
Health condition	Intermittent

Trusted communication to the patients in real time and efficient routing mechanism is to be addressed in Wireless Body Sensor Network. Trust of the network varies with its location (i.e.) Trust worthiness varies when the patient is monitored indoor and outdoor environment. Trust worthiness of the patients data is not to be sacrificed with respect to his mobility. Increase in resource would increase the size of the nodes making it unsuitable for implanted and to be sticked in the subject (Otta, Jovanov, & Milenkovic, 2006; Anlinker, et al., 2004; Srinivasan, Stankovic, & Whitehouse, 2008; Casas, et al., 2008; Jasemian, 2008; Atallah, Lo, Yang, & Siegemund, 2008; Hori & Nishida, 2000; Zhang, Poon, Li, & Zhang, 2009; Bao, Zhang, & Shen, 2005). The Nodes should exhibits it Trust without sacrificing its resource like size of processor, memory and battery size. Security to data is mainly determined by encryption key size directly determined by the memory and processors capability, increase in capability of processor increases the other constraints sacrificing energy and size oriented constraints. Figure 4 represents the Sensor Node of a typical Wireless Body Sensor Network.

The physiological sensing unit senses the signals from our body. The data from the body is processed by the processing unit and transmitted by Transceiving unit (Pereira, et al., 2007; Misic, 2008; Renaud, et al., 2008; Lauterbach, Strasser, Jung, & Weber, 2002; Marinkovic & Popovici, 2009; Schoellhammer, et al., 2004; Sadler & Martonosi, 2006; Marcelloni & Vecchio, 2010).

Figure 4. Sensor node architecture

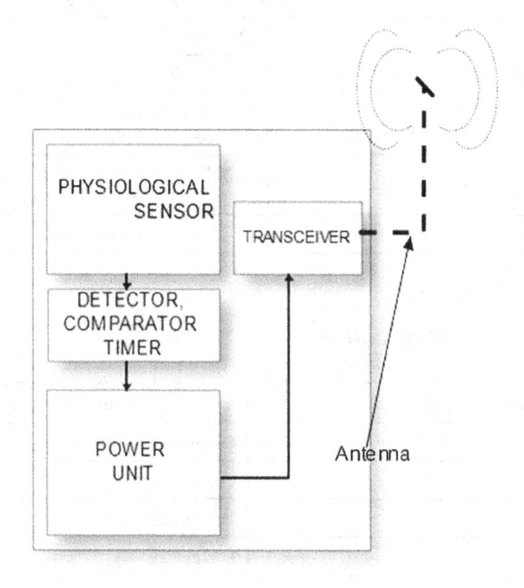

Figure 5. Wireless Body Sensor Network scenario

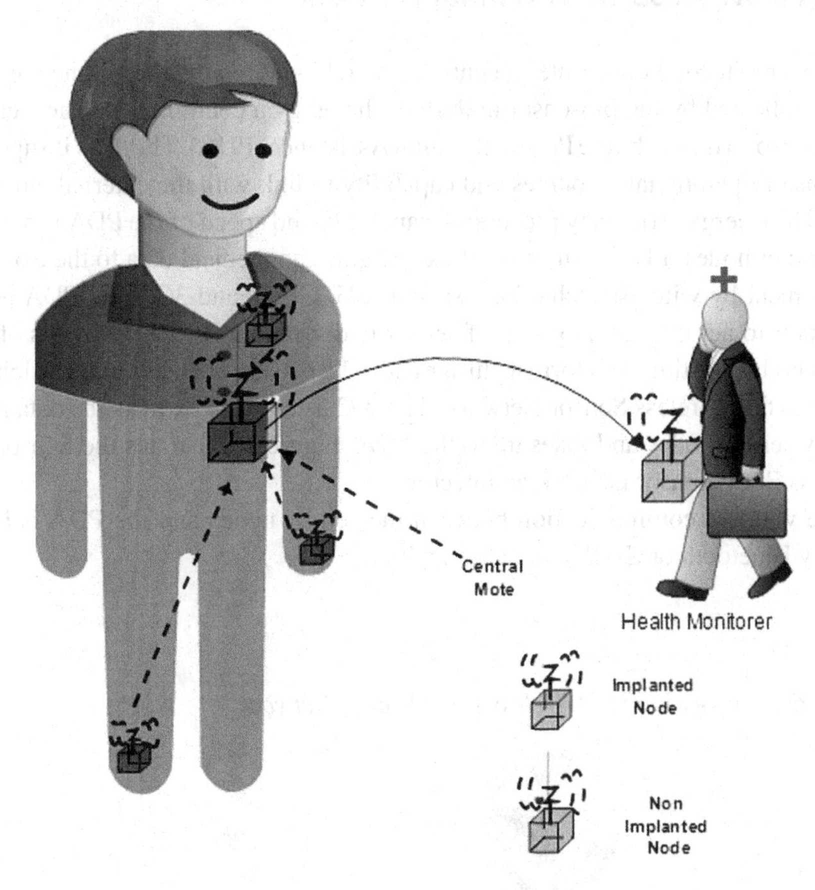

Figure 5 represents the Classical Wireless Body Sensor Network, Implanted and Non Implanted nodes together sense the physiological signal and transmitting wirelessly to the sink node.

1.2. Sensor Deployment

Sensor properties, size sink placement, battery specification influences the location of the sensor node to be deployed. In some cases the parameters to be sensed also influences the sensor node placement based on the availability of the physiological signals. As these factors influences the bio heat, compatibility, mobility of the user the deployment is to be done carefully to the subject.

1.3. Star Wireless Body Sensor Network

The star architecture constitutes a centralized architecture at which all the medical data is collected by the biosensor nodes is collated by a central node. The Central Node is also termed to be the Personal Digital Assistance (PDA). The PDA is superior in terms of operational resources and capability to link with the external environment. The Energy efficiency processing capability and speed of the PDA is n times to the other nodes. PDA is the sole of exchanging the medical data to the external environment by wireless technology such as 2G, GPRS and 3G. This PDA helps the doctor to act in time. A group of sensor nodes monitoring the motions of the human body constitutes to form a cluster and selects Cluster Heads and participate similar to the Wireless Sensor Network. These Cluster Head extracts the data from the tiny sensor nodes and fuses up to the PDA. Figure 6 illustrates the Star based Wireless Body Sensor network architecture.

The wireless communication between the sensor nodes and the PDA is basically by Bluetooth or ZigBee in nature.

Figure 6. Star topology of Wireless Body Sensor Network

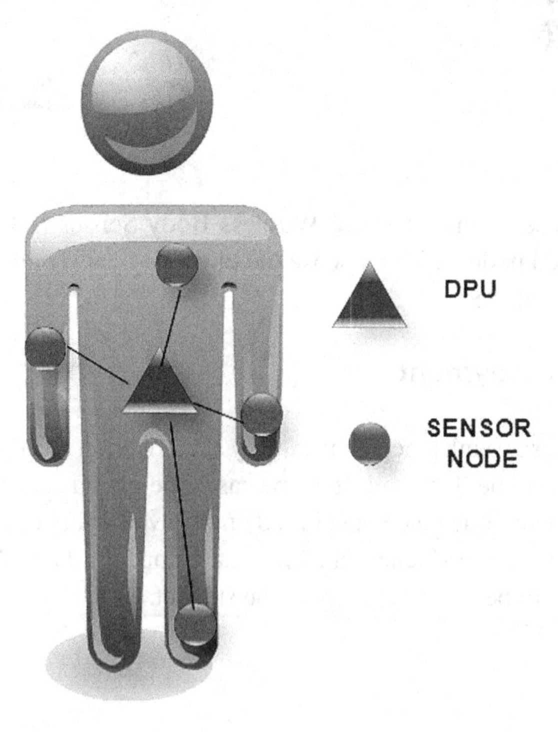

1.4. Mesh Wireless Body Sensor Network

The mesh architecture refers to the peer to peer based distributed network without having a centralized master participant. Here the biosensor nodes communicate with each other, capable of executing intelligent treatment. Mesh based architecture basically suits automated drug delivery system and other actuator type medications. Figure 7 illustrates the Mesh based Wireless Body Sensor Network.

All the nodes in the Wireless Body Sensor Network are of similar capability. Implanted node naturally lag with resources when compared to the wearable one. Mesh topology becomes inappropriate in case of the Implanted nodes. Making implanted node as a cluster head and routing data via implanted nodes makes the sensor node to dissipate more heat causing energy decay. Providing lengthy key, complex keying mechanism makes the implanted node to overload thereby creating a massive tissue damage.

Figure 7. Mesh topology of Wireless Body Sensor Network

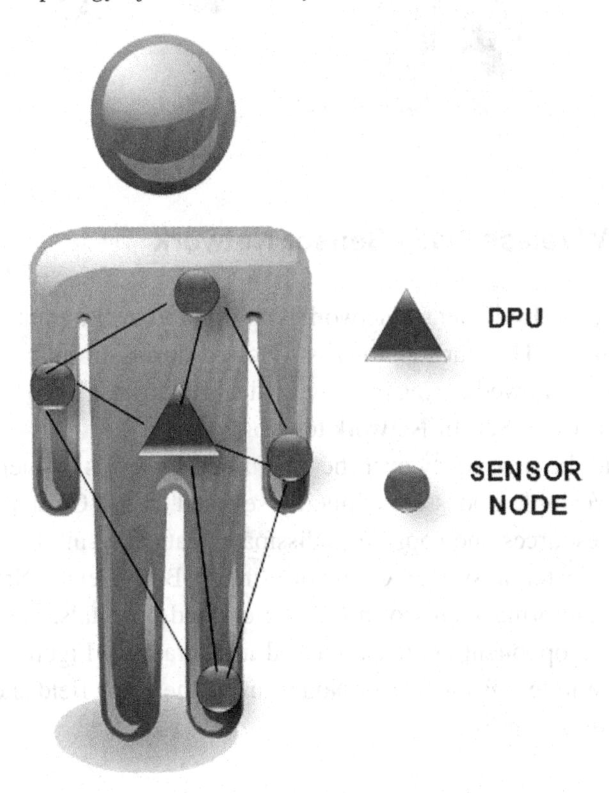

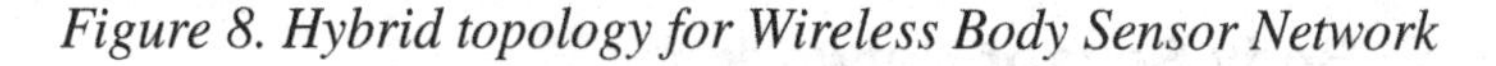

Figure 8. Hybrid topology for Wireless Body Sensor Network

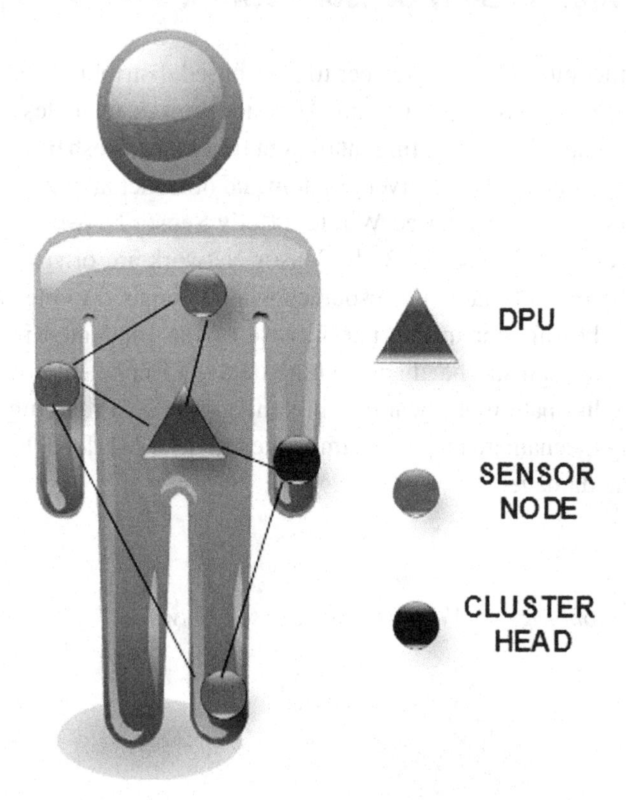

1.5. Hybrid Wireless Body Sensor Network

The hybrid Wireless Body Sensor network is configured by the combination of Star and Mesh topology. The star topology is followed during normal conditions and Mesh topology is followed during extreme critical situation. Figure 8 illustrates the Hybrid Wireless Body Sensor Network topology.

The CH hold the superiority over the other nodes in terms of energy and memory capacity. Wearable nodes are explicitly selected as the Cluster Head due its availability of resources and capable to dissipate heat. The sink node is wearable and connected to external system. Group of Wireless Body Sensor Network creates Group based monitoring scenario in Game field and hospitals. Passing personal health data to an opponent could be hacked and attacked. Figure 9 exhibits the deployment of Wireless Body Sensor Nodes inside the game field and other group based monitoring scenarios.

Figure 9. Deployment inside the game field

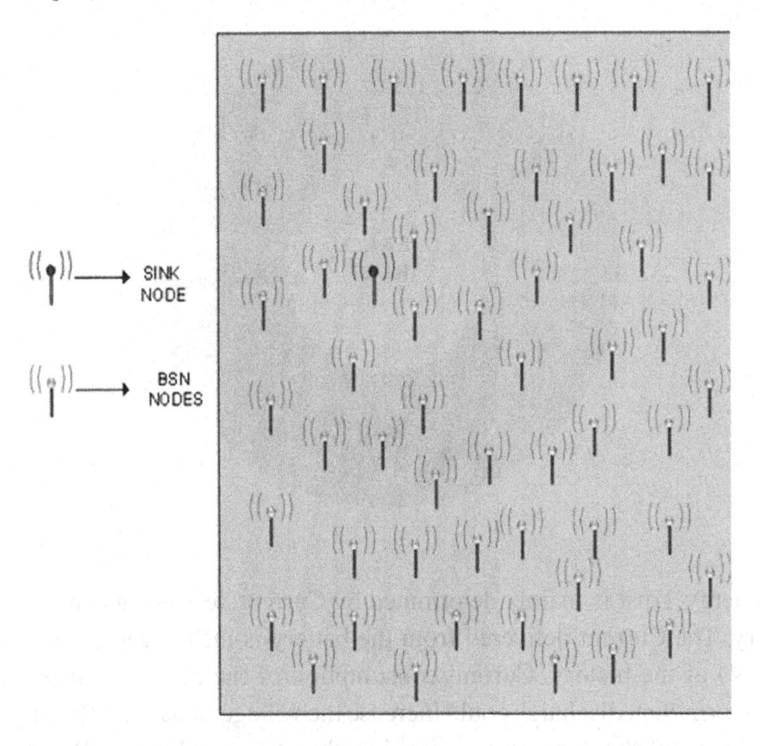

Here the sink represents the gateway to monitor all the Wireless Body Sensor Network status. Each Wireless Body Sensor Network relays data to the sink via another Wireless Body Sensor Network forming a Group based monitoring system.

2. TRUST FOR WIRELESS BODY SENSOR NETWORKS

Trust promises the delivery of data to the Sink in a secured way ensuring patients lifestyle. Trust for nodes is time varying and varying with positions and based on its individuality. Trust to the node is determined based on the Security-Encryption key size and residual Energy of the Sensor Nodes.

2.1. Trust Overview

The values of Trust mainly depends on the residual energy, energy harvesting capability, encryption of data. Over all Trust to the network is combination of Security and Energy Trust of the nodes. Figure 10 exhibits the Trust model incorporating Security and Energy Trust.

Figure 10. Trust model for Wireless Body Sensor Network

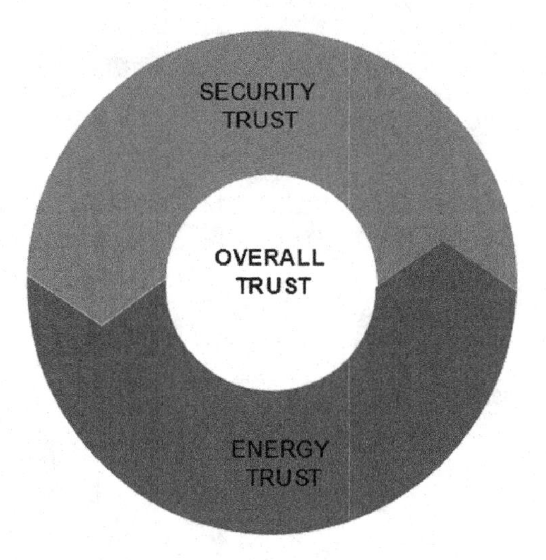

The Energy Trust is mainly determined by Current delivering capacity (Ah) of the battery. The Current delivered from the battery is influenced by the Terminal Voltage (V) of the battery. Current consumption of the battery reduces terminal voltage; intermittent discharge could increase the voltage of battery, thereby extending the lifetime of the network to a considerable manner. Trust in Wireless Body sensor network includes confidentiality, integrity, authentication, and non repudiation, all these things are included by the cryptographic key implementation and energy of the Wireless Body Sensor Node. Architecture of the Wireless Body Sensor Network mainly influences the Trust of network (Marcelloni & Vecchio, 2010; Ting & Liao, 2010; Demirkol & Ersoy, 2006; Durmus & Ozgovde, 2009; Pandian, et al., 2008; Pandian, et al., 2007).

2.2. Energy Trust and Thermal Aware

Wireless Body Sensor Network works in a tedious environment surrounded by bones, tissues, water, blood and blood vessels. The Wireless Channel in the human body composed of tremendous water content leading high path loss and exhibits more amount of heat causing more resistivity. Range of transmission and total bits transmitted across the surface size of the node, energy capacity, range of transmission influences the heat dissipation of the node. Fading due to high path loss complicates the design of the antenna and energy consumed and heat dissipated by the node. Heat generation affects the tissue and damages the communication channel.

2.3. Thermal Factors

2.3.1. Heat by RF Power Supply

Implanted nodes are normally powered by high frequency RF inductive power supply, which dissipates massive heat damaging the surrounding tissue due to RF path loss.

2.3.2. Radiation-Implanted Node

Implanted device communicates the data to sink causing heat dissipation by the antennas due to propagation loss.

2.3.3. Thermal Dissipation from Circuitry

Sensor node process the data, the microwave emitted by the processor dissipated heat on contact with the high density medium making it to damage.

The heat dissipation is directly related with the number of bits transmitted majorly influencing the security trust of the network. Specific Absorption Rate (SAR) of the human varies with person to person based on the salinity of the human blood.

SAR relationship with radiation is given as

$$SAR = \frac{\sigma\left[E^2\right]}{\rho} \tag{4}$$

where E is the induced electric field by radiation, ρ is the density of the tissue and σ is the conductivity of the tissue. The power dissipation of the circuitry in turn raises the temperature of the wireless Body sensor Node.

$$\gamma = \alpha + \beta i \tag{5}$$

$$\beta = \omega\sqrt{\frac{\mu e}{2}}\left[\sqrt{1+\left(\frac{\sigma}{\omega e}\right)^2}+1\right]^{0.5} \quad (rad\ /\ m) \tag{6}$$

The power dissipation in turn dominates the current density. Equation 4, 5 and 6, the SAR with respect to different person. SAR signifies that the trust depends with person to person.

Table 3. Body parts and its characteristics

Tissue Type	Relative Permittivity *Er*	Conductivity σ (S/m)
Bone	0.28	0.0144
Liver	9.8–14	0.15–0.16
Spleen	3.3	0.62
Blood	2.7–4.0	0.55–0.68
Kidney	10.9–12.5	0.24–0.25
Retina	4.75	0.52
Bone (cancellous)	0.47	0.09
Bone (cortical)	0.23	0.02
Bone (marrow)	0.11	0.003
Cartilage	2.57	0.18
Skeletal muscle	14.4–27.3	0.38–0.65
Fat	0.09	0.02
Cerebrospinal fluid	0.1	2
Brain (grey matter)	3.8	0.17
Brain(white matter)	1.9–3.4	0.12–0.15

Table 3 illustrates the various conductivity and relative permittivity of various body parts (Halin, Junnila, Loula, &Aarnio, 2005; Mundt, 2005; Gopalsamy, Park, Rajamanickam, & Jayaraman, 2005; Yazicioglu, et al., 2009; Shankar, Natarajan, Gupta, & Schwiebert, 2001).

2.4. Trust Design

The Trust of the Wireless Body Sensor Network is been dispersed between -1 <x <1 for classifying its state of trust. Trust classification is done by High Trust, Low Trust and Distrust. The Trust value between 0.5 to 1 is said to be High Trust Zone, 0 to 0.5 is termed as Low Trust Zone and -1 to 0 is termed to be Distrust Zone.

2.4.1. Security Trust

Many Timestamp based key management scheme has been proposed which schedules and distribute key among the sensor nodes. The Wireless Body Sensor Network need to broadcast and receive the key management schedule messages which would reduce the energy consumption of the Wireless Body Sensor Network.

Security trust is greatly determined by

- Key size;
- Key sharing;
- Key rotation;
- Key management.

The Security trust are influenced by the Cryptographic key parameters and its management process. The Size of the key influences the strength of the cryptographic algorithm. The key sharing between trusted node member should be done in a secured way. Key rotation should be done frequently in order to avoid attacks to the Wireless Body Sensor Network. The key should be managed between the other monitoring Health monitoring nodes in order to have an effective trusted communication between the doctor and the subject. Key management scheme should based on the topology of the network. Effective key management scheme should be implemented to serve the implanted node to be aware of the heat dissipation. Transmission range and capacity of the implanted node is very much lesser than that of wearable. Low energy based keying mechanism and topology is the effective solution to the problem for implanted nodes in Wireless Body Sensor Network. Effective usage of keying mechanism could be implemented by multihop key sharing in hybrid networks in turn increases the overhead of the network, resulting higher energy cost to the nodes. Low energy based key mechanism is advised in Wireless Body Sensor Network. Key management is the prime part of the Security trust in the Wireless Body Sensor Network. Frequent key sharing, Key rotation and Key management would increase the number of data transmitted rather increases the Energy consumed by Wireless Body Sensor Network.

The key size greatly determines the Time complexity to decode the data for the hacker. Increase in key size increases the data size which greatly increases the constraints demanding a high end processor and memory RAMs. Figure 11 represents the relation between the Encryption key and time complexity of the encryption mechanism.

The Trust values is dispersed between -1 to 1; higher the value, higher the Trust.

$$S_t = mx + c \tag{7}$$

Equation 7 describes the Security Trust distribution, the security Trust is modeled between $-1<S_t<1$.

Figure 11. Encryption time complexity

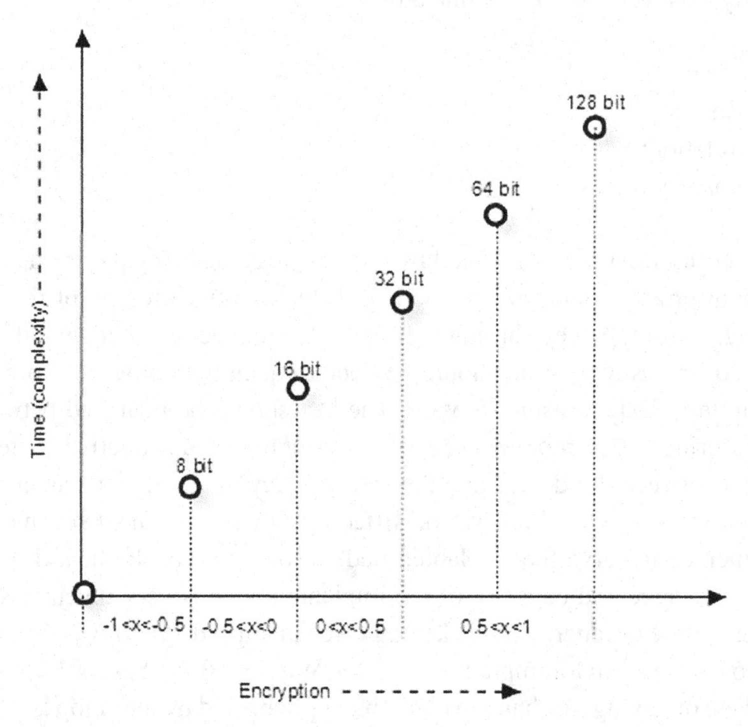

2.4.2. Energy Trust

The Energy Trust of the Node is greatly determined by the number of bits, distance and the terminal voltage. The transmitted number bit by uplink and downlink from the Wireless Body Sensor Nodes varies. The total circuit power consumption is the sum of power consumed due to uplink and downlink of data.

$$P_t = P_d + P_u \quad (8)$$

$P_t \rightarrow$ Total power consumed
$P_d \rightarrow$ Power consumed by downlink
$P_u \rightarrow$ Power consumed by uplink

$$E_t = \left(P_d + P_u\right) \times n \times f^{-1} \quad (9)$$

$n \rightarrow$ Number of bits transmitted
$f \rightarrow$ Frequency of antenna operation

Equation 8 and 9 represents the total energy consumed by the node as a function of power dissipation, total number of bits transmitted and frequency of antenna operation. The frequency of antenna operation denotes the speed of data transfer and higher frequency transmission induces high path loss to the implanted nodes. Wireless Body sensor node consumes a fraction of energy during the idle state of antenna.

$$E_I = P_I \times T_I \tag{10}$$

$T_I \rightarrow$ Idle time period

Total uplink power consumption is given as for L- bits is

$$E_L = \left(P_d + P_u\right) f^{-1} + P_I + T_I \tag{11}$$

Now IEEE 802.15 standards enhances the low power based protocols for Wireless Body Sensor Network and Future Nano Body Sensor Network.

The Battery capacity of the implanted node and its lifetime varies with type of implanted node. Table 4 represents the implanted node, its type of battery, power requirement, Energy Density and lifetime of the battery.

The specification and working values varies with the environment temperature and subject work style. Voltage depression happens due to irregular loading of Wireless Body Sensor Node, operating in drastic region of the battery curve. Load to the Wireless Body Sensor Node should spread out based on battery curve. Node operation as Full Function Device (FFD) (i.e. as Cluster Head is to operated in the high energy trust regions). Frequent operation charging, recharging of batteries greatly affects the lifetime of the battery, its energy holding capacity, varies the cut

Table 4. Implanted device characteristics

Type of Implanted Node	Type of Battery	Power Requirement (mW)	Energy Density $(Wh / I)^{-1}$	Lifetime
Pacemaker	Li/ I2	0.030–0.1	700	>10 years
Defibrillator	Li/SVO	10,000	780	Several years
Neurological stimulator	Li/SOCl2	0.3 to several	680	>5 years
Drug pump	Li/SOCl2	0.1–2	680	>5 years

off voltage making a distrust to the Wireless Body Sensor Node. The operating Voltage of the implanted node is operated in between 1.4 V to 3.7 V for safe operation. Operating Voltages for the battery constitutes have three various values, Charging voltage is a voltage at which the battery is charged. Discharging Voltage is a voltage at which the battery dissipates its energy by powering the implanted nodes circuitry and radio. Floating voltage is given to battery to reduce the unnecessary power wastage due to long term resting of batteries without usage.

2.5. Battery Considerations

The factors to be considered to a battery for an implanted node are Amp-Hour Capacity (C), Cell Voltage, Multiple Cell configuration, Weight and Volume, Charge Characteristics, Discharge Characteristics, Cost and complexity of Fast Charging circuit. The time of charging majorly influences the charging circuit complexity and methodology to charge. Fast charging circuits should be matched with the battery significantly in order to provide a smooth reliable charging and charge termination. Overcharging to batteries could cause the reduced lifespan of battery and it also reduces the amount to storing capacity of the battery. Environmental concerns are to be noted on proper disposal battery in the human body. The hazardous metal content might significantly affect the human tissue and cause allergic reaction. Battery capacity, C is expressed in Amp- Hours or mA hours. Battery current is described in units of C- Rate.

C-RATE is nothing but the normalized Battery charge and Discharging Currents,

$$C - RATE = \frac{c}{1\ hour} \tag{12}$$

- C is the battery capacity expressed in A- Hour.
- Self discharge is the rate at which a battery discharges with no load.

Figure 12 illustrates the discharging characteristics of the three batteries Li-Ion, SLA, NiCd and NiMH.

Table 5 illustrates the slow charging characteristics of the Battery.

Slow charging mechanism to the battery makes it unsuitable for implanted nodes. Fast charging to batteries requires much more complicated techniques. Under charging the batteries will reduce energy content whereas the over charged batteries might damage the battery, causing catastrophic out gassing and even may explode the battery. Batteries are acutely sensitive to operating temperature, both Surface tem-

Figure 12. Discharge characteristics of batteries

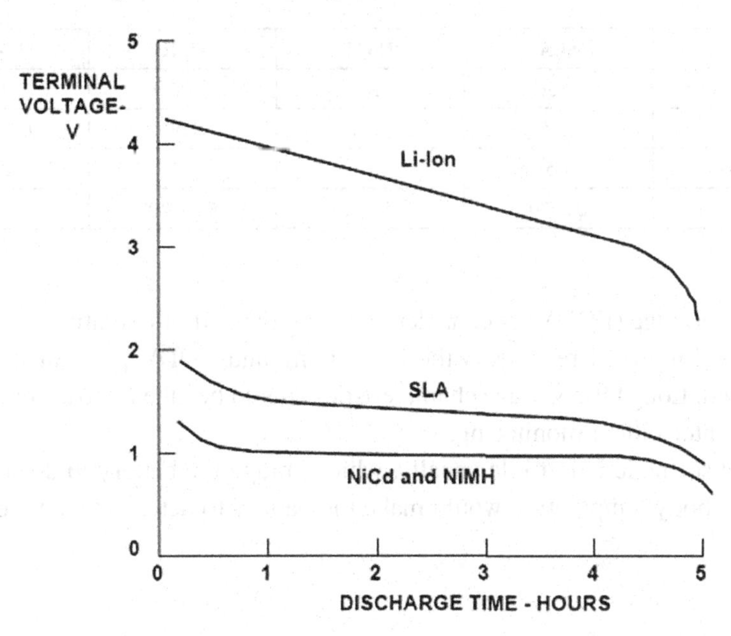

perature and temperature due to heating of Wireless Body Sensor Node. Design of Charging Circuit should also keep temperature sensor to monitor its temperature while charging and discharging the battery.

Table 6 represents the typical battery characteristics for SLA, NiCd, NiMH and Li-Ion. Proper charging and termination is mandatory to batteries in Wireless Body Sensor Nodes.

This can be properly monitored using battery voltage, voltage change vs. time, temperature change, temperature change vs. time, minimum current at full voltage, charge time. Figure 13 represents the terminal voltage to discharge relation of the battery, the terminal Voltage gets reduced on increased discharge.

The MPV (Mid Point Voltage) is the nominal voltage of the implanted battery that is measured when the battery completes 50% of its total energy. The End of

Table 5. Battery slow charging characteristics

	SLA	NiCd	NiMH	Li-Ion
Current	0.25C	0.1C	0.1C	0.1C
Voltage (V/cell)	2.27	1.50	1.50	4.1 or 4.2
Time (hr)	24	16	16	16
Temp. Range	0°/45°C	5°/45°C	5°/40°C	5°/40°C

Table 6. Battery characteristics for SLA, Ni-Cd, NiMH and Li-Ion

	SLA	NiCd	NiMH	Li-Ion
Current	1.55C	1C	1C	1C
Voltage (V/cell)	2.45	1.50	1.50	4.1 or 4.2
Time (hr)	1.5	3	3	2.5
Temp. Range	0 to 30	15 to 40	15 to 40	10 to 40

Discharge Voltage (EODV) is considered during the end of operating life cycle of the battery. Figure 14 represents the MPV pong and EODV point of the typical battery used. Long term storage characteristics should be taken into account for the battery in intermittent monitoring.

The sensor node is left idle usually, which enrage the battery to deterioration. Higher the body temperature would make the battery to deteriorate and loss more

Figure 13. Battery characteristics of the sensor node

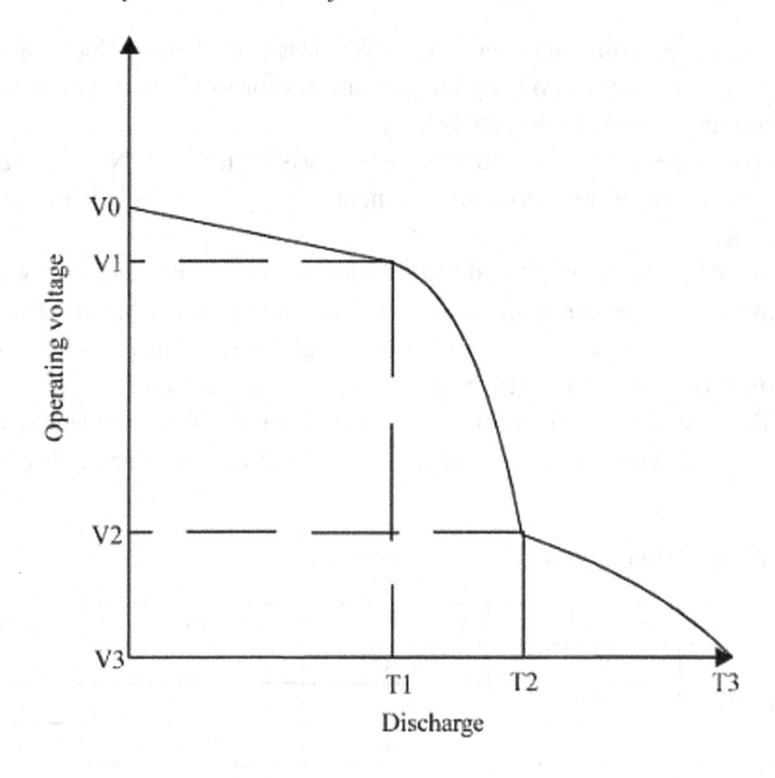

Figure 14. Charge and discharge characteristics of rechargeable battery

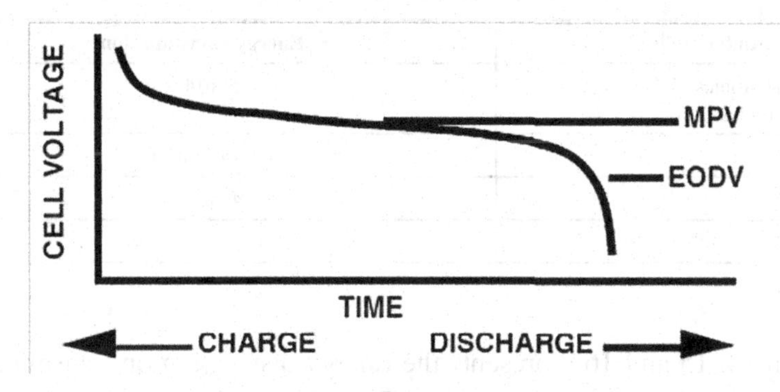

energy making the Wireless Body Sensor node unavailable during most extreme situations. Higher the storage temperature greater the deterioration.

$$F(a) = x_1 \sin\left(y_1 a + z_1\right) + x_2 \sin\left(y_2 a + z_2\right) \tag{13}$$

where

x_1, y_1, z_1, x_2, y_2 and z_2: Curve constants.

Equation 13 represents the characteristics curve of the battery.

$$E_{tx}(k,d) = E_{elec}k + E_{fs}kd^2; d < d_0 \tag{14}$$

$$E_{tx}(k,d) = E_{elec}k + E_{mp}kd^4; d > d_0 \tag{15}$$

$$E_{rx}(k) = E_{elec}k \tag{16}$$

where *k*: No. of bits,

d: Distance,
E_{elec}: Energy dissipated per bit to run the transmitter or the receiver circuit,
E_{rx}: Energy dissipated during receiving data.
E_{fs}(pJ/(bit-$^{m-2}$)), E_{mp}(pJ/(bit-m $^{-2}$)): Energy dissipated per bit to run the transmit amplifier based on the distance between the transmitter and receiver.

Table 7. Wireless sensor node radio energy consumption

Radio Mode	Energy Consumption
Transmitter Electronics Receiver Electronics	50nJ/bit
Transmit Amplifier	$100pJ/bit/m^2$
Idle	40 *nJ/bit*
Sleep	0

Equation 14, 15 and 16 represents the Energy dissipation on transmitting and receiving a bit in wireless environment. Table 7 explains the typical parameter values in Joules for transmitting, receiving and operating amplifier for data sharing operation.

$$E = V_b \times I_N \times t_d \tag{17}$$

where

V_b: Battery end Voltage,
I_N: Current drawn from the battery,
t_d: Time of data transmission which depends on the speed of operation of the radio.

Equation 17 represents the Terminal Voltage and Energy relation of the battery, the current taken by the sensor node increases on decreased terminal Voltage. The widely used solution to avoid power wastage is duty cycling. Turning off the radio during non transmission conditions saves the major part of energy of the Wireless Body Sensor Node. Duty cycling allows switching on the radio front- ends only when the data is transmitted and received, which would significantly reduce the major amount of power dissipation due to radio operation. Duty cycling request the bandwidth to be greater than the symbol rate. Wake up radio scheme considerable reduces the energy consumption of the Wireless Body Sensor Nodes. The Wake up radio are very low power consuming, low data rate, low bandwidth radios, which are mainly operated to share the I am Alive packet to the sink. Under extreme situations these radio helps in waking up the main radio which is high power consuming and capable to handle large bandwidth data and start sharing its data towards the sink. Figure 15 illustrates the typical architecture of the two radio-based Wireless Sensor node mounted with Wake up transceiver and Main transceiver.

Figure 15. Wireless Body Sensor node with wake up transmitter

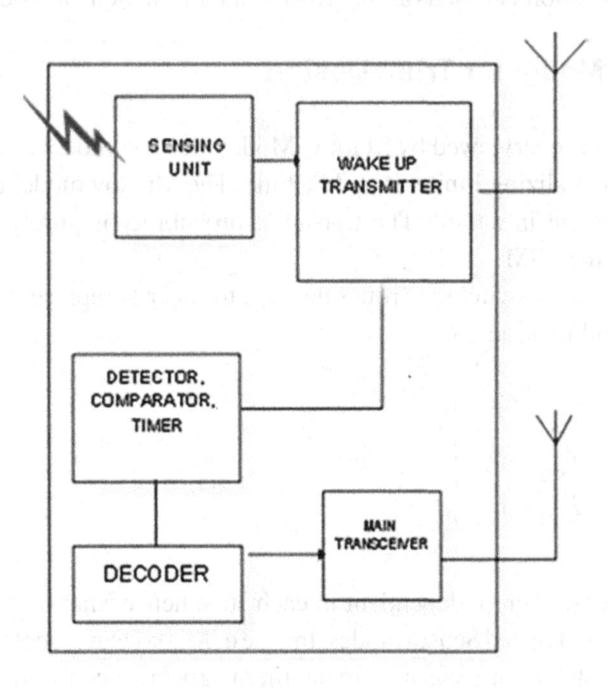

Figure 16. Finite state machine of trust

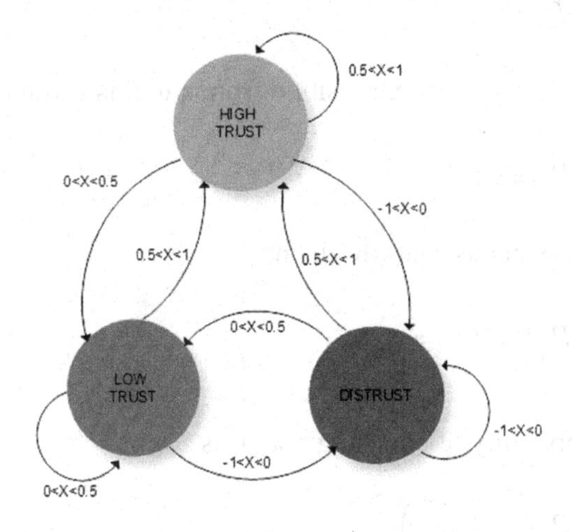

3. FINITE STATE MACHINE (FSM) ANALYSIS

Trust of the Wireless Body Sensor Network is realized as Finite State Machine with three State HIGH Trust, LOW Trust and DISTRUST zones. Figure 16 represents the Finite State Machine based realization of Trust of the Wireless Body Sensor Network.

The state transition is based on the Trust value of the Sensor Node.

3.1. Markov Model for Trust Design

The Trust model is overviewed by Markov Model. Markov model is a memory less model meant for realizing Finite State Machine. The Markov model uses variables that are independent in nature. The transition probability is purely meant by the current state of the FSM.

The probability of switching from one state to other is represented in a matrix form as given in Equation 18.

$$P_{Si\rightarrow i+1} = \begin{pmatrix} P_{S11} & P_{S12} & P_{S13} \\ P_{S21} & P_{S22} & P_{S23} \\ P_{S31} & P_{S32} & P_{S33} \end{pmatrix} \quad (18)$$

Data from sensors are independent in each state hence Markov model helps in routing the data via Trusted Sensor Nodes. In case of Markovian model the probability of selection of r steps from one state to another is given by conditional approach.

Probability of choosing x state to y state for n steps is given by

$$P_{xy} = P_r\left(P_n = y \mid P_0 = x\right) \quad (19)$$

The probability of single-step transition from x to k is given by

$$P_{xk} = P_r\left(P_1 = q \mid P_0 = x\right) \quad (20)$$

For a time-homogeneous Markov chain:

$$P_r\left(P_n = y\right) = \sum_{r \in s} P_{ry} P_r\left(P_{n-1} = r\right) \quad (21)$$

Generalized probability of choosing r steps is

$$P_r\left(P_n = y\right) = \sum_{r \in s} P_{ry} P_r\left(P_0 = r\right) \quad (22)$$

Equations 19, 20, 21, and 22 represent the probability of choosing the next state by the node in the system model.

$$P = \begin{pmatrix} P_{r11} & P_{r12} & P_{r13} \\ P_{r21} & P_{r22} & P_{r23} \\ P_{r31} & P_{r32} & P_{r33} \end{pmatrix} \tag{23}$$

Equation 23 represents the probability transition matrix of the Finite State Machine. The data transmission can be done based on the Trust state as proposed in Equation 23.

3.2. Trustworthy Architecture

The data from the implanted node should be more confidential, attack to an implanted node causes fatal conditions than an attack to the non-implanted node. The data from the implanted node should be transmitted via a more trusted node. Figure 17 represents the Trustworthy architecture of the Wireless Body Sensor Network, transmitting confidential packet stream via more trusted path.

Figure 17. Trustworthy architecture

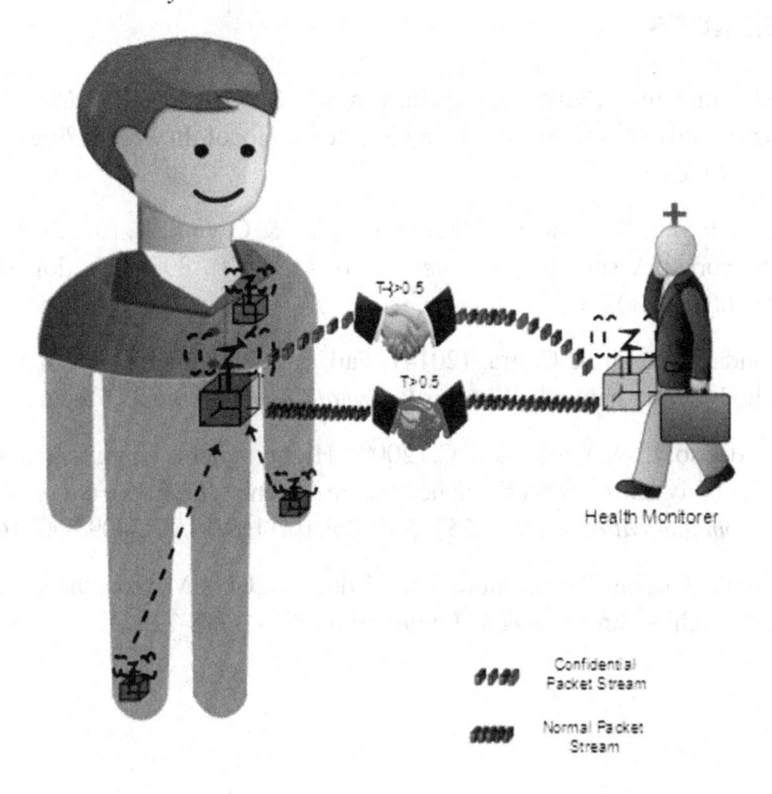

Data from implanted node is termed to be Confidential packet stream and Data from non-implanted node is named as Normal packet stream. The Normal packet stream is transmitted via high trust or low trust zone.

CONCLUSION

The essentiality of Trust in Wireless Body Sensor Network is discussed in this chapter. The trust to Wireless Body Sensor Network ensures the novel delivery of data to the sink. The Trust is designed based on the encryption key size and Residual energy of the battery. The encryption key size increases the size of the data, compromising some of the basic constraints of the node. Increasing the battery size of the node increases the size of the node, making it unsuitable to implant inside the human body. The proposed trustworthy architecture gives suitable key to solve the issue without compromising the basic constraints of the Wireless Body Sensor Network. The Overall trust depends on the individual trust (i.e. Security Trust and Energy Trust). Markov Model helps in getting a peer view about State machine analysis of Trust. The transition from one state to other is viewed as Markov Model.

REFERENCES

Kanagachidambaresan, SarmaDhulipala, Vanusha, & Udhaya. (2011). Matlab based modeling of body sensor network using ZigBee protocol. In *Proceedings of CIIT* (pp. 773-776). CCIS.

Akyildiz, I. F., Su, W., Sankarasubramaniam, Y., & Cayirci, E. (2002). Wireless sensor networks: A survey. *Computer Networks*, *38*(4), 393–422. doi:10.1016/S1389-1286(01)00302-4

Kanagachidambaresan & Chitra. (2014). Fail safe fault tolerant mechanism for wireless body sensor network. *Wireless Personal Communications*, *78*(2), 247–260.

Otal, B., Alonso, L., & Verikoukis, C. (2009). Highly reliable energy-saving MAC for wireless body sensor networks in healthcare systems. *IEEE Journal on Selected Areas in Communications*, *27*(4), 553–565. doi:10.1109/JSAC.2009.090516

Kanagachidambaresan, SarmaDhulipala, & Udhaya. (2011). Markovian model based trustworthy architecture. *Procedia Engineering, 38*(4), 718-725.

SarmaDhulipala, Kanagachidambaresan, & Chandrasekaran. (2012). Lack of power avoidance: A fault classification based fault tolerant framework solution for lifetime enhancement and reliable communication in wireless sensor network. *Information Technology Journal, 11*(6), 247–260.

Shaikh, J., Brian, A., & Heejo, L. (2009). Group-based trust management scheme for clustered wireless sensor network. *IEEE Transactions on Parallel and Distributed Systems, 20*(11), 1698–1718. doi:10.1109/TPDS.2008.258

Zhang, Das, & Liu. (2006). A trust based framework for secure data aggregation in wireless sensor networks. In *Proceedings of IEEE SECON* (vol. 1, pp. 60-72). IEEE.

Momani, M., Challa, S., & Alhmouz, R. (2008). Can we trust trusted nodes in wireless sensor networks? In *Proceedings of International conference on Computer and Communication Engineering* (ICCCE 2008), (vol. 1, pp. 37-45). Kuala Lumpur, Malaysia: ICCCE.

Boukerche, A., Li, X., & Khatib, K. (2006). Trust based framework for secure data aggregation in wireless sensor networks. In *Proceedings of IEEE SECON* (pp. 718-725). IEEE.

Marsh, S. (1994). *Formalizing trust as a computational concept.* (Unpublished PhD Thesis). University of Stirling, Stirling, UK.

Hoffman, L. J., Lawson-Jenkins, K., & Blum, J. (2006). Trust beyond security: An expanded trust model. *Communications of the ACM, 49*(7), 95–101. doi:10.1145/1139922.1139924

Ng, H. S., Sim, M. L., & Tan, C. M. (2006). Security issues of wireless sensor networks in healthcare applications. *BT Technology Journal, 24*(2), 138–144. doi:10.1007/s10550-006-0051-8

Pirzada, A. A., & McDonald, C. (2004). Establishing trust in pure ad-hoc networks. In *Proceedings of 27th Australasian Computer Science Conf.* (ACSC '04), (*vol. 26*, pp. 47-54). ACSC.

Sun, Y. L., Yu, W., Han, Z., & Liu, K. J. R. (2006). Information theoretic framework of trust modeling and evaluation for ad hoc networks. *IEEE Journal on Selected Areas in Communications, 24*(2), 305–317. doi:10.1109/JSAC.2005.861389

Shaikh, R. A., Jameel, H., Lee, S., Rajput, S., & Song, Y. J. (2006). Trust management problem in distributed wireless sensor networks. In *Proceedings of 12th IEEE Intl Conf Embedded Real-Time Computing Systems and Applications*. IEEE.

Momani, M. S. Challa, & Aboura, K. (2007). Modeling trust in wireless sensor networks from the sensor reliability prospective. In Innovative algorithms and techniques in automation, industrial electronics, and telecomm (pp. 317-321). Springer.

Liu, Z., Joy, A. W., & Thompson, R. A. (2004). A dynamic trust model for mobile ad hoc networks. In *Proceedings of 10th IEEE Int Workshop Future Trends of Distributed Computing Systems*. IEEE. doi:10.1109/RTCSA.2006.61

Grandison, T., & Sloman, M. (2000). A survey of trust in internet applications. *IEEE Communications Surveys and Tutorials*, *3*(4), 2–16. doi:10.1109/COMST.2000.5340804

Shi, E., & Perrig, A. (2004). Designing secure sensor networks. *IEEE Wireless Comm*, *11*(6), 38–43. doi:10.1109/MWC.2004.1368895

Walters, J. P., Liang, Z., Shi, W., & Chaudhary, V. (2006). Wireless sensor network security: A survey. *Security in Distributed Grid and Pervasive Computing*, *43*(5), 367–410.

Buchegger, S., & Boudec, Y. L. (2005). Self-policing mobile ad hoc networks by reputation systems. *IEEE Communications Magazine*, *43*(7), 101–107. doi:10.1109/MCOM.2005.1470831

Heinzelman, W. B., Chandrakasan, A. P., & Balakrishnan, H. (2002). An application-specific protocol architecture for wireless microsensor networks. *IEEE Transactions on Wireless Communications*, *1*(4), 660–670. doi:10.1109/TWC.2002.804190

Olariu, S., Xu, Q., Eltoweissy, M., Wadaa, A., & Zomaya, A. Y. (2005). Protecting the communication structure in sensor networks. *International Journal of Distributed Sensor Networks*, *1*(2), 187–203. doi:10.1080/15501320590966440

Misra, S., & Xue, G. (2006). Efficient anonymity schemes for clustered wireless sensor networks. *International Journal of Sensor Networks*, *1*(1/2), 50–63. doi:10.1504/IJSNET.2006.010834

Cherukuri, S., Venkatasubramanian, K. K., & Gupta, S. K. S. (2003). Biosec: A biometric based approach for securing communication in wireless networks of biosensors implanted in the human body. In *Proceedings of 32nd Int Conf Parallel Processing* (pp. 432-439). Academic Press. doi:10.1109/ICPPW.2003.1240399

Ruzzelli, A. G., Jurdak, R., O'Hare, G. M., & van Der Stok, P. (2003). Energy-efficient multi-hop medical sensor networking. In *Proceedings of the 1st ACM SIGMOBILE International Workshop on Systems and Networking Support for Healthcare and Assisted Living Environments*. ACM.

Kim, K., Lee, I.-S., Yoon, M., Kim, J., Lee, H., & Han, K. (2009). An efficient routing protocol based on position information in mobile wireless body area sensor networks. In *Proceedings of the 1st International Conference on Networks and Communications (NETCOM)*. Chennai, India: Academic Press. doi:10.1109/NetCoM.2009.36

Maskooki, A., Soh, C. B., Gunawan, E., & Low, K. S. (2011). Opportunistic routing for body area network. In *Proceedings of Consumer Communications and Networking Conference* (CCNC). Las Vegas, NV: Academic Press.

Sayrafian-Pour, K., Wen-Bin, Y., Hagedorn, J., Terrill, J., & Yazdandoost, K. Y. (2009). A statistical path loss model for medical implant communication channels. In *Proceedings of the 20th IEEE International Symposium on Personal, Indoor and Mobile Radio Communications*. Tokyo, Japan: IEEE. doi:10.1109/PIMRC.2009.5449869

Liang, X., Balasingham, I., & Byun, S.-S. (2008). A reinforcement learning based routing protocol with QoS support for biomedical sensor networks. In *Proceedings of the 1st IEEE International Symposioum on Applied Sciences on Biomedical and Communication Technologies* (ISABEL). Aalborg, Denmark: IEEE.

Djenouri, D., & Balasingham, I. (2009). New QoS and geographical routing in wireless biomedical sensor networks. In *Proceedings of the 6th IEEE International Conference on Broadband Communications, Networks, and Systems* (BROADNETS). Madrid, Spain: IEEE. doi:10.4108/ICST.BROADNETS2009.7188

Razzaque, M. A., Hong, C. S., & Lee, S. (2011). Data-centric multiobjective QoS-aware routing protocol for body sensor networks. *Sensors (Basel, Switzerland), 11*(12), 917–937. doi:10.3390/s110100917 PMID:22346611

Tang, Q., & Tummala, N. (2005). TARA: Thermal-aware routing algorithm for implanted sensor networks. In *Distributed computing in sensor systems*. Springer. doi:10.1007/11502593_17

Bag, A., & Bassiouni, M. A. (2006). Energy efficient thermal aware routing algorithms for embedded biomedical sensor networks. In *Proceedings of the IEEE International Conference on Mobile Adhoc and Sensor Systems* (MASS). Vancouver, Canada: IEEE. doi:10.1109/MOBHOC.2006.278619

Takahashi, D., Xiao, Y., & Hu, F. (2007). LTRT: Least total-route temperature routing for embedded biomedical sensor networks. In *Proceedings of Global Telecommunications Conference* (GLOBECOM). Washington, DC: IEEE. doi:10.1109/GLOCOM.2007.125

Bazaka & Mohan, V. (2013). Jacob: Implantable devices: Issues and challenges. *Electronics*, *2*, 1–34.

Bag, A., & Bassiouni, M. A. (2007). Hotspot preventing routing algorithm for delay-sensitive biomedical sensor networks. In *Proceedings of the IEEE International Conference on Portable Information Devices* (PORTABLE). Orlando, FL: IEEE. doi:10.1109/PORTABLE.2007.30

Ullah, S., Higgins, H., Braem, B., Latre, B., Blondia, C., Moerman, I., & Kwak, K. S. et al. (2012). A comprehensive survey of wireless body area networks. *Journal of Medical Systems*, *36*(3), 1065–1094. doi:10.1007/s10916-010-9571-3 PMID:20721685

Hirata, A., & Shiozawa, T. (2003). Correlation of maximum temperature increase and peak SAR in the human head due to handset antennas. *IEEE T. Microw. Theory*, *51*(7), 1834–1841. doi:10.1109/TMTT.2003.814314

Havenith, G. (2001). Individualized model of human thremoregulation for the simulation of heat stress response. *Journal of Applied Physiology (Bethesda, Md.)*, *90*, 1943–1954. PMID:11299289

Natarajan, A., de Silva, B., Kok-Kiong, Y., & Motani, M. (2009). To hop or not to hop: Network architecture for body sensor networks. In *Proceedings of the 6th Annual IEEE Communications Society Conference on Sensor, Mesh and Ad Hoc Communications and Networks* (SECON). Rome, Italy: IEEE. doi:10.1109/SAHCN.2009.5168978

Braem, B., Latre, B., Moerman, I., Blondia, C., & Demeester, P. (2006). The wireless autonomous spanning tree protocol for multihop wireless body area networks. In *Proceedings of the 3rd IEEE Annual Conference on Mobile and Ubiquitous Systems: Networking & Services*. San Jose, CA: IEEE.

Latre, B., Braem, B., Moerman, I., Blondia, C., Reusens, E., Joseph, W., & Demeester, P. A. (2007). Low-delay protocol for multihop wireless body area networks. In *Proceedings of the 4th IEEE Annual International Conference on Mobile and Ubiquitous Systems: Networking & Services* (MobiQuitous). Philadelphia, PA: IEEE. doi:10.1109/MOBIQ.2007.4451060

Quwaider, M., & Biswas, S. (2009). On-body packet routing algorithms for body sensor networks. In *Proceedings of the 1st IEEE International Conference on Networks and Communications* (NETCOM). Chennai, India: IEEE. doi:10.1109/NetCoM.2009.54

Culpepper, J., Dung, L., & Moh, M. (2003). Hybrid indirect transmissions (HIT) for data gathering in wireless micro sensor networks with biomedical applications. In *Proceedings of the 18th IEEE Annual Workshop on Computer Communications* (CCW). Dana Point, CA: IEEE. doi:10.1109/CCW.2003.1240800

Gosalia, Weiland, Humayun, & Lazzi. (2004). Thermal elevation in the human eye and head due to the operation of a retinal prosthesis. *IEEE Tran. Biomedical Eng.*, *51*(8), 1469–1477.

Moneda, I., Ioannidou, M. P., & Chrissoulidis, D. P. (2003). Radio-wave exposure of the human head: Analytical study based on a versatile eccentric spheres model including a brain core and a pair of eyeballs. *IEEE Transactions on Bio-Medical Engineering*, *50*(6), 667–676. doi:10.1109/TBME.2003.812222 PMID:12814233

Gao, T., Greenspan, D., Welsh, M., Juang, R. R., & Alm, A. (2005). Vital signs monitoring and patient tracking over a wireless network. In *Proceedings of IEEE-EMBS 27th Annual International Conference of the Engineering in Medicine and Biology*. Shanghai, China: IEEE.

Otto, C. A., Jovanov, E., & Milenkovic, E. A. (2006). WBAN-based system for health monitoring at home. In *Proceedings of IEEE/EMBS International Summer School, Medical Devices and Biosensors*. Boston, MA: IEEE. doi:10.1109/ISSMDBS.2006.360087

Anliker, U., Ward, J. A., Lukowicz, P., Troster, G., Dolveck, F., Baer, M., & Vuskovic, M. et al. (2004). AMON: A werable multiparameter medical monitoring and alert system. *IEEE Transactions on Information Technology in Biomedicine*, *8*(4), 415–427. doi:10.1109/TITB.2004.837888 PMID:15615032

Srinivasan, V., & Stankovic, J. (2008). Protecting your daily in home activity information from a wireless snooping attack. In *Proceedings of the 10th International Conference on Ubiquitous Computing*. Seoul, Korea: Academic Press.

Casas, R., Blasco, M. R., Robinet, A., Delgado, A. R., Yarza, A. R., & McGinn, J., … Grout, V. (2008). User modelling in ambient intelligence for elderly and disabled people. In *Proceedings of the 11th International Conference on Computers Helping People with Special Needs*. Linz, Austria: Academic Press. doi:10.1145/1409635.1409663

Jasemian, Y. (2008). Elderly comfort and compliance to modern telemedicine system at home. In *Proceedings of the Second International Conference on Pervasive Computing Technologies for Healthcare*. Tampere, Finland: Academic Press. doi:10.4108/ICST.PERVASIVEHEALTH2008.2516

Atallah, L., Lo, B., Yang, G. Z., & Siegemund, F. (2008). Wirelessly accessible sensor populations (WASP) for elderly care monitoring. In *Proceedings of the Second International Conference on Pervasive Computing Technologies for Healthcare*. Tampere, Finland: Academic Press. doi:10.4108/ICST.PERVASIVEHEALTH2008.2777

Hori, T., Nishida, Y., Suehiro, T., & Hirai, S. (2000). SELF-network: Design and implementation of network for distributed embedded sensors. In *Proceedings of IEEE/RSJ International Conference on Intelligent Robots and Systems*. Takamatsu, Japan: IEEE. doi:10.1109/IROS.2000.893212

Zhang, G. H., Poon, C. C. Y., Li, Y., & Zhang, Y. T. (2009). A biometric method to secure telemedicine systems. In *Proceedings of the 31st Annual International Conference of the IEEE Engineering in Medicine and Biology Society*. Minneapolis, MN: IEEE. doi:10.1109/IEMBS.2009.5332470

Bao, S., Zhang, Y., & Shen, L. (2005). Physiological signal based entity authentication for body area sensor networks and mobile healthcare systems. In *Proceedings of the 27th Annual International Conference of the IEEE EMBS*. Shanghai, China: IEEE.

Pereira, P., Grilo, A., Rocha, F., Nunes, M., Casaca, A., & Chaudet, C., ... Johansson, M. (2007). End-to-end reliability in wireless sensor networks: Survey and research challenges. In *Proceedings of EuroFGI Workshop on IP QoS and Traffic Control*. Lisbon, Portugal: Academic Press.

Misic, J. (2008). Enforcing patient privacy in healthcare WSNs using ECC implemented on 802.15.4 beacon enabled clusters. In *Proceedings of the Sixth Annual IEEE International Conference on Pervasive Computing and Communications*. Hong Kong: IEEE.

Renaud, M., Karakaya, K., Sterken, T., Fiorini, P., Hoof, C. V., & Puers, R. (2008). Fabrication, modelling and characterization of MEMS piezoelectric vibration harvesters. *Sens. Actuat. A*, 380-386.

Lauterbach, C., Strasser, M., Jung, S., & Weber, W. (2002). Smart clothes self-powered by body heat. In *Proceedings of Avantex Symposium*. Frankfurt, Germany: Avantex.

Marinkovic, S., & Popovici, E. (2009). Network coding for efficient error recovery in wireless sensor networks for medical applications. In *Proceedings of International Conference on Emerging Network Intelligence*. Sliema, Malta: Academic Press. doi:10.1109/EMERGING.2009.22

Schoellhammer, T., Osterweil, E., Greenstein, B., Wimbrow, M., & Estrin, D. (2004). Lightweight temporal compression of microclimate datasets. In *Proceedings of the 29th Annual IEEE International Conference on Local Computer Networks*. Tampa, FL: IEEE. doi:10.1109/LCN.2004.72

Sadler, C. M., & Martonosi, M. (2006). Data compression algorithms for energy-constrained devices in delay tolerant networks. In *Proceedings of the 4th ACM International Conference on Embedded Networked Sensor Systems*. Boulder, CO: ACM. doi:10.1145/1182807.1182834

Marcelloni, F., & Vecchio, M. (2009). An efficient lossless compression algorithm for tiny nodes of monitoring wireless sensor networks. *The Computer Journal*, *52*(8), 969–987. doi:10.1093/comjnl/bxp035

Marcelloni, F., & Vecchio, M. (2010). Enabling energy-efficient and lossy-aware data compression in wireless sensor networks by multi-objective evolutionary optimization. *Inf. Sci*, *180*(10), 1924–1941. doi:10.1016/j.ins.2010.01.027

Ting, C. K., & Liao, C.-C. (2010). A memetic algorithm for extending wireless sensor network lifetime. *Inf. Sci.*, *180*(24), 4818–4833. doi:10.1016/j.ins.2010.08.021

Demirkol, I., Ersoy, C., & Alagoz, F. (2006). MAC protocols for wireless sensor networks: A survey. *IEEE Communications Magazine*, *44*(4), 115–121. doi:10.1109/MCOM.2006.1632658

Durmus, Y., Ozgovde, A., & Ersoy, C. (2009). Event based queueing for fairness and on-time delivery in video surveillance sensor networks. In *Proceedings of IFIP Networking*. Aachen, Germany: IFIP.

Pandian, P. S., Safeer, K. P., Gupta, P., Shakunthala, D. T., Sundersheshu, B. S., & Padaki, V. C. (2008). Wireless sensor network for wearable physiological monitoring. *J. Netw.*, *3*, 21–28.

Pandian, P. S., Mohanavelu, K., Safeer, K. P., Kotresh, T. M., Shakunthala, D. T., Gopal, P., & Padaki, V. C. (2007). Smart vest: Wearable multi-parameter remote physiological monitoring system. *Medical Engineering & Physics*, *30*(4), 466–477. doi:10.1016/j.medengphy.2007.05.014 PMID:17869159

Halin, N., Junnila, M., Loula, P., & Aarnio, P. (2005). The life shirt system for wireless patient monitoring in the operating room. *Journal of Telemedicine and Telecare*, *11*(8), 41–43. doi:10.1258/135763305775124623 PMID:16375793

Mundt, W., Montgomery, K. N., Udoh, U. E., Barker, V. N., Thonier, G. C., Tellier, A. M., & Kovacs, G. T. A. et al. (2005). A multiparameter wearable physiologic monitoring system for space and terrestrial applications. *IEEE Transactions on Information Technology in Biomedicine*, *9*(3), 382–391. doi:10.1109/TITB.2005.854509 PMID:16167692

Gopalsamy, C., Park, S., Rajamanickam, R., & Jayaraman, S. (2005). The wearable motherboard TM: The first generation of adaptive and responsive textile structures (ARTS) for medical applications. *Virtual Reality (Waltham Cross)*, *4*(3), 152–168. doi:10.1007/BF01418152

Yazicioglu, R. F., Torfs, T., Merken, P., Penders, J., Leono, V., Puers, R., & VanHoof, C. et al. (2009). Ultra-low-power biopotential interfaces and their applications in wearable and implantable systems. *Microelectronics Journal*, *40*(9), 1313–1321. doi:10.1016/j.mejo.2008.08.015

Gupta & Schwiebert. (2001). Energy efficient protocols for wireless communication in biosensor networks. In *Proceedings of 12th IEEE Int'l Symp. Personal, Indoor and Mobile Radio Comm.* San Diego, CA: IEEE. doi:10.1109/PIMRC.2001.965503

KEY TERMS AND DEFINITIONS

Energy: The network is Energy Starving in nature powered by small battery based energy sources.

Markov Model: Mathematical model which is memory less in nature.

Security: Makes the network secure from attacks and hackers.

Thermal Aware: Implanted and Sticked nodes causes tissue damage due to heat dissipation.

Trustworthy: Trusted architecture capable of communicating confidential packets.

Wireless Body Sensor Network: Tiny sensor nodes sticked or implanted to the human body for sensing physiological signals.

Wireless Sensor Network: Small embedded machine forms a network to achieve a common goal.

Chapter 3
Design of Logistic Map-Based Spreading Sequence Generation for Use in Wireless Communication

Katyayani Kashayp
Gauhati University, India

Kandarpa Kumar Sarma
Gauhati University, India

Manash Pratim Sarma
Gauhati University, India

ABSTRACT

Spread spectrum modulation (SSM) finds important place in wireless communication primarily due to its application in Code Division Multiple Access (CDMA) and its effectiveness in channels fill with noise like signals. One of the critical issues in such modulation is the generation of spreading sequence. This chapter presents a design of chaotic spreading sequence for application in a Direct Sequence Spread Spectrum (DS SS) system configured for a faded wireless channel. Enhancing the security of data transmission is a prime issue which can better be addressed with a chaotic sequence. Generation and application of chaotic sequence is done and a comparison with Gold sequence is presented which clearly indicates achieving better performance with simplicity of design. Again a multiplierless logistic map sequence is generated for lower power requirements than the existing one. The primary blocks of the system are implemented using Verilog and the performances noted. Experimental results show that the proposed system is an efficient sequence generator suitable for wideband systems demonstrating lower BER levels, computational time and power requirements compared to traditional LFSR based approaches.

DOI: 10.4018/978-1-4666-8687-8.ch003

INTRODUCTION

Over the last decades there has been an exponential growth in wireless communication systems. Among many techniques develop for communication through the wireless medium, spread spectrum modulation (SSM) finds an important place in wireless communication due to many striking features like robustness to noise and interference, low probability of intercept, application to Code Division Multiple Access (CDMA) and so on. The idea behind SSM is to use more bandwidth than the original message while maintaining the same signal power. A spread spectrum signal does not have a clearly distinguishable peak in the spectrum. This makes the signal more difficult to distinguish from noise and therefore more difficult to jam or intercept. There are two predominant techniques to spread the spectrum, one is the frequency hopping (FH) technique, which makes the narrow band signal jump in random narrow bands within a larger bandwidth. Another one is the direct sequence (DS) technique which introduces rapid phase transition to the data to make it larger in bandwidth (Rappaport, 1997). Pseudo noise (PN) sequence, Gold code etc are the spreading codes which play a prominent role in SSM techniques. There are used as spreading code in SSM. A PN code is one that has a spectrum similar to a random sequence of bits but is determinately generated (Tse & Viswanath, 2005). A Gold code is a type of binary sequence, used in telecommunication primarily in CDMA and satellite navigation system like GPS (Tse & Viswanath, 2005). Gold codes have bounded small cross-correlations within a set, which is useful when multiple devices are broadcasting in the same frequency range. A set of Gold code sequences consists of 2^n-1 sequences each one with a period of 2^n-1. But PN and Gold codes are limited to fixed sequence lengths with a system configuration. Again flexibility is also poor because for same sequence length we cannot generate multiple numbers of sequences. Traditionally, to generate PN sequence, linear feedback shift register (LFSR) and certain sum-store blocks are used. Since Gold code is generated by doing exclusive-OR of two PN sequences, here also LFSR is required. Therefore, to generate PN or Gold code, a definite physical structure is required which consume significant power. The fixed length of LFSRs improve further constraints. The PN sequence length become confined within the LFSR size. In fading situations or in conditions where there are variations in the propagation medium, a varying length PN sequence shall be more suitable than a fixed length one primarily to use the advantages of SSM to counter detrimental effects observed in wireless channels. Continuous researches are going on to design devices that save power and demonstrate dynamic behaviour with respect to channel conditions required. Therefore, in this chapter an efficient spreading sequence generation method is presented using chaotic logistic map. The sequence thus generated is used as part of a SSM system designed for application method in faded environment. A logistic map is a polynomial mapping having a degree of 2. It gives the idea of how complex the generated sequence is. The chaotic behaviour arises from simple non-linear dynamical equations. Chaos happens when a small difference initially in the system leads to very big difference in the final state. A small error initially could lead to a big one in the final state. Prediction thus becomes impossible, and then the system behaves randomly (Reddy, 2007). Therefore, logistic map sequences have several advantages over Gold code. First, flexibility is more because the period of logistic map sequence is no longer limited to 2^n-1 like Gold code For same spreading code length we can generate extended number of spreading sequences, which is not possible in case of Gold code. Next, bit error rate (BER) performance in Direct Sequence Spread Spectrum System (DS SS) for different spreading code lengths and for different modulation schemes (BPSK, QPSK, DPSK) using logistic map code is better than Gold code. Further, computational times using different spreading code lengths and also using BPSK and QPSK modulation schemes are lesser than Gold code as observed in implementations. Also, mutual information (for different spreading code lengths and for different modulation schemes) are better than Gold code. Performance of the proposed sequence with SSM is compared with Gold code. The comparison parameters are bit BER, mutual information and computational time. The generated dynamic and recursive logistic map sequences have moderate correlation property. Therefore, the proposed sequences can be effectively used as spreading sequences in high data rate modulation schemes. Again the general form of logistic map equation contains two multiplication terms which may yield more area and power. We further modify the logistic map sequence generator to a multiplierless system which lead to less area and less power consumption than the existing logistic map systems. Thus, the proposed system is an efficient sequence generator suitable for wideband systems demonstrating lower BER levels, computational time and power requirements compared to traditional LFSR based approaches.

The chapter is organized as follows: Section I provides an introduction. Section II describes mathematical analysis of Logistic map and related works (an overview of the existing binary sequences using logistic map). Section III states about the proposed method and simulation. Section IV describes results. Finally, conclusion and discussions are summarized in section V.

THEORETICAL BACKGROUND

Here we briefly cover different related concepts.

Spreading the Spectrum

Spread spectrum (SS) was originally developed for military applications, to offer secure communications by spreading the signal over a large frequency band. The idea behind spread spectrum is to use more bandwidth than the original message while maintaining the same signal power and cover the signal with noise. Noise like signals are used as carriers. This makes the signal hard to distinguish from noise and therefore more difficult to jam or intercept (Tse & Viswanath, 2005).

Benefits of spread-spectrum transmission over fixed-frequency transmission are as follows:

- Narrowband interference can be tolerated by SS.
- SS signals are hard to intercept. If the pseudorandom sequence is not familiar to user, then in real time it is hard to intercept a transmission.

There are two major techniques to spread the spectrum:

- **Frequency Hopping (FH):** Which ensures the narrow band signal make rapid and random transitions within a larger bandwidth. Again, overall bandwidth requirement is less using only one carrier frequency to transmit the same information than the bandwidth which is required for frequency hopping. However, in transmission process small portion of the bandwidth is used at any given time, therefore bandwidth of effective interference is same. Degradation created by narrowband interference sources can be reduced by frequency hopping approach.

Synchronization of transmitter and receiver is the one the challenging issues of frequency hopping technique. In a fixed time period transmitter uses all channels. Then the transmitter can be found by the receiver by taking a random channel and picking for valid data on the used channel. The data of transmitter is recognized by a set of special sequence that is improbable to take place over the section of data for this channel and the section (segment) can have a checksum for reliability and further recognition. Again, Adaptive Frequency-hopping (AFH) SS is a technique which is used in Bluetooth. This technique avoids the hopping sequence crowded by frequency and improves resistance to interference. Implementation of adaptive transmission is easier with FHSS than with DSSS. In AFH, corrupted frequency channels are avoided and only the higher quality frequencies are used. Those corrupted frequency channels are possibly affected by frequency selective fading, or maybe some other factors attempts to communicate on same bands etc. Hence, in AFH technique detection of proper or corrupted channel detection is done (http://en.wikipedia.org/wiki/Frequency-hopping_spread_spectrum).

- **Direct Sequence (DS):** Introduces rapid phase transitions to the data making it capture the entire bandwidth and hence, become hard to recover. In DSSS a sine wave is phase modulated with an unbroken string of PN code termed as chips. The chip has shorter period (duration) than a message sequence i.e., with the help of sequences having faster chips, message bits are modulated. Hence, bit rate of message signal is lower than that of chip rate. In DSSS, receiver already knows the chip sequence of transmitter. Therefore receiver can apply the same PN sequence that is used by the transmitter to reconstruct the message signal by counteracting the result of PN sequence. In DSSS system, transmitted data is multiplied by a noise signal. The noise that is multiplied with the data is a pseudorandom sequence having values 1 and -1, which have frequency greater than that of message signal. This noise signal i.e the pseudorandom sequence is also used in receiver side to reconstruct the original message signal by multiplying the transmitted data with the same noise signal. This technique is termed as de -spreading. Therefore this process comprises a correlation of the already known received sequences and the transmitted sequences. Therefore the effect of the outcome improves the channel's signal to noise ratio and is termed as process gain. This effect can be increased by giving more chips per bit along with the longer PN sequence. If using dissimilar PN sequence, on the same channel, an undesired transmitter transmits, for that, signal processing gain is reduced. Therefore CDMA technique is used to allow more than one transmitter to share the same channel. But in FHSS technique, uniform frequency distribution is done rather than PN sequence to the original message bit sequence (http://en.wikipedia.org/wiki/Direct-sequence_spread_spectrum).

Coding

In wireless communication system the main aim is to ensure error free transmission of data and provide zero physical contact. With the help of analog and digital signals information can be broadcasted in communication system. For analog signal amplitude of the signal represent the source information, but for digital signal, the information is first converted to bit of binary stream of '1' and '0'and then transmitted. The benefit of using digital signal in transmission process is that error detection and correction is easy. Therefore for error detection and correction coding is necessary. Again for data compression and for cryptography application also coding is used. There are basically two types of coding (http://en.wikipedia.org/wiki/Coding_theory).

- Source coding (data compression),
- Channel coding (error coding).

Source encoding compresses the source data so as to transmit it more efficiently with more bandwidth. Creating a zipped data file is one of the common example. In zip data, compression files are make smaller to decrease the network load. On the whole, source code attempts to reduce the source redundancy and create small amount of bits which carry more information and require less storage. Data compression attempts to reduce the average length of information according to a probability model termed as entropy encoding. Usually there are many methods to convert to binary (http://en.wikipedia.org/wiki/Coding_theory).

Let a word 'Zebra' has to be sent. But before transmission the information should be converted to binary bit stream ('1' and '0'). This technique is termed as source coding. ASCII is one of that methods. In this method each alphabet is represented with the help of 7 binary bits, therefore, called code word. The alphabets in the word 'Zebra' are coded as

Z = '1010101'; e = '0110110'; b = '0010110'; r = '0010111'; a = '0001110';

ASCII is an illustration of code having fixed length, because each alphabet of the word 'Zebra' is represented by same code length i.e. 7 bits. However, in resourceful communication, happening of 'a' and 'e' is more than that of 'Z'. In that case, there is a method of encoding the alphabets such as the alphabets having greater probability of happening are denoted by shorter code words, and alphabets having lower probability of happening are denoted by longer code words. One of the examples of such a code is the Huffman code (http://en.wikipedia.org/wiki/Coding_theory).

Error Coding

In digital communication, channel coding is a process that connects transmitter and receiver with less probability of error. In channel coding, some extra bits (parity bits) are used, which consumes additional bandwidth. This makes correlation between transmitter and receiver more reliable. Strategies dealing with channel coding are

- Forward error correction (FEC), and
- Automatic repeat request (ARQ).

FEC attempts to find out the errors and makes corrections, but ARQ just find out the error and transmit a retransmit request towards the transmitter. FEC based techniques are more difficult than ARQ methods. But for greater benefit, combinations of both of the above are used. Therefore, it can be said that error coding or error control coding is a technique to detect and to correct errors by adding some redundancy bits to the data stream and then send it to the receiver. Encoder of the channel adds the extra bits to the information bits i.e. the message bits and then are transmitted. The receiver or decoder receive the transmitted bits and first detect the error. After that attempts are made to correct it. For example, let "000" has to be sent to the channel in place of sending '0' (only one bit). But because of the noise after receiving the bits to the receiver, it may turn out to be "001". But for majority of '0' bits, the received bit is decoded as "000" i.e finally, received message bit is '0' (http://en.wikipedia.org/wiki/Coding_theory).

Channel codes are of two types:

1. Linear block codes, and
2. Convolutional codes.

- **Linear Block Code:** Generally block code is a type having input bits k and forwards n bits at the output and often known as (n, k) codes. r = n-k is called as check bits. In block codes these check bits are added to the coded block.

Linear codes have linearity property, that means addition of any two code words is again a code word and they can be applied to the source blocks, therefore they are called linear block codes. Symbol alphabets that defines the linear block codes are (n, m, d_{min}), where,

n is codeword length, m the number of source symbol used for encoding and d_{min} is the minimum Hamming distance of the code.

There are various types of linear block codes. These are

- Cyclic codes (e.g. Hamming codes),
- Repetition codes,
- Parity codes,
- Polynomial codes (e.g. BCH codes),
- Reed Solomon codes,
- Algebraic geometric codes,
- Reed Muller codes and
- Perfect codes.

- **Cyclic Code:** A code C is called cyclic if C has linearity property and also, if a codeword after cyclic shifting is also a codeword. The code is called cyclic code, i.e. if $a_0 \ldots a_{n-1} \in C$, then $a_{n-1}a_0 \ldots a_{n-2} \in C$. Cyclic codes are generated using shift registers, and they have rich algebraic structure (http://en.wikipedia.org/wiki/Coding_theory).
- **Repetition Code:** Out of the error correcting codes, one of the basic code is repetition code. When a message signal is transmitted in a noisy channel, in some places the transmission may be corrupt. In the repetition code method the codes are repeated so many times because the channel corrupts some of the repeated bits. In that way the receiver will observe that a transmission error take place since the received data stream is not the recurrence of a single message, and furthermore, the receiver can detect the original message signal by observing at the received message in the stream of data that happens frequently (http://en.wikipedia.org/wiki/Coding_theory).
- **Parity Code:** It is a sedimentary form of detection of error code. Parity bit also known as check bit is a type of bit which is added to the end of binary bit steam that specify whether with the value of one, number of bits in the bit stream is even or odd. Parity bits are of two types:
    - Even parity bit, and
    - Odd parity bit.

In even parity, if the total number of one's (1's) are found to be odd in the bit stream then the parity bit is set to 1 (one) so that total number of 1's in the bit stream are even but if the bit stream contain even numbers of 1's then the parity bit is set to 0.

But for odd parity, the condition is opposite i.e. if the total number of one's (1's) are found to be odd in the bit stream then the parity bit is set to 0 (zero), just the reverse for the other case. The parity bit is said to be mark parity (bit is always 1) or space parity (bit is always 0) if the parity bit is not used but is presented (http://en.wikipedia.org/wiki/Coding_theory).

- **Polynomial Code:** Polynomial code is a kind of a linear code. The code words of the polynomial code composed of fixed length polynomials which are divisible by a known fixed polynomial having shorter length known as generator polynomial. In digital codes, at first to detect the error and then correction capacities of polynomial codes are calculated by minimum Hamming distance of that code. Here, minimum Hamming distance is the weight of non-zero codeword because the polynomial codes are linear codes. BCH code is an example of polynomial code (http://en.wikipedia.org/wiki/Coding_theory).

- **Reed Solomon Codes:** In 1960, Irving S. Reed and Gustave Solomon developed Reed Solomon codes. These codes are cyclic and non-binary error correcting codes which can detect and correct many symbol errors. If *t* check symbols are added in the data, Reed Solomon code can detect any arrangement of up to *t* erroneous codes and correct up to *t*/2 codes. Reed Solomon codes are transparent code like convolutional code i.e. if the channel symbols are inverted somewhere in the line, decoder still works. When these codes are shortened then transparency of these codes are lost. Depending on that the data bits are complemented or not, missing bits in the shortened code can be filled up by '1' or '0' (http://en.wikipedia.org/wiki/Coding_theory).
- **Algebraic Geometric Code:** Algebraic geometry code is also known as Goppa code. It is a general type of linear code.
- **Reed Muller Code:** In communication system Reed Muller codes are linear error correcting codes. Hadamard codes, Reed Solomon codes and Walsh Hadamard codes fall under Reed Muller code. Reed Muller code can be written as RM (*d, r*), where *r* is code length, $n = 2^r$ and *d* determines the order of the code. RM codes are connected to binary functions over the elements $\{1, 0\}$ (http://en.wikipedia.org/wiki/Coding_theory).
    - RM (0, *r*) codes are called repetition codes having length $n = 2^r$, minimum distance $d_{min} = n$ and rate $R = \frac{1}{n}$.
    - RM (1, *r*) codes are called parity check codes having length $n = 2^r$, minimum distance $d_{min} = \frac{n}{2}$ and rate $R = \frac{r+1}{n}$.
    - RM (*r*-2, *r*) codes are called family of extended Hamming codes having length $n = 2^r$ and minimum distance $d_{min} = 4$.
- **Perfect Code:** A code is said to be perfect if it attains the Hamming bound. Hamming bound is a boundary on the parameters of a random block code. Hamming bound is also known as volume bound or sphere packing bound (http://en.wikipedia.org/wiki/Coding_theory).
- **Convolutional Codes:** Generation of convolutional code can be done by transmitting the message sequence through a linear finite state shift register. Convolutional encoder (n, k, m) encodes 'k' bit input blocks to 'n' bit output block, which is dependent on current input block and preceding input blocks 'm'. Block codes are convolutional codes without memory. Convolutional code gives better simplicity of implementation but does not offer more protection against noise than block code (http://en.wikipedia.org/wiki/Coding_theory).

Spreading Codes

There are several codes used for spreading in communication systems. The spreading codes create space or buffer zones between signals or channels thereby minimizing distortions. The ability to mitigate distortions effects vary from code to code. Hence we discuss a few such codes used as spreading sequences. Broadly codes can be orthogonal and non orthogonal. Orthogonal codes are the codes which have zero cross-correlation and auto-correlation is a delta function. Similarly non orthogonal codes have non zero cross-correlation. Non orthogonal codes can be converted to orthogonal forms by using Walsh-Hadamard code, zero padding etc. Some advantages of orthogonal schemes are

- Intra-cell interference can be avoided by assigning orthogonal signal subspaces to other users.
- They are not distorted by near-far effect.

On the other hand, some common disadvantages of orthogonal schemes are

- They are affected by cross-cell interference.
- Frame synchronization is essential to maintain orthogonality.

Again, spectral-power efficiency is the advantage of non orthogonal approaches over orthogonal ones in fading environments for delay-sensitive applications, but intra-cell interference is more than orthogonal signal (Tse & Viswanath, 2005).

Some of the spreading codes (orthogonal and non orthonal) are as follows

- **PN Code:** PN sequence is one that has binary sequence of 1's and 0's and it is periodic in nature. Some of its characteristics are similar to binary random sequences having same numbers of 0's and 1's. They have very low correlation property between any two shifted version of the sequence and cross correlation is also low for any two sequences. Pseudo-random sequence is deterministic i.e not random but for user who does not know the code it behaves randomly (Rappaport, 1997). A linear-feedback shift register (LFSR) is used in PN Sequence Generator block which generates binary pseudorandom sequence. LFSR is implemented using simple shift register generator (SSRG, or Fibonacci series) configuration. PN sequence can be used in DS SS system, pseudorandom scrambler and descrambler etc. Generally, three factors are used to determine the generated PN sequence. Factors are shift register length (m), feedback logic and initial state of flip flop's. For m flip flop's there is at most 2^m number of achievable state of the shift register. Therefore, if the generated PN sequence have at most 2^m period, then the sequence is periodic. If the feedback logic comprises of exclusive-OR gates, then the shift register is termed as linear and in that case, the state zero is not allowed. Therefore, the PN sequence period of a linear m-stage shift register cannot be more than $2^m - 1$. The sequence having period $2^m - 1$ is called maximal length (ML) sequence (Rappaport, 1997), (Tse & Viswanath, 2005), (Dinan & Jabbari, 1998).

- **Gold Code:** The m sequence do not have proper cross correlation property although they have appreciable autocorrelation property. Therefore, a particular type of PN sequences are used and they are termed as Gold sequences. A Gold code or Gold sequence is a binary sequence, having application in telecommunication (CDMA) and satellite navigation (GPS). Gold codes have small cross-correlations properties but better than PN sequences, which is necessary when transmission from multiple devices are propagating through a channel in the same frequency range. The Gold sequences are described using a pair of sequences u and v, having period $N = 2^n - 1$, termed as preferred pair, which is defined by following Preferred Pairs of Sequences.

The set G (u, v) of Gold sequences is defined by

$$G(u,v) = \{u, v, u \oplus v, u \oplus Tv, u \oplus T^2v, \ldots, u \oplus T^{N-1}v\}$$

where, *T* represents the shifts vectors cyclically one place to the left, and $\oplus$ represents addition modulo 2. G (u, v) contains $N + 2$ sequences having period *N*. Generation of Gold sequence can be done by doing modulo 2 addition of two maximum length sequences having same length. The new generated code have the same length as of maximum length sequence because the generation is done by doing the modulo 2 addition operator (Dinan & Jabbari, 1998).

- **Walsh Code:** Orthogonal codes are easily generated by starting with a seed of 0, repeating the 0 horizontally and vertically, and then complementing the 1 diagonally. This process is to be continued with the newly generated block until the desired codes with the proper length are generated. Sequences created in this way are referred as Walsh code. The Walsh code is used to separate the user in the forward CDMA link. In any given sector, each forward code channel is assigned a distinct Walsh code (Tse & Viswanath, 2005).
- **Kasami Code:** Kasami codes are binary sequences of length $2^N - 1$, where N is an even integer. Kasami sequences have better cross-correlation values. There are two classes of Kasami sequences - the small set and the large set (Tse & Viswanath, 2005).

MODULATION (BPSK, QPSK AND DPSK)

Modulation is a process of mixing a signal with a sinusoid to produce a new signal. This new signal have several advantages over un modulated signal. Modulation allows us to send a signal over a band pass frequency range. If every signal gets its own frequency range, then we can transmit multiple signals simultaneously over a single channel, all using different frequency ranges. Another reason to modulate a signal is to allow the use of a smaller antenna. A baseband (low frequency) signal would need a huge antenna because in order to be efficient, the antenna needs to be about 1/10th the length of the wavelength. Modulation shifts the baseband signal up to a much higher frequency, which has much smaller wavelengths and allows the use of a much smaller antenna (Molisch, 2011).

Phase shift keying (PSK) is a large class of digital modulation schemes. PSK is widely used in the communication industry. It conveys data by changing, or modulating, the phase of a reference signal (the carrier wave). Basically PSK are of two types -

- Coherent PSK, and
- Non coherent PSK.

Coherent PSK

Any digital modulation scheme uses a finite number of distinct signals to represent digital data. PSK uses a finite number of phases, each assigned a unique pattern of binary digits. Usually, each phase encodes an equal number of bits. Each pattern of bits forms the symbol that is represented by the particular phase. The demodulator, which is designed specifically for the symbol-set used by the modulator, determines the phase of the received signal and maps it back to the symbol it represents, thus recovering the original data. This requires the receiver to be able to compare the phase of the received signal to a reference signal, such a system is termed coherent PSK (and referred to as CPSK) (Molisch, 2011). BPSK, QPSK are examples of coherent PSK.

Binary Phase Shift Keying (BPSK)

BPSK modulation is the simplest modulation method, the carrier phase is shifted by ± /2, depending on whether a +1 or -1 is sent (Molisch, 2011). The general form for BPSK follows the equation

$$s_0(t) = \sqrt{\frac{2E_b}{T_b}} \cos 2\pi f_c t + \pi = -\sqrt{\frac{2E_b}{T_b}} \cos 2\pi f_c t$$

$$s_1(t) = \sqrt{\frac{2E_b}{T_b}} \cos 2\pi f_c t$$

$$Basis\ Function = \varphi(t) = \sqrt{\frac{2}{T_b}} \cos 2\pi f_c t$$

$$s_0(t) = \varphi(t)\sqrt{E_b}$$

$$s_1(t) = -\varphi(t)\sqrt{E_b}$$

Quadrature Phase Shift Keying (QPSK)

QPSK modulated signal is a PAM where the signal carries 1 bit per symbol interval on both the in-phase and quadrature-phase component. The original data stream is split into two streams, b_{1i} and b_{2i} (Molisch, 2011).

$$b_{1i} = b_{2i}$$

$$b_{2i} = b_{2i+1}$$

Non Coherent PSK

Alternatively, instead of using the bit patterns to set the phase of the wave, it can instead be used to change it by a specified amount. The demodulator then determines the changes in the phase of the received signal rather than the phase itself. Since this scheme depends on the difference between successive phases, it is termed non coherent phase-shift keying (e.g. DPSK). DPSK can be significantly simpler to implement than ordinary PSK since there is no need for the demodulator to have a copy of the reference signal to determine the exact phase of the received signal. In exchange, it produces more erroneous demodulations. The exact requirements of the particular scenario under consideration determine which scheme is used (Molisch, 2011). DPSK is the example of non coherent PSK.

Differential Phase Shift Keying (DPSK)

The basic idea is that the transmitted phase is not solely determined by the current symbol; rather, we transmit the phase of the previous symbol plus the phase corresponding to the current symbol (Molisch, 2011). In DPSK, the data bits are encoded according to the following expression

$$\tilde{b}_i = b_i b_i - 1$$

Fading Channel

Fading is a phenomenon observed in wireless channels. The signal transmitted reach the destination using various paths giving rise to phase shifts and path delays. The received signal is a combination of several wave streams with add destructively and constructively leading to variation in signal quality and magnitude. This generates fading and is a severe problem in wireless channels through with mobile communication take place. The easiest probabilistic model for the channel filter taps is based on the assumption that there are a large number of statistically independent reflected and scattered paths with random amplitudes in the delay window corresponding to a single tap. Phase of the ith path is $2\pi f_c \tau_i$ modulo 2π. Now,

$$f_c \tau_i = \frac{d_i}{\lambda}$$

where λ is the distance travelled by the ith path and ¸ is the carrier wavelength. Since the reflectors and scatterers are far away relative to the carrier wavelength, i.e., $d_i \gg \lambda$, it is reasonable to assume that the phase for each path is uniformly distributed between 0 and 2π and that the phases of different paths are independent. The involvement of each path in the tap gain $h_l[m]$ is

$$a_i\left(\frac{m}{W}\right) e^{-j2\pi f_c \tau_i \left(\frac{m}{W}\right)} \sin c\left(l - \tau_i\left(m / W\right)W\right)$$

and this can be modelled as a circular symmetric complex random variable. Each tap h_l $[m]$ is the sum of a large number of such small independent circular symmetric random variables. It follows that $R\left(h_l\left[m\right]\right)$ is the sum of many small independent real random variables, and so by the Central Limit Theorem, it can reasonably be modelled as a zero-mean Gaussian random variable. Similarly, because of the uniform phase, $R\left(h_l\left[m\right]e^{j\varphi}\right)$ is Gaussian with the same variance for any fixed φ. This assures us that $h_l\left[m\right]$ is in fact circular symmetric. It is assumed here that the variance of $h_l\left[m\right]$ is a function of the tap l, but independent of time m (there is little point in creating a probabilistic model that depends on time). With this assumed Gaussian probability density, we know that the magnitude $\left|h_l\left[m\right]\right|$ of the lth tap is a Rayleigh random variable with density

$$\frac{1}{\sigma_l^2}\exp\left\{\frac{-x^2}{2\sigma_l^2}\right\}, x \geq 0;$$

and the squared magnitude $\left|h_l\left[m\right]\right|^2$ is exponentially distributed with density

$$\frac{1}{\sigma_l^2}\exp\left\{\frac{-x}{\sigma_l^2}\right\}, x \geq 0;$$

This model, which is called Rayleigh fading, is somewhat reasonable for scattering mechanisms where there are many small reflectors, but is adopted mainly for its simplicity in typical cellular situations with a relatively small number of reflectors. There is no direct line of sight (LOS) component and is observed frequently in city areas. The word Rayleigh is universally used for this model, but the supposition is that the tap gains are circularly symmetric complex Gaussian random variables. There is a commonly used alternative model in which the line of- sight path (often called a specular path) is large and has a well-known magnitude, and that there are also a huge number of independent paths. In this case, $h_l\left[m\right]$, at least for one value of l, can be modelled as

$$h_l\left[m\right] = \sqrt{\frac{\kappa}{\kappa+1}}\sigma_l e^{j\theta} + \sqrt{\frac{\kappa}{\kappa+1}}CN\left(0,\sigma_l^2\right)$$

with the first term corresponding to the specular path arriving with uniform phase θ and the second term corresponding to the aggregation of the large number of reflected and scattered paths, independent of θ. The parameter κ (so-called K-factor) is the ratio of the energy in the specular path to the energy in the scattered paths; the larger κ, the more shall be the deterministic component. Magnitude of such a random variable is said to have a Rician distribution. Its density has quite a complex form, it is often a better model of fading than the Rayleigh model (Molisch, 2011). It has a strong LOS component and is useful for rural areas and satellite channels.

Bit Error Rate

Bit Error Rate or Bit Error Ratio (abbreviated as BER) is a quality measure used to judge digital transmissions. When a data stream is transmitted via a communication channel, the received data gets changed due to various factors like distortion, noise, interference etc. So this changed number of received bits of the transmitted data stream is also called the number of bit errors. Thus in a time interval the ratio between the number of bit errors and the total number of transmitted bits is referred to as Bit Error Rate (BER). BER is expressed in percentages (http://en.wikipedia.org/wiki/Bit_error_rate).

For example, in a 10 bit transmission system if the transmitted bits are: 0 1 0 1 1 1 0 1 0 1 and the received bits are: 0 1 1 1 0 1 1 0 0 1 then the BER will be 4 altered bits divided by 10 transmitted bits, resulting BER equals to 0.4 or 40%.

Mutual Information

Mutual Information (also formerly known as Transinformation) is the quantity which measures (between two random variables) how much we may know about one random variable from another. Mutual information is defined in terms of bits and it is a dimensionless quantity. If the mutual information between two random variables is high then the reduction in uncertainty is higher, if it is low then reduction in uncertainty is lower and if it is zero then both the random variables are independent. Thus mutual information is the measure of reduction in uncertainty between two random variables (http://www.scholarpedia.org/article/Mutual_information).

Mathematically, if X and Y are two discrete random variables the Mutual Information between X and Y is given as

$$I\left(X;Y\right)=\sum_{y\in Y}\sum_{x\in X}p\left(x,y\right)\log\left(\frac{p\left(x,y\right)}{p\left(x\right)p\left(y\right)}\right)$$

where, $p\left(x,y\right)$ – Joint probability distribution function of X and Y.

$p(x)$ *and* $p(y)$ – Marginal probability distribution function of X and Y respectively (http://en.wikipedia.org/wiki/Mutual_information).

Logistic Map

The logistic map was first projected to explain the dynamics of insect populations in 1976 by biologist Robert May. Chaotic behavior appears in its simplest form in a noninvertible system. Also, logistic map was formulated with an objective to use it in dynamical systems. The logistic map is a polynomial mapping of degree 2, it gives the idea of how complex, chaotic behaviour can be while derived from simple non-linear dynamical equations. Chaos happens with small differences in the initial state of the system which can lead to large differences in its final state. A small error in the former could then produce an enormous one in the latter. Prediction becomes impossible, and the system appears to behave randomly (Reddy, 2007). The basic equation of logistic map is written as

$$x_{n+1} = r * x_n \left(1 - x_n\right)$$

where x_n is a number between zero and one, and represents the ratio of existing population to the maximum possible population at year n, and hence x_0 represents the initial ratio of population to maximum population (at year 0). Also, r is a positive number, and represents a combined rate for reproduction and starvation.

Chaotic behaviour is dependent on *r* such that:

- With r between 0 and 1, the population will eventually die, independent of the initial population.
- With r between 1 and 2, the population will quickly approach the value r-1/r, independent of the initial population.
- With r between 2 and 3, the population will also eventually approach the same value r-1/r, but first will fluctuate around that value for some time. The rate of convergence is linear, except for r=3, when it is dramatically slow, less than linear.
- With r between 3 and approximately 3.44949, from almost all initial conditions the population will approach permanent oscillations between two values. These two values are dependent on r.
- With r between 3.44949 and 3.54409 (approximately), from almost all initial conditions the population will approach permanent oscillations among four values. The latter number is a root of a 12th degree polynomial.

- With r increasing beyond 3.54409, from almost all initial conditions the population will approach oscillations among 8 values, then 16, 32, etc. The lengths of the parameter intervals which yield oscillations of a given length decrease rapidly.

The relative simplicity of the logistic map makes it suitable for generation of chaos. A rough description of chaos is that chaotic systems exhibit a great sensitivity to initial conditions-a property of the logistic map for most values of r between about 3.57 and 4. A common source of such sensitivity to initial conditions is that the map represents a repeated folding and stretching of the space on which it is defined (Reddy, 2007).

Mathematically logistic map is explained as follows:

Let time be discrete rather than continuous for a new class of dynamical system . These systems possess recursion relations, difference equations, iterated maps which are otherwise known simply maps. An example of 1-dimensional map is $x_{n+1} = \cos\left(x_n\right)$, where, the sequence x_0, x_1, x_2, ... is called the orbit starting from x_0.

Maps arise in various way such as

- Tools for analysing differential equations (e.g. Poincare and Lorenz),
- Models of natural phenomena (in economics and finance), and
- Simple examples of chaos.

Maps are capable of much wilder behaviour than differential equations because the points x_n hop discontinuously along their orbits rather than flow continuously.

For Fixed Point

An equation $x_{n+1} = f\left(x_n\right)$ is considered, where $f\left(x_n\right)$ is a smooth function from the real line onto itself. Suppose x^* satisfies $f\left(x^*\right) = x^* \Rightarrow x^*$ is a fixed point of the map. Its stability is determined by considering a nearby orbit $x_n = x^* + \eta_n$. Thus

$$x^* + \eta_{n+1} = x_{n+1} = f\left(x^*\right) f'\left(x^*\right)\eta_n + O\left(\eta_n^2\right)$$

Since $f\left(x^*\right) = x^*$,

$\eta_{n+1} = f'\left(x^*\right)\eta_n$ is the linearized map and $\lambda = f'\left(x^*\right)$ is the eigen value or multiplier.

- If $|\lambda| = \left|f'\left(x^*\right)\right| < 1$ then $\eta_n \to 0$ as $n \to \infty \Rightarrow x^*$ is linearly stable.
- If $|\lambda| = \left|f'\left(x^*\right)\right| > 1$ then x^* becomes unstable.
- If $|\lambda| = \left|f'\left(x^*\right)\right| = 1$ then we have to consider the terms $O\left(\eta_n^2\right)$.

Similarly if we consider $x_{n+1} = x_n^2$, no for fixed points at $x^* = (x^*)^2 \Rightarrow x^* = 0,1$.

$\lambda = f\left(x^*\right) = 2\left(x^*\right) \Rightarrow x^* = 0$ is stable and $x^* = 1$ instable.

Considering the equation of logistic map, let $0 \le r \le 4, 0 \le x \le 1 \Rightarrow$ be a parabola with maximum value of $r/4$ at $x = 0.5$.

- For $r < 1, x_n \to 0$ as $n \to \infty$
- For $1 < r < 3$, x_n grows as n increases, reaching a non-zero steady state.
- For larger r (e.g. $r. = 3.3$), x_n eventually oscillates about the former steady state $\Rightarrow$ period 2 cycle.
- At still larger r (e.g. $r = 3.5$), x_n approaches a cycle which repeats every 4 generations $\Rightarrow$ period 4 cycle.
- Further period doublings to cycles of period 8, 16, 32 etc occur when r increases. Therefore,
  - $r\ 1 = 3$ (period 2 is born);
  - $r\ 2 = 3.449$ (period 4 is born);
  - $r\ 3 = 3.54409$ (period 8 is born)
  - $r\ 4 = 3.5644$ (period 16 is born);
  - ....;
  - $r_\infty = 3.569946$ (period r is born).
- Successive bifurcations become faster and faster as r increases.
- The r_n converge to a limiting value r_∞.
- For large n, the distance between successive transitions shrinks by a constant factor

$$\delta = \lim_{n \to \infty} \frac{r_n - r_{n-1}}{r_{n+1} - r_n} = 4.669 .$$

When $r > r_\infty$, for many values of *r*, the sequence $\{x_n\}$ never settles down to a fixed point or a periodic orbit, the long term behaviour is aperiodic.

One might think that the system would become more and more chaotic as *r* increases, but in fact the dynamics are more subtle.

At *r* = *3.4* the attractor is a period 2 cycle. As r increases, both branches split, giving a period 4 cycle i.e. a period-doubling bifurcation has occurred. A cascade of further period-doublings occurs as r increases, until at $r = r_\infty \cong 3.57$, the map becomes chaotic and the attractor changes from a finite to an infinite set of points. When $r > r_\infty$, the orbit reveals a mixture of order and chaos, with periodic windows interspersed with chaotic behaviour. When, $r \cong 3.83$, then there is a stable period 3 cycle.

Now, considering the logistic map equation $x_{n+1} = r * x_n (1 - x_n)$; $0 \le r \le 4, 0 \le x_n \le 1$

For fixed point, $x^* = f(x^*) = rx^*(1 - x^*)$

$$\Rightarrow x^* = 0 \text{ or } 1 - 1/r ,$$

where $x^* = 0$ is a fixed for all *r*, and $x^* = 1 - 1/r$ only if $r \ge 1$, (since $0 \le x_n \le 1$)

Stability depends on $f'(x^*) = r - 2rx^*$.

$x^* = 0$ is stable for $r < 1$ and unstable for $r > 1$

$x^* = 1 - 1/r$ is stable for $-1 < (2 - r) < 1$, i.e. $1 < r < 3$ and unstable for $r > 3$.

Therefore, at $r = 1$, x^* bifurcates from the origin in a transcritical bifurcation. As r increases beyond 1, the slope at x^* gets increasingly steep. The critical slope $f'(x^*) = -1$ is attained when $r = 3$, the resulting bifurcation is called a flip bifurcation that means two cycle.

Now, logistic map has a 2-cycle for all r > 3, which is clear from the following description:

A 2 cycle exists if and only if there are two points *p* and *q* such that *f (p)* = *q* and *f (q)* = *p*. Equivalently, such a *p* must satisfy *f(f(p))* = *p,* where *f(x)* = *r x(1 - x)*.

Hence, p is a fixed point of the second iterate map $f^2(x) = f(f(x))$. Since $f(x)$ is a quadratic map and $f^2(x)$ is a quadratic polynomial.

Now, to solve $f^2(x) = x$, $x^* = 0$ and $x^* = 1 - 1/r$ are trivial solutions. The other two solutions are

$$p, q = \frac{r + 1 \pm \sqrt{(r-3)(r+1)}}{2r}$$

which are real for $r > 3$. Hence, a 2 cycle exists for all $r > 3$ for logistic map (May & Oster, 1980).

RELATED WORK

The study relating to logistic map sequence generating method is prepared after carefully going through other relevant works, a jist of which are presented serially according to the year of their publication.

In (Reddy, 2007), a work is related to generation of chaotic sequences for DS-CDMA system using various receiver techniques is presented.

In (Xueyi, Lu, Kejun, & Dianpu, 2000), a typical chaotic system, logistic map, under linear transformation is studied. A model, linear transform (LT) logistic map, is proposed, and its statistical property is derived. The result shows that the statistical properties of the model are identical with white noise. LT-logistic map sequences are used to the spreading spectrum address codes. Using this model, large number of chaotic sequences can be generated and the privacy of transmission in spreading spectrum communication system can be enhanced.

In (Bateni & McGillem, 1992), a method for generating chaotic sequences appropriate for use in DS/SS system has been established . In this generation process, the spreading sequences transform from one bit to the next in a very random-like nature, causing undesired interception are very difficult. The result shows that the generated sequence have low auto- and cross-correlations with spectral density similar to that of wideband noise. Again, the generated sequence out-perform PN sequence in low probability of intercept Therefore, interception and detection becomes much harder when chaotic sequences are used instead of binary PN sequences in direct sequence spread spectrum systems.

In (Phatak & Rao, 1994), a chaos base pseudorandom sequence generator is developed. Then some statistical testing related to randomness are also performed.

The result presented in this paper shows that the newly generate sequence can be used as spreading sequence in communication system.

In (Lawrance & Wolff, 2003), comparison is done between conventional spread spectrum techniques and chaotic spread spectrum technique and the comparison factors are power spectral density, autocorrelation sequence.

In (Netto & Eisencraft, 2008), comparison of conventional spread spectrum techniques and spread spectrum using chaotic signals in terms of power spectral density, autocorrelation sequence and bit error rate is done.

In (Patidar, Sud, & Pareek, 2009), based on two chaotic logistic maps a pseudo random bit generator is proposed. The outputs of both the newly generated chaotic logistic maps are compared and a new pseudo random bit sequence is generated.

In (Kadirl & Maarof, 2009), chaos based pseudorandom bit sequence and conventional pseudorandom bit sequence are examined. In this paper several methods such as Linear Congruential method, Mersenne Twister method and Marsaglia's Ziggurat algorithm are used to generate the chaos based pseudorandom bit sequence and to verify the randomness of the generated bit, mono bit test, run test and long run test are also done.

In (Mandi, Haribhat, & Murali, 2010), a new algebraic structure for generation of chaotic sequences defined over finite field GF (2^8) is projected here. The results indicate that, a periodic sequence of period ($2^8 - 1$) can be achieved for appropriate choice of initial values and bifurcation parameters both from GF (2^8). These sequences over GF (2^8) are changed to binary using three methods first one is the expressing every element in GF (2^8) of the sequence as binary 8 tuple. Second one is selecting a particular binary bit from each element of the sequence over GF (2^8). Third one is the mapping every element in GF (2^8) to GF (2) using Trace function. The cross correlation and Linear Complexity properties of generated binary sequences are shown here. It is clear from the result that they have good cross correlation values and large linear complexity property and therefore can be used as spreading sequences in CDMA applications.

In (Guyeux, Wang, & Bahi, 2010), a new chaotic pseudo random number generator (PRNG) is described. It joins the famous ISAAC and XORshift generators with chaotic iterations. This PRNG have important properties of topological chaos and successfully passes NIST tests. An application in the field of watermarking is described here. Again it also offers a satisfactory level of security for a range of applications in computer science. Not only that, an application in data hiding is also presented in this paper.

In (Krishna, 2011), the use of fractional order logistic equation to generate radar sequences is presented. A binary code is generated using fractional order logistic map equation. From the result it is shown that sequences with proper merit factor can be generated by varying fractional order.

In (Lakhtaria, 2011), a novel method based on radial basis function (RBF) for both blind and non-blind multiuser detection in chaos-based DS-CDMA systems is proposed. Again, a new method for optimizing generation of binary chaotic sequences using Genetic Algorithm is also proposed. Simulation results show that the proposed nonlinear receiver with optimized chaotic sequences outperforms in comparison to other conventional detectors such as a single-user detector, de correlating detector and minimum mean square error detector, particularly for underloaded CDMA condition.

In (Lakhtaria, 2012), a novel noise-resisting ciphering method resorting to a chaotic multi-stream pseudorandom number generator (Cms-PRNG) is proposed. This Cms-PRNG co-generates an arbitrarily large number of uncorrelated chaotic sequences. These cogenerated sequences are used in several steps of the ciphering process. Here, noisy transmission conditions are considered. The efficiency of the proposed method for ciphering and deciphering is illustrated through numerical simulations based on a Cms-PRNG involving ten coupled chaotic sequences.

In (Shaerbaf & Seyedin, 2011), an equation is proposed and proved that it is chaotic in the imaginary axis. And based on the chaotic equation a pseudorandom number generator is done. The alteration of the definitional domain of the chaotic equation from the real number field to the complex one provides a new approach to the construction of chaotic equations, and a new technique to generate pseudorandom number sequences accordingly. Both theoretical study and experimental results show that the sequences generated by the proposed pseudorandom number generator possess good properties.

In (Lozi & Cherrier, 2011), speech scrambling methods are used to scramble clear speech into unintelligible signal in order to neglect eavesdropping. Here, residual intelligibility of the speech signal is reduced by dropping correlation between the speech samples. Security level enhancement in speech encryption system using large set of Kasami sequence is described here. By using chaotic map speech signal is divided into segments, the polynomial is generation of the polynomial is done. Here using this polynomial large set of Kasami sequences are generated. Using this sequence as a key the speech signal is encrypted with AES-128 bit algorithm. The evaluation is carried out with respect to noise attack.

In (Yang & Jun, 2012), a study of possibility and usefulness of chaotic sequence in spread spectrum system based applications like wireless communication and watermarking . The results of this paper shows that availability of large number of variety of sequences of given length comparing with Gold sequences and m-sequence.

In (Kohad, Lngle, & Gaikwad, 2012), a chaos based secure direct-sequence spread-spectrum (DS SS) communication system is proposed, which is based on a novel combination of the conventional DS SS and chaos methods. In the proposed system, bit duration is varied according to a chaotic behaviour but is always equal to a multiple of the fixed chip duration in the communication process. Structure and operation of the proposed system are analyzed in detail. Theoretical evaluation of bit-error rate (BER) performance in presence of additive white Gaussian noise (AWGN) is provided. Parameter choice for different cases of simulation is also considered. In this paper, simulation and theoretical results are shown to verify the reliability and feasibility of the proposed system. Security of the proposed system is also discussed.

In (Mankar, Das, & Sarkar, 2012), chaotic spreading sequences is generated with arbitrary period, which optimizes the binary-quantization thresholds with different fractal parameters and initial values. The result of this paper shows that the method can generate a huge amount of high-performance chaotic spreading sequences with arbitrary period. The article supports the application of the chaotic sequences in spread spectrum communications.

In (Quyen, Yem, & Hoang, 2012), using three chaotic logistic maps a novel pseudo-random bit sequence generator (PRBG)is proposed here. Starting from randomly chosen initial seeds, algorithm generates at each iteration sequences of 32 bit-blocks. IEEE 754-2008 floating-point representation format is also taken into account for the generator. The performance of the generator is estimated through various statistical tests. The results demonstrate that the generated sequences possess properties of high randomness and good security level which make it appropriate for cryptographic applications.

In (Chengji & Bo, 2012), improvement of pseudo random number generation algorithm based on the logistic map is done. Here, the seed of the random generation algorithm is replaced with a sequence of numbers which is generated by the original logistic map and the improved logistic map. And then the two improved algorithms is described in detail including the essence of two algorithms and the processes. Also a series of experiments is done to prove the improved algorithms have a good randomness and they are efficient. The results shows that the two improved pseudo-random number generation algorithms take advantage of the chaotic characteristics of the logistic map which makes it become more efficient.

In (Francois & Defour, 2013), schemes for generating binary sequences from logistic maps are proposed. Using same length but different initial conditions of logistic map several binary sequences are generated. A comparison between maximum length sequences, Gold sequences and proposed sequences has been established, and demonstrate that the generated sequences are comparable and even better than maximum length sequences, Gold sequences.

In (Han, Wang, Liu, & Li, 2013), a improved logistic map sequence is used. Using this newly generated sequence, binary and ternary sequences are developed. Again results of the paper states that binary sequence have lower discrimination factor than that of ternary and from that it is clear that ternary the high resolution codes are good in performance.

In (Lakhtaria, 2009), chaotic logistic map sequences and Tent map sequences are used to generate one new sequence by EXORing the output bits of both chaotic logistic map sequences and Tent map sequences.

Proposed Approach

Here, we describe the proposed system. The primary focus is to generate a spreading sequence for a DS SS system. Initially we generate a binary sequence from the logistic map generator and use it as part of a DS SS system. The performance of the system is compared by varying the modulation types and varying the spreading sequence length. Figure 1 shows the flow logic system suitable for a statically defined S SS system.

First the data is generated from a random source. It consists of a series of ones and zeros. Data input bits are converted into symbol vector using modulation. Modulation scheme used to map the bits to symbols used in this work is BPSK, QPSK and DPSK. The modulated data is spreaded by a chaotic code in the transmitter side. Chaotic code is generated using logistic map as described in Figure 2. Thus a new chaotic sample is generated.

Figure 1. Flow diagram of static system

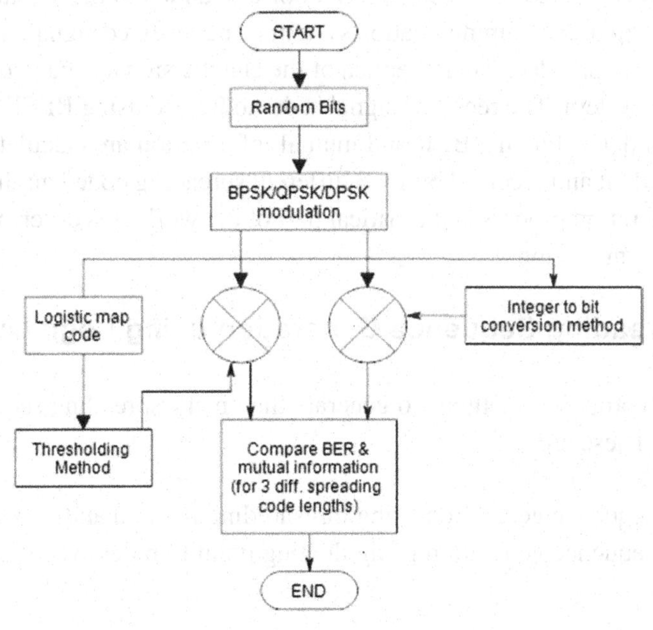

Figure 2. Flow diagram of both thresholding method and integer to bit conversion method

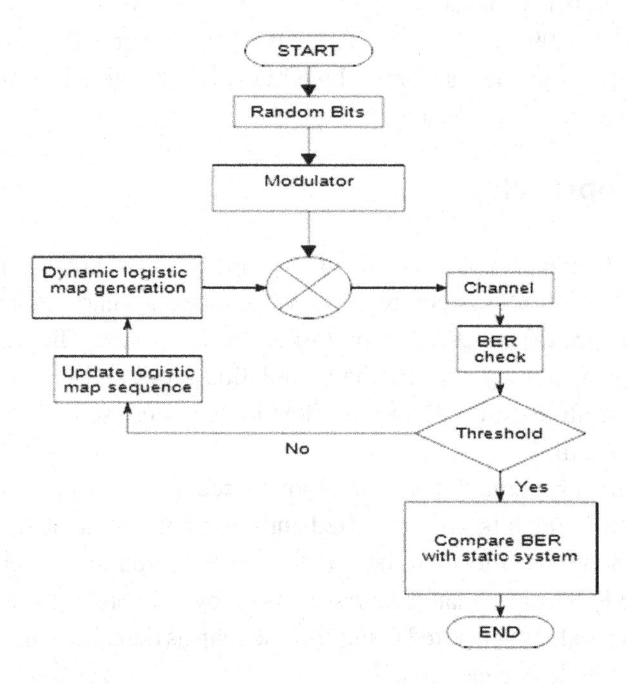

Next, the signal is passed through a channel, which has Rician fading characteristics. In Rician fading, there is only one direct component, and all other signals reaching the receiver are reflected. AWGN noise to added to the signal.

The final step of the communication system is the received signal. The received signal is first de-spreaded using a replica of the chaotic signal used at the transmitter side of the system. The received signal is demodulated using BPSK, QPSK and DPSK demodulator. Finally, BER and mutual information are calculated between the transmitted bit and received bit for 3 different spreading code lengths. Since the sequence generation process is the critical part of the work, it is described in detals in the subsequent sections.

Binary Spreading Sequence Generation using Logistic Map

There are two proposed methods to generate the binary spreading sequence using logistic map. These are

- Binary sequence generation using thresholding method and
- Binary sequence generation using floating point to bit conversion method.

Binary sequence is generated using a logistic map based approach as shown in the Figure 2.

Let x be the floating point valued chaotic sequence. For transforming this floating point valued x sequence to binary sequence, we define a threshold function as shown below.

$$f\left(x\right)=\begin{cases}1 & if\ x\geq 0.5\\ 0 & if\ x<0.5\end{cases}$$

Using this function we can generate binary sequence from logistic map.

Again, the Figure 2 shows the binary sequence generation using floating to binary converter. Here, floating point values are converted to integer and then integer to binary conversion is done. Here to convert integer to binary, frame based sequence is used. Properties of both the methods are explained in detail in the following.

Property of Generated Sequences for Both (Thresholding and Integer to Bit Conversion) Methods

Property of Method 1

- **Mono Bit Test:** The focus of the test is the proportion of zeroes and ones for the entire sequence. The purpose of this test is to determine whether the number of ones and zeros in a sequence are approximately the same as would be expected for a truly random sequence. The test assesses the closeness of the fraction of ones to 1/2, that is, the number of ones and zeroes in a sequence should be the same. But in this method number of ones are 66 and number of zeros are 34 out of 100 bits.
- **Run Length Test:** The focus of this test is the total number of runs in the sequence, where a run is an uninterrupted sequence of identical bits. A run of length k consists of exactly k identical bits and is bounded before and after with a bit of the opposite value. The purpose of the runs test is to determine whether the number of runs of ones and zeros of various lengths is as expected for a random sequence. In particular, this test determines whether the oscillation between such zeros and ones is too fast or too slow. Run in this method is 5.
- **Correlation Property:** This method shows moderate correlation property and which is clear from the Figure 3.
- **Computational Complexity:** Computational complexity is less in this method. Since this method requires 13.88 second computational time.

Figure 3. Correlation property of thresholding method

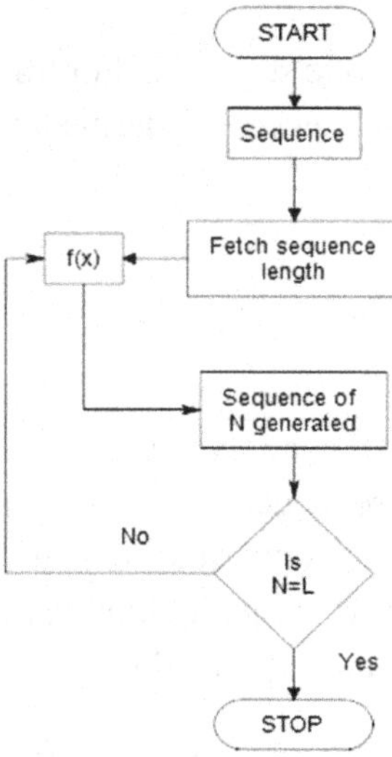

Property of Method 2

Similarly for mono bit test, number of ones are 59 and number of zeros are41 out of 100 bits.

For the run length test, run in this method is 5. The outcomes of correlation property is as shown in the Figure 4.

- **Computational Complexity:** Computational complexity is more in this method. This method requires 20.99 second computational time.

From the properties of both methods it is clear that mono bit test gives better result in method 2, run length test and the correlation property almost give same results but computational complexity is more in the second method, but for frame based sample the second method is more suitable. The chaotic behaviour is observed for sequences generated by logistic map with r taking values between 3.57 and 3.83. We show the chaotic nature in a few binary sequences generated by a logistic map

Table 1. Sequences generated by logistic map for r=3.61, 3.65 and 3.69

Value of r	Binary Sequence
3.61	00111000011010010100
3.65	01001001010010010101
3.69	01101001001110000110

Figure 4. Correlation property of integer to bit conversion method

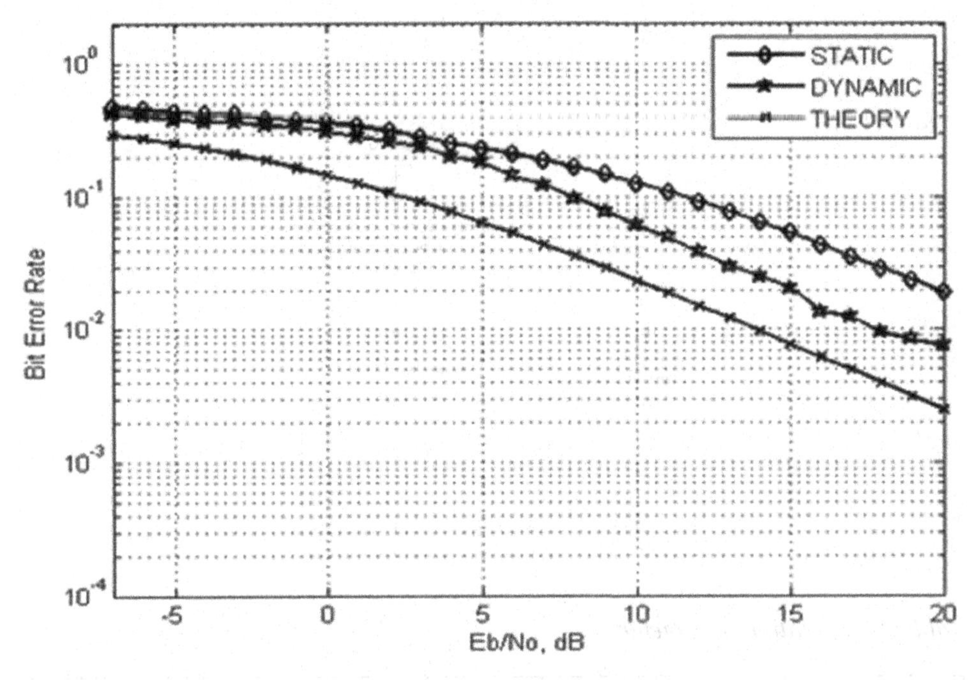

function with values like 3.61, 3.65 and 3.69 given to *r*. Table 1 shows that three different binary sequences for three different values of *r*.

System Model

The overall system model is shown in the Figure 5. The entire system is implemented first in MATLAB and then in Verilog. This is a DS SS system using logistic map code as spreading sequence for three different modulation schemes (BPSK, QPSK, and DPSK) and for three different spreading code length also (63, 127 and 511). In this model, Rician fading channel is used. After implementing the system model in MATLAB, BER is calculated using BPSK, QPSK, and DPSK modulation. Then BER is calculated for the same system using Gold code. The computational time of the system and the mutual information using both logistic map and Gold codes are also calculated. The simulation parameters are summarized in Table 2.

RESULTS AND DISCUSSIONS

The results obtained from the algorithmic performance and hardware implement ability is included. These are described in terms of BER, computational time and

Figure 5. System model

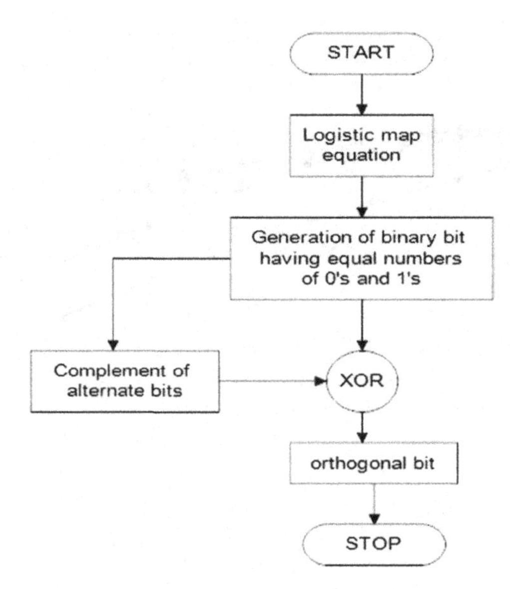

Table 2. Simulation parameter

Sl. No.	Item	Description
1	Modulation Type	BPSK, QPSK, DPSK
2	Data block Size	Between 1000 to 10,000
3	Gold code and logistic map sequence length	63, 127, 511
4	Channel Type	AWGN, Rician
5	No. Of trials per sequence length	10

mutual information. Also the transmitted and received bit stream from the hardware platform is presented.

Figure 6 shows that BER curves using BPSK and QPSK modulation for both logistic map code and Gold code almost give same results and their performances are also better than DPSK modulation, i.e BER curves for BPSK and QPSK modulation are closer to theoretical value than DPSK modulation. Again, Figure 7 shows that as the number of logistic map sequence increases the BER curves approach closer to the theoretical one i.e. for spreading code length 511 the BER curve is more close than spreading code length 127 and 63. The most noticeable aspect is that the BER curves of logistic map code are better than Gold code in the considered case. Computational time of the system model using logistic map code and Gold code is calculated for three different modulation schemes. Table 3 shows that the system

Figure 6. BER in fading channel for gold and logistic map code in DS SS systems using BPSK, QPSK and DPSK modulation schemes

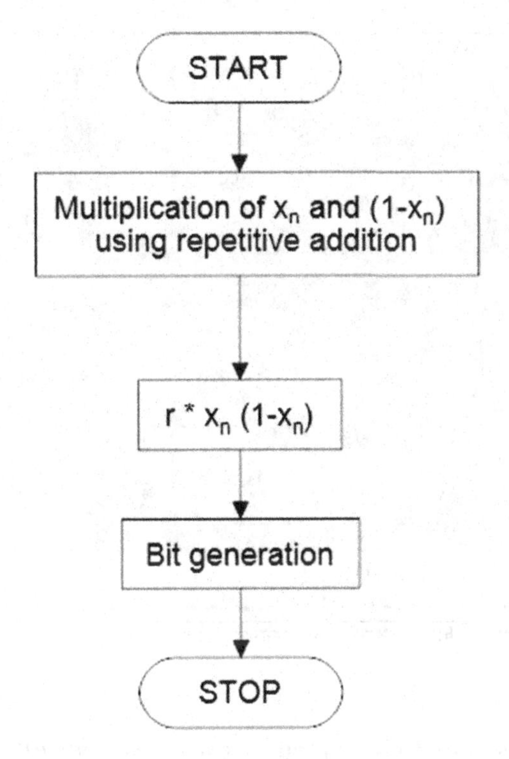

model using logistic map requires less computational time than using Gold code as spreading code in BPSK and QPSK modulation. But for DPSK modulation, system model using Gold code requires less computational time than system model using logistic map code. Again, Table 4 shows that the system model using logistic map requires less computational time than system using Gold code for different spreading code lengths. But for more spreading code length the system requires more computational time. Hence, lower BER can be achieved at the cost of computational time.

Figure 8 shows that mutual information using BPSK and QPSK modulation give almost same results but better than DPSK modulation. But when we consider both logistic map and Gold codes, the mutual information of logistic map code gives better results than Gold code for all type of modulation schemes i.e BPSK, QPSK and DPSK modulation. Again, Figure 9 shows that as the spreading code length increases the mutual information also increases for both Gold code and logistic map code. But for same spreading code length the logistic map code gives better result in comparison to Gold code in case of mutual information. Again signal power is an important factor in communication system. Gold code require more signal power than logistic map code.

Figure 7. BER in fading channel for gold and logistic map code in DS SS systems for spreading code length 63, 127, 511

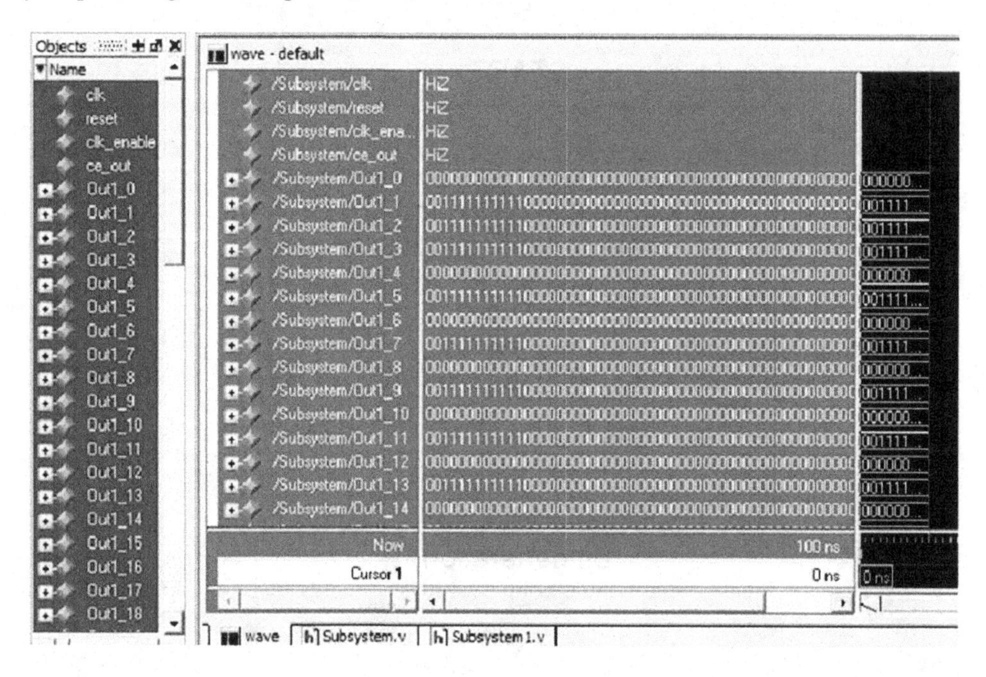

Table 3. Computational time (CT) of the system model using logistic map code and gold code in seconds for different modulation schemes

Modulation Type	CT (in Second) using Chaotic Code	CT (in Second) using Gold Code
BPSK	1.66	2.53
QPSK	3.22	3.90
DPSK	2.20	2.08

Table 4. Computational time (ct) of the system model using logistic map code and gold code in seconds for different spreading code lengths

Spreading Code Length	CT (in Second) using Chaotic Code	CT (in Second) using Gold Code
63	1.20	1.92
127	1.66	2.53
511	2.00	2.59

Figure 8. Mutual information for gold and logistic map code in DS SS systems using BPSK, QPSK and DPSK modulation schemes

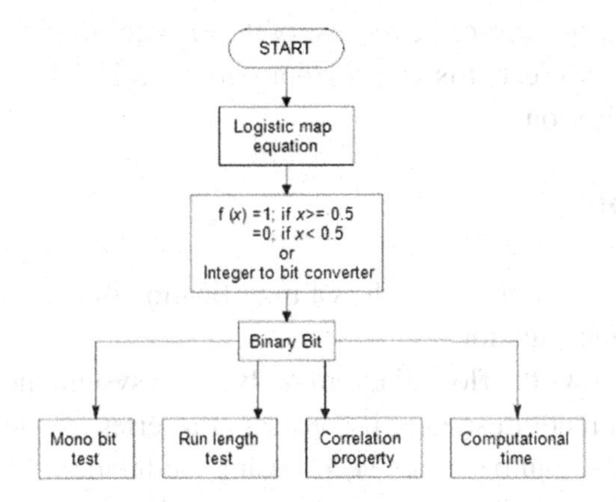

Figure 9. Mutual information for gold and logistic map code in DS SS systems for spreading code length 63, 127, 511

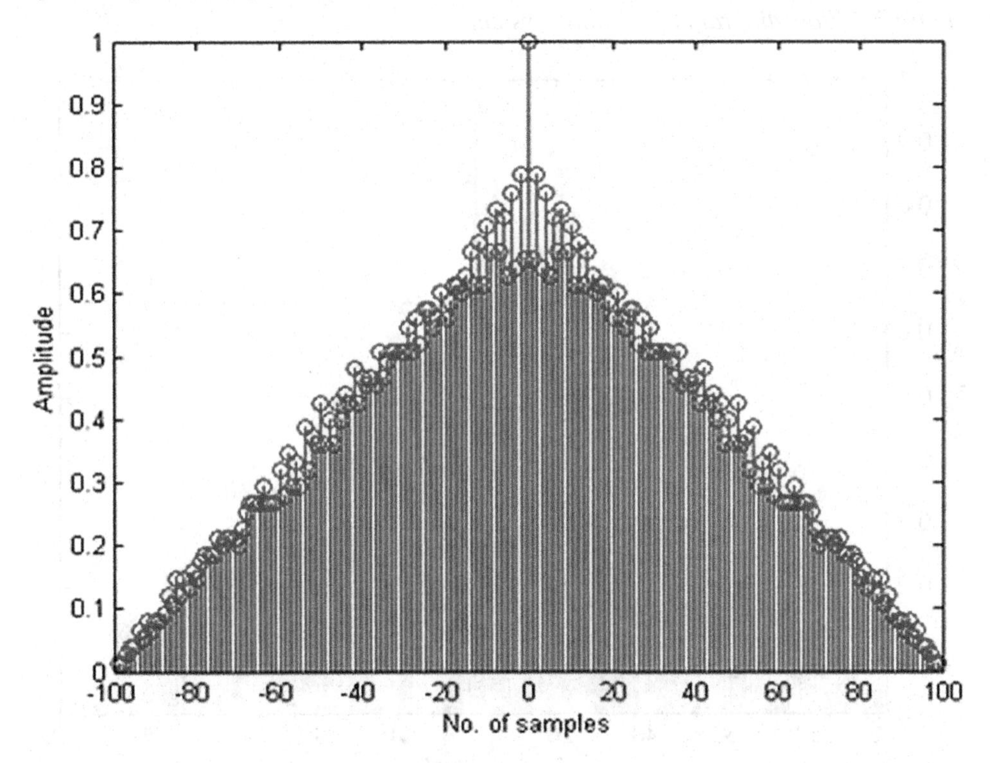

The system described until now based on the model shown in figure 5 is suitable for slowly varying channel conditions and is a prototype used for validating the concepts. The next section describes the use of the logistic map based chaos sequence for a dynamic situation created as a representation of a real life like scene as observed in mobile communication.

Dynamic System

Here, we describe an approach which is a modification of the earlier method configured for a dynamic situation.

The Figure 10 shows the flow diagram of dynamic system model. First the data is generated from a random source. It consists of a series of ones and zeros. Data input bits are converted into symbol vector using modulation. Modulation scheme used to map the bits to symbols used in this work is BPSK. The modulated data is spreaded by a dynamic chaotic code in the transmitter side. The word dynamic is used here because the logistic map sequence is not fixed here, length is varied with the BER value. Dynamic chaotic code is generated using logistic map as described

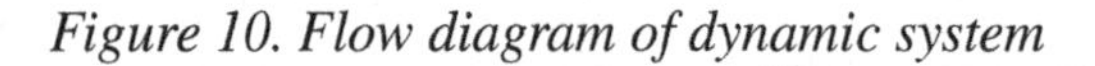

Figure 10. Flow diagram of dynamic system

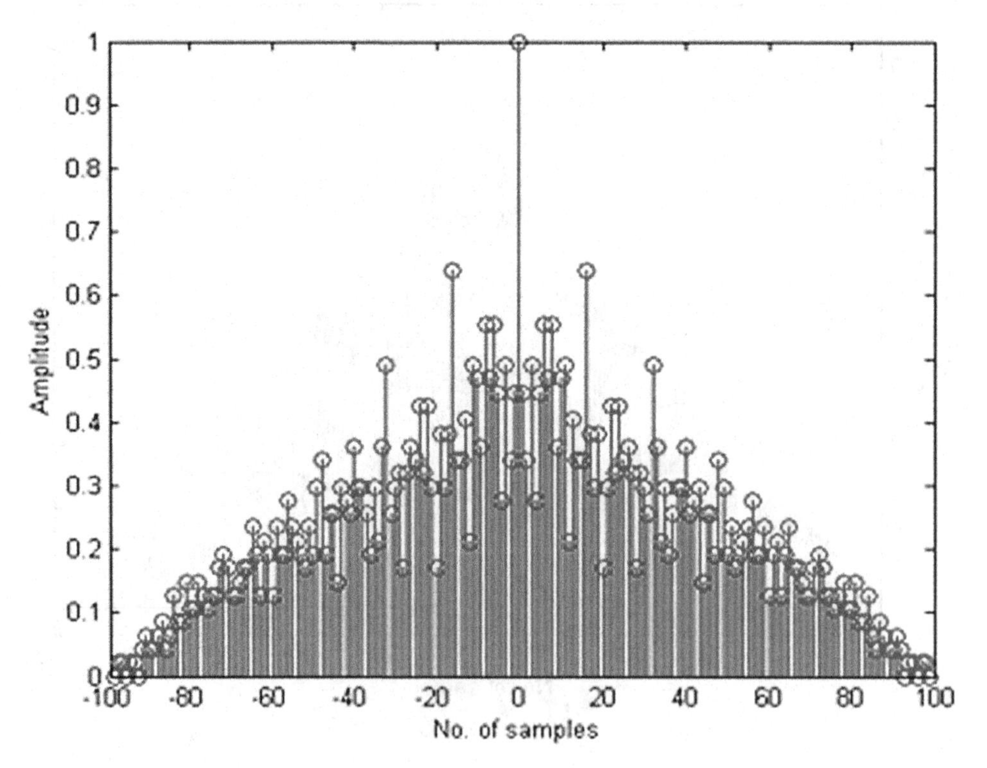

Figure 11. Flow chart to update logistic map sequence

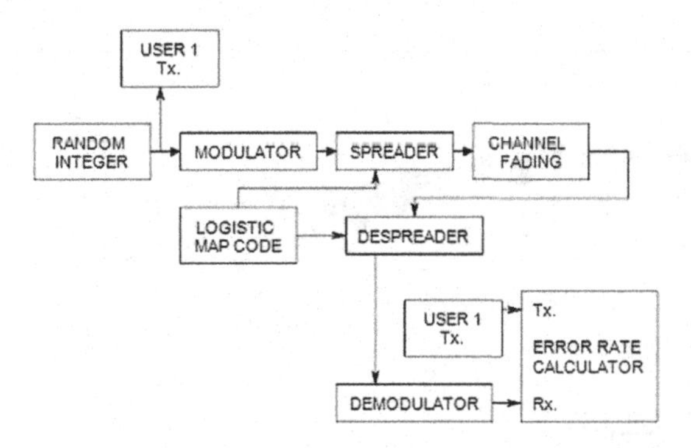

in Figure 11. Thus a new chaotic sample is generated. Next, the signal is passed through a channel, which has Rician fading characteristics. In Rician fading, there is only one direct component, and all other signals reaching the receiver are reflected. AWGN noise added to the signal. The final step of the communication system is the received signal. The received signal is first de-spreaded using a replica of the chaotic signal used at the transmitter side of the system. The received signal is demodulated using BPSK demodulator. After demodulation if the BER value crosses the threshold value then automatically length of logistic map change. The length of logistic map change according to the flow diagram shown in the Figure 11. Finally, BER is calculated between the transmitted bit and received bit.

For dynamic system BER curve is as shown in the Figure 12. From the Figure 12 it is clear that BER curve for dynamic system is more close to the theoretical value than the static one. Therefore dynamic system of logistic map gives better result than the static logistic map. But signal power and computational time requirement are more than that of static system.

ORTHOGONAL CHAOTIC SEQUENCE

The flow diagram to generate orthogonal chaotic sequence is shown in the Figure 13.

To generate orthogonal chaotic sequence the first step is to make equal numbers of 0's and 1's. After that alternate bits are complemented and then the alternate complemented bits and the bits having equal numbers of 0's and 1's are Exclusive-ORed. After Exclusive-ORing the generated bit sequence is orthogonal in nature. Signal power is calculated for both orthogonal and non orthogonal code since it is a important factor in communication system. To calculate the signal power, at first MATLAB function is written for both orthogonal and non orthogonal codes, then root mean square (RMS) value is calculated after that it is converted to dBm using

Figure 12. BER curve for static and dynamic logistic map

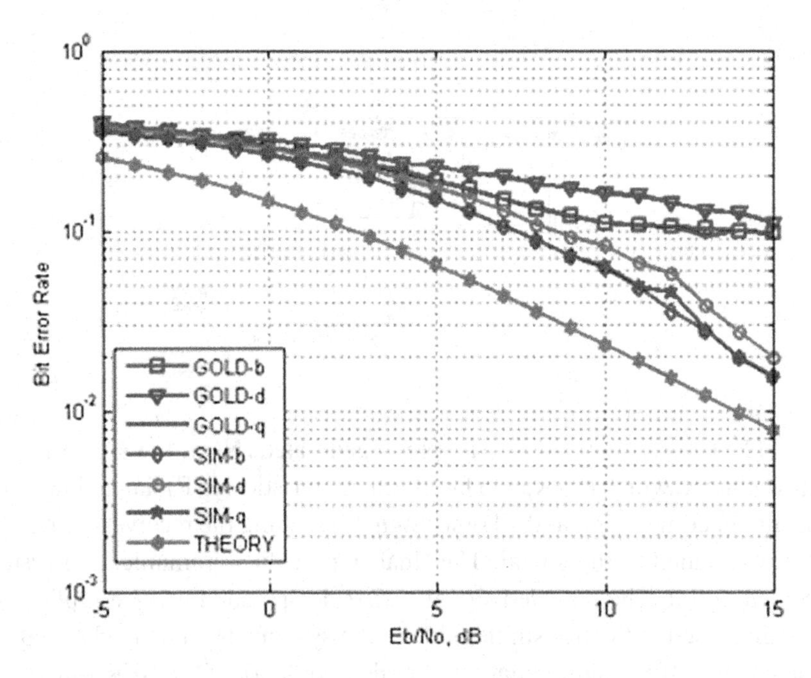

Figure 13. Flow diagram of orthogonal chaotic sequence generation

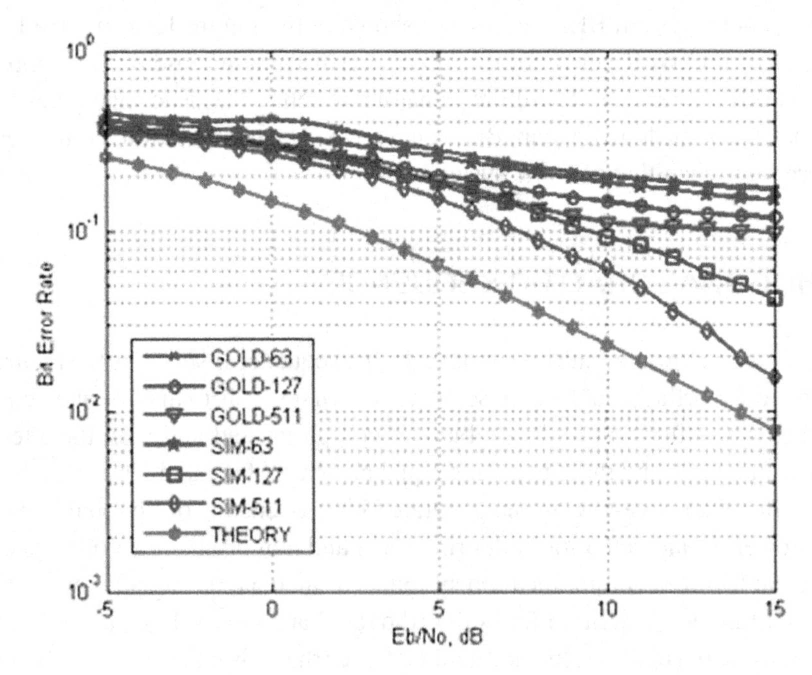

dBW converter. In dBW converter a load resistance is required. Here, for both orthogonal and non orthogonal system the load resistance is assumed to be 50 ohm. For non orthogonal DS SS system signal power is found to be 7.87 dBm and for orthogonal DS SS system the signal power is 5.94 dBm. Therefore, the orthogonal system requires less signal power than the non orthogonal system.

MULTIPLIERLESS CHAOTIC SEQUENCE GENERATOR

Flow diagram of multiplier less system is shown in the Figure 14. In the equation of logistic map there are two multiplication factors, in multiplierless chaotic sequence generator one of the multiplication is replaced by repetitive addition to reduce the amount of power required. The generated sequence have the same property as the existing one because the generated sequence is same as that of the existing one, just one of the multiplication is replaced by repetitive addition for lower power requirement. By using XPE (xilinx power estimator) power is measured and it is found that multiplierless logistic map required 1.448 W and existing logistic map required 1.579 W.

Figure 14. Flow diagram of multiplierless system

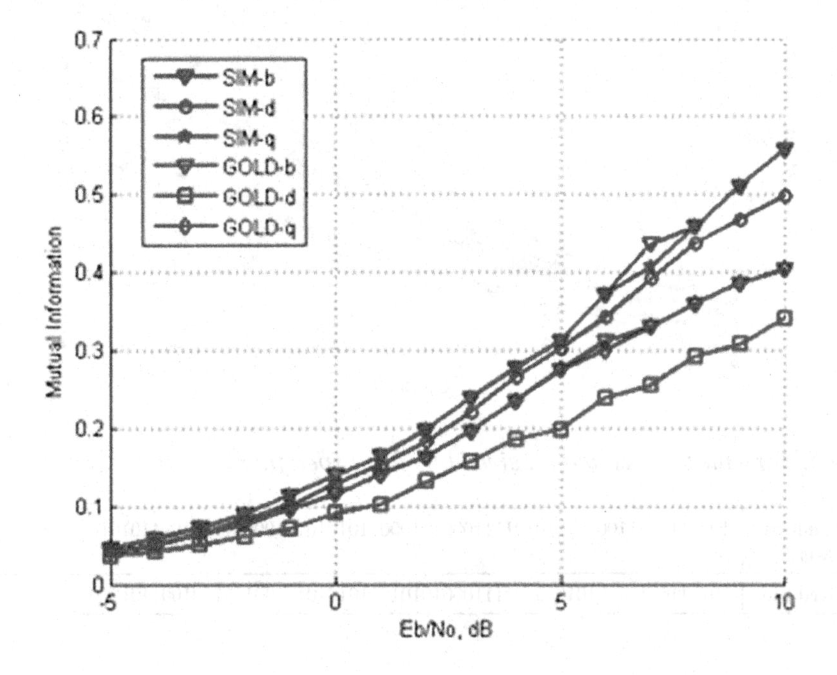

FPGA Approach

In order to perform the signal processing of the DS SS system in FPGA, HDL code (VHDL or Verilog) needs to be generated for the Simulink model. The Mathworks HDL Coder Toolbox will be used to generate this HDL code. Prior to use HDL Coder Toolbox on the Simulink model, some modifications to the model must be made which can be done with the help of HDL Workflow Advisor. Using the Mathworks HDL Workflow Advisor, which is part of the HDL Coder Toolbox, an implementability check of this FPGA model has been completed. Therefore, the generated Verilog code can be used for FPGA implementation.

The generated sequence is as shown in Figure 15. The generated chaotic sequence is then applied in DS SS transmitter as spreading sequence using BPSK modulation and Rician fading channel and the transmitted sequence is as shown in the Table 5. After that the transmitted sequence is received in DS SS receiver and the received sequence is as shown in the Table 5.

Figure 15. Logistic map spreading sequence

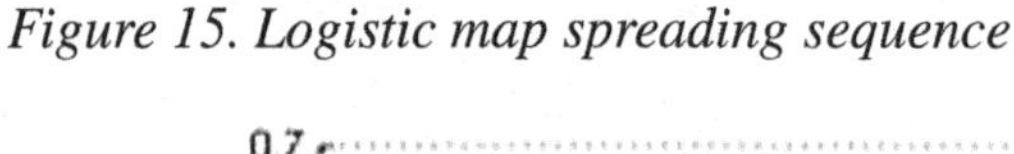

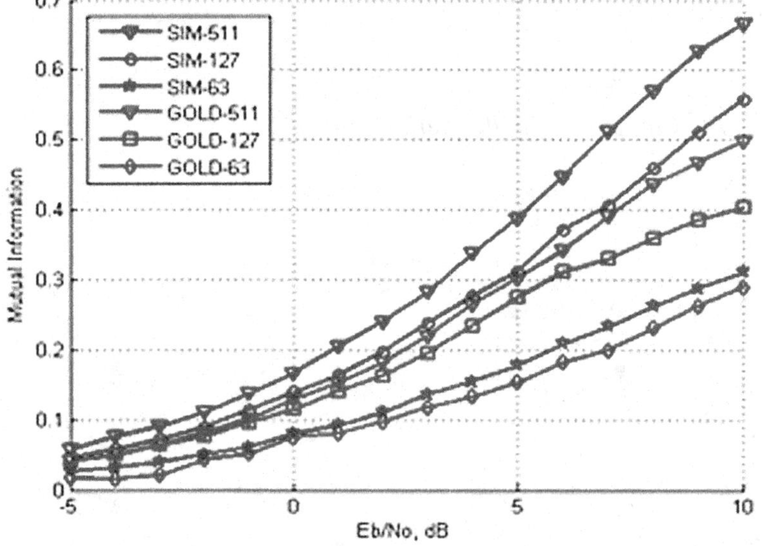

Table 5. Transmitted bits and received bits for a specific simulation

Transmitted Bits	101011001110010101010111100010100011010101010000111010011010101
Received Bits	101011001110010101010111100010101011010101010010111010011010101

SUMMARY

It has been observed that logistic map sequences have several advantages over Gold code. First, flexibility is more because the period of logistic map sequence is no longer limited to 2^n-1 like Gold code and for same spreading code length we can generate extended number of spreading sequences, which is not possible in case of Gold code. Second, the BER performance is more close to the theoretical value than Gold code in faded environment. Third, computational time is lesser than Gold code in BPSK and QPSK modulation and for different spreading code lengths as observed in implementations. Fourth, mutual information is also better than Gold code. Again from the experimental results it is clear that dynamic chaotic sequence gives better result than static one. Also for multiplierless logistic map equation consume less power than existing one. Hence, logistic map sequence can be used as efficient spreading code. Hardware implementation using Verilog makes the system efficient to use in practical situation. But simulations are constrained to baseband only and bandwidth, capacity are also not considered. In future, simulations can be extended to some nonlinear receivers like neural network receivers and ASIC-FPGA implementation of the nonlinear receiver using logistic map can also be investigated.

REFERENCES

Bateni, G. H., & McGillem, C. D. (1992). Chaotic Sequences for Spread Spectrum: An Alternative to PN – Sequences. In *Proceedings of IEEE International Conference on Wireless Communications*, (pp. 437 – 440). IEEE.

Chengji, P., & Bo, W. (2012). New Optimal Design Method of Arbitrary Limited Period Spreading Sequences based on Logistic Mapping. *4th International Conference on Signal Processing Systems*, (vol. 58, pp. 246 – 250). doi:10.1109/ICCIS.2012.29

Dinan, E. H., & Jabbari, B. (1998, September). Spreading codes for direct sequence CDMA and wideband CDMA cellular networks. *IEEE Communications Magazine*, *36*(9), 48–54. doi:10.1109/35.714616

Francois, M., & Defour, D. (2013, February). A Pseudo-Random Bit Generator Using Three Chaotic Logistic Maps. *Hyper Articles en Ligne*, *1*, 1–22.

Guyeux, C., Wang, Q., & Bahi, J. M. (2010). A Pseudo Random Numbers Generator Based on Chaotic Iterations. Application toWatermarking. *International Conference on Web Information Systems and Mining*, 6318, 202-211. doi:10.1007/978-3-642-16515-3_26

Han, Wang, Liu, & Li. (2013). Two Improved Pseudo-Random Number Generation Algorithms Based on the Logistic Map. *Research Journal of Applied Sciences, Engineering and Technology*, 2174 – 2179.

Kadir & Maarof. (2009). Randomness Analysis of Pseudorandom Bit Sequences. *International Conference on Computer Engineering and Applications, 2*, 390 – 394.

Kohad, H., Lngle, V. R., & Gaikwad, M. A. (2012, July – August). Security Level Enhancement In Speech Encryption Using Kasami Sequence. *International Journal of Engineering Research and Applications*, *2*, 1518–1523.

Krishna, B. T. (2011, April 7). Binary Phase Coded Sequence Generation Using Fractional Order Logistic Equation. *Circuits, Systems, and Signal Processing*, *31*(1), 401–411. doi:10.1007/s00034-011-9295-8

Lakhtaria, K. I. (2009). Enhancing QOS and QOE in IMS enabled next generation networks. In *Networks and Communications, 2009. NETCOM'09. First International Conference on* (pp. 184-189). IEEE. doi:10.1109/NetCoM.2009.29

Lakhtaria, K. I. (2011). Protecting computer network with encryption technique: A Study. In *Ubiquitous Computing and Multimedia Applications* (pp. 381–390). Springer Berlin Heidelberg. doi:10.1007/978-3-642-20998-7_47

Lakhtaria, K. I. (2012). *Technological Advancements and Applications in Mobile Ad-hoc Networks: Research Trends*. Information Science Reference.

Lawrance, A. J., & Wolff, R. C. (2003, June). Binary Time Series Generated by Chaotic Logistic Maps. *International Journal of Bifurcation and Chaos in Applied Sciences and Engineering*, *3*, 529–544.

Lozi, R., & Cherrier, E. (2011). Noise-resisting Ciphering based on a Chaotic Multi-stream Pseudo-random Number Generator. *IEEE International Conference for Internet Technology and Secured Transactions (ICITST)*, (pp. 91 – 96). IEEE.

Mandi, Haribhat, & Murali. (2010). Generation of Large Set of Binary Sequences Derived from Chaotic Functions Defined Over Finite Field GF (2^8) with Good Linear Complexity and Pair wise Cross - correlation Properties. *International Journal of Distributed and Parallel Systems, 1*, 93 – 112.

Mankar, V. H., Das, T. S., & Sarkar, S. K. (2012). Discrete Chaotic Sequence based on Logistic Map in Digital Communications. *National Conference on Emerging Trends in Electronics Engineering & Computing (E3C 2010)*, 1, 1016 – 1020.

May, R. M., & Oster, G. F. (1980, July 7). Chaos from Maps. *Journel of Physics Letters A*, *78*, 1–124. doi:10.1016/0375-9601(80)90788-4

Molisch, F. (2011). Wireless Communications (2nd ed.). Wiley and IEEE.

Netto, F. S., & Eisencraft, M. (2008). Spread Spectrum Digital Communication System Using Chaotic Pattern Generator. *The 10th Experimental Chaos Conference*.

Patidar, V., Sud, K. K., & Pareek, N. K. (2009). A Pseudo Random Bit Generator Based on Chaotic Logistic Map and its Statistical Testing. *International Journal of Modern Physics*, *251*, 441–452.

Phatak, S. C., & Rao, S. S. (1994, July 15). Logistic Map: A Possible Random Number Generator. *Physical Review. Statistical, Nonlinear and Soft matter Physics*, *51*, 3670–3678.

Quyen, N. X., Yem, V. V., & Hoang, T. M. (2012, November 21). A Chaos-Based Secure Direct-Sequence/Spread-Spectrum Communication System. *Journal of Abstract and Applied Analysis*, *2013*, 1–11. doi:10.1155/2013/764341

Rappaport, T. S. (1997). *Wireless Communications - Principles and Practice* (2nd ed.). Pearson Education.

Reddy, G. V. (2007). Performance Evaluation of Different DS-CDMA Receivers Using Chaotic Sequences. *International Conference on RF and Signal Processing Systems*, 32, 49-52.

Shaerbaf, S., & Seyedin, S. A. (2011, September). Nonlinear Multiuser Receiver for Optimized Chaos-Based DSCDMA Systems. *Iranian Journal of Electrical & Electronic Engineering*, *7*, 149–160.

Tse, D., & Viswanath, P. (2005). Fundamentals of Wireless Communications (2nd ed.). Cambridge. doi:10.1017/CBO9780511807213

Xueyi, Z., Lu, J., Kejun, W., & Dianpu, L. (2000). Logistic-Map Chaotic Spreading Spectrum Sequences Under Linear transformation. *The 3rd World Congress on Intelligent Control and Automation*, *4*, 2464 - 2467.

Yang, L., & Jun, T. X. (2012, May 2). A new pseudorandom number generator based on complex number chaotic equation. *Chinese Physics B*, *21*.

Chapter 4
Security Aspect in Multipath Routing Protocols

Prasanta K. Manohari
Silicon Institute of Technology, India

Niranjan K. Ray
Silicon Institute of Technology, India

ABSTRACT

In the absence of central authority, dynamic topology, limited bandwidth and the different types of vulnerabilities secured data transmission is more challenging in Mobile Ad hoc Network (MANET). A node in MANET acts as a host as well as a router. Routing is problematic due to node mobility and limited battery power of node. Security mechanisms are required to support secured data communication. It also requires mechanism to protect against malicious attacks. In recent times multipath routing mechanisms are preferred to overcome the limitation of the single path routing. Security in routing dealt with authentication, availability, secure linking, secure data transmission, and secure packet forwarding. In this chapter, we discuss different security requirements, challenges, and attacks in MANETs. We also discuss a few secured single path and multipath routing schemes.

1. INTRODUCTION

Mobile ad hoc network (MANET) is a self-configuring, infrastructure-less network. The nodes in this network are mobile in nature and they do not depend on any fixed

DOI: 10.4018/978-1-4666-8687-8.ch004

infrastructure to support communication among themselves (Mohapatra, & Krishnamurthy, 2005). Network topology changes rapidly due to the node characteristics. The network is easy to deploy in any environments. MANETs lead themselves to countless applications such as battlefield communication, disaster recovery management, environmental monitoring, etc. In the same time, it suffers with different challenges due to its physical constraints.

MANET performs many tasks and routing is major among them. The path between a source and destination is established through many intermediate nodes. The packet is forwarded by many nodes and passes through multi-hops to reach at destination. Nodes can move, leave and join the network at any point of times. As a results, path between source-destination changes rapidly. It causes link-failure, frequent modification at routing table, increased delay and congestion in the network. Subsequently it is very easy for hackers to eavesdrop and gain access to confidential information. Attacker inserts erroneous routing information, changes routing updates and transmits incorrect routing information.

The security objectives in the mobile ad hoc networks are also affected by different attacks. Such objectives are: *authentication, confidentiality, integrity, availability, and non-repudiation* (Yang, Haiyun, Fan, Songwu, & Zhang, 2004). Excluding these parameters security of MANETs cannot be complete. In literature, numerous security mechanisms are discussed. But, they do not give any concrete solutions to protect against the different attacks. In this chapter, we focus on different security aspects in single and multipath routing (Tarique, Tepee, Adibi, & Erfani, 2009). Single-path routings are not enough to provide higher data rate transmission. However, that can be achieved by establishing multiple paths between a source-destination pair. Therefore in many applications multipath routing are preferred. Multipath routing provides many benefits and at the same time, security in multipath is also more challenging. Several attacks or vulnerabilities may affect the route discovery of multipath routing protocols. Any malicious node or misbehaving node can create unfriendly attack or remove all other nodes from providing any service. Attacks or vulnerabilities are allowing a small set, or even a single node, to control the routing paths of critical nodes. The use of multiple paths in MANETs could diminish the effect of unreliable wireless links and the frequent topological changes. Nodes in an ad hoc network are usually powered by carefully distributing traffic load into multiple paths.

The remaining of this chapter is coordinated as follows: In Section 2, different security challenges and requirements are discussed. In Section 3, different secured routing mechanisms are discussed. Conclusions are given in Section 4.

2. BACKGROUND AND PRELIMINARIES

Secured linking, secured data transmission and packet forwarding are major security areas in MANETs. Identification and privacy are two basic security requirements in MANET routing. MANETs have several security requirements such as: *authentication, confidentiality, integrity, availability, and non-repudiation.* There is also numerous challenges due to dynamic topology, wireless links, scalability, memory and computational power limitation, and limited resources. In the presence of these challenges security is more challenging. In this section we briefly discuss a few security challenges in MANETs.

2.1 Security Issue and Requirements in MANETs

- **Identification Issue:** Nodes use common radio link to setup ad-hoc infrastructure. A communication should be secure between nodes. To establish a secure communication, one node should be able to detect the other nodes. For this the node has to provide its identity as well as associated identifications to others nodes. The identity and credentials provided by the node need to be authenticated and protected. So that, the receiver node cannot be questioned for a received authenticate message.
- **Privacy Issue:** The identification issues all together leads the privacy issues for MANETs. The mobile node uses different types of identities and that are varied from the link level to the operational level. In privacy point of view, a common mobile node is not ready to expose its identity or credentials to another mobile node. In many cases revealing identity is expected to generate communication link. Hence, a whole privacy protection is required in ad hoc networking.
- **Authentication Issue:** Authentication means the identity of the source information of mobile nodes. Without authentication, an unauthorized node could easily attack the network, and in the same time the unauthorized node may harm to the data resources. So it is necessary to have a mechanism for preventing malicious nodes. Authentication may be mutual authentication, it is crucial because each node needs to know the information of another node. Another authentication is pairwise authentication. This is required to protect the privacy of the communication between the two participating nodes. Efficient authentication needs due to the physical limitation of the MANETs.
- **Integrity:** It gives assurance that a packet is not tampered even when it passes through multiple numbers of nodes before reaching at destination. It helps

to maintain the genuineness of transmitting messages or data (Singh, Kaur, & Das, 2012).

- **Non-Repudiation:** It is a security service that provides protection against false denial of involvement in a communication. If a node with some undesired function is compromised, then it helps to discriminate. It ensures that the sender cannot deny later that packet is forwarded by that node. The receiver also cannot be denied that it has not received that packet forwarded by the sender node.
- **Confidentiality:** It is the process where unauthorized readers cannot read the information sent between the users. Network transmission of sensitive information such as strategic or tactical military information requires confidentiality. Confidentiality gives more privacy to network, where authorized systems or persons are able to access the protected data or programs. It is also known as privacy.
- **Availability:** It is the ability of the network to allow or provide services. It is very important in many applications. Resources should be available to genuine users whenever required.
- **Authorization:** It defines the activity of each network node. It provides rules by which a node not allowed to perform a specific task. It determines which node can access resources or information across the network. This requirement can be the result of the network organization, or the supported application. Authorization needs robust methods to check correctness of protocols and proper utilization of resources.

MANETs have also several challenges due to its physical characteristics and constraints (Raghavendran, Satish, & Varma, 2013). Topology is mostly dynamic in MANET, so the security proposal should have mechanism to adjust the topology. Updating information of dynamic links among nodes is a major challenge. Scalability is another key issue in many routing mechanisms. Security mechanisms should be support for large as well as small networks. Nodes have restricted storage and computational capabilities. These limitations should be consider while designing a secured protocol for MANETs.

2.2 Attacks in MANETs

Security attacks in MANET can be classified as (i) Passive attacks, and (ii) Active attacks (Sevil, Clark, & Tapiador, 2010). They are briefly discussed below:

Figure 1. Example of attacks

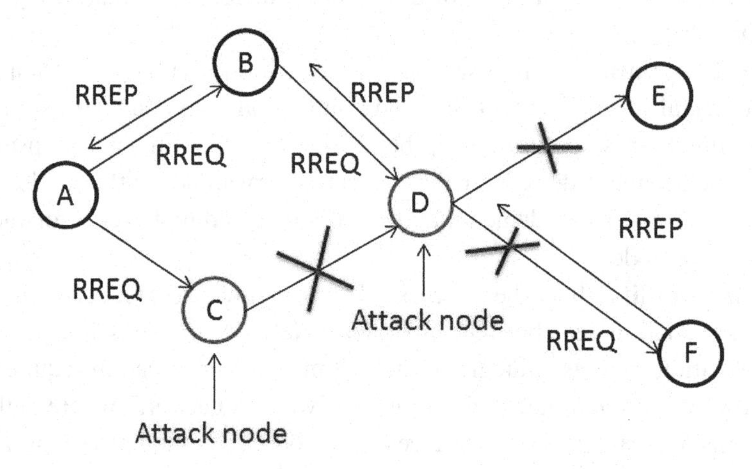

2.2.1 Passive Attacks

In passive attacks (Sevil, Clark, & Tapiador, 2010), data are not affected by attackers. It analyzes network traffic to identify communicating nodes, and monitors data which is exchanged between them. It only attempts to discover valuable information by listening to the routing traffic. Attacker looks and watches the transmission and does not modify the data packets. It is very difficult to detect this type of attacks. Figure 1 shows an example of attack scenario in a typical on-demand routing protocol. Some passive attacks are: *traffic analysis*, *eavesdropping*, *traffic monitoring*, and *snooping, etc*. Two passive attacks are discussed briefly below.

1. **Traffic Analysis Attack:** In this approach, attackers employ different techniques such as RF direction finding, traffic rate analysis, and time-correlation monitoring. For example, by timing analysis, it can be revealed that two packets in and out of an explicit forwarding node at time t and (t+δ) are likely to be from the same packet flow. Traffic analysis in ad hoc networks may reveal: the existence and location of nodes; the communications network topology; the roles played by nodes; the current sources and destination of communications; and the current location of specific individuals or functions.
2. **Eavesdropping Attack:** This is also known as disclosure attack. The attack is executed by external or internal nodes. The attacker can analyze broadcast messages to reveal some useful information about the network.

2.2.2 Active Attacks

Active attacks (Sevil, Clark, & Tapiador, 2010) modify user data, such as message modifications, message replays and message fabrications. It disrupts normal functionality of the network. Some active attacks are: *modification attacks, fabrication attacks, Denial of Service (DOS) Attacks, etc.* A few active attacks are discussed below:

1. **Modification Atta cks:** Attackers modify packets to disrupt the network. For example, in the sinkhole attack, the attacker tries to attract nearly all traffic from a particular area through a compromised node. It is especially effective in routing protocol which advertises information such as remaining energy and nearest node to the destination in the route discovery process, etc. A sinkhole attack can be used as a basis for further attacks like dropping and selective forwarding attacks. A black hole attack is like a sinkhole attack that attracts traffic through itself and uses it as the basis for further attacks. The aim is to prevent packets being forwarded to the destination. If the black hole is a virtual node or a node outside the network, it is hard to detect (Yadav, Jain, & Faisal, 2012).
2. **Fabrication Attacks:** Here the attacker forges network packets. Fabrication attacks (Yadav, Jain, & Faisal, 2012) are classified into "active forge" in which attackers send fake messages without receiving any related messages. The attacker also sends a "forge reply" message. In the forge reply attack, the attacker forges a route reply (RREP) message after receiving a route request message (RREQ). The reply message contains falsified routing information showing that the node has a fresh route to the destination node. It causes route disruption.
3. **Denial of Service Attacks:** Denials of Service (DOS) attacks are most concerning problems for network managers (Mueller, Tsang, & Ghosal, 2004). In this attack, a malicious node makes an attempt to prevent legitimate nodes from accessing information or services. DOS attacks target resources in three scenarios. First, it attacks point to storage and processing resource. It directly attacks to memory, storage space or CPU of the service provider and preventing the node from sending or receiving packets from other legitimate nodes. Second, it attacks energy resources, especially the battery power of the service provider. A malicious node may continuously send a fake packet to a node. Due to flooding of fake packet, a malicious node consuming the victim's battery energy and keeping out from communicating with the other node. Finally attackers waste the precious bandwidth. An attacker placed between the multiple

communications nodes and wastes the network bandwidth and disrupts the connectivity.

4. **Interception Attacks:** Attackers might introduce the interception attacks to get an unauthorized access to the routing message that are not sent to them. These kinds of attacks expose the integrity of the packets because such packet might be reformed before being forwarded to the next-hop. Further, the intercepted packets strength is analyzed before passing that to the destination. Thus, it violates the confidentiality. Some examples of interception attacks are: wormhole attacks (Yadav, Jain, & Faisal, 2012), blackhole attacks (Mahdi, Othman, Ibrahim, Desa, & Sulaiman, 2013), rushing attacks, spoofing attacks etc.
5. **Wormhole Attack:** In this attack, nodes create a tunnel, using encapsulation out-of bound channel. In MANET the route discovery is based on peer-to-peer information passing. So a node heavily relies on neighborhood information. False information sends by a neighboring node may cause severe affects. If the routing protocol uses the number of hop count to establish the shortest path, it computes a path through the wormhole (Yadav, Jain, & Faisal, 2012). The attacker receives packets at one point in the network, tunnels them to an attacker at another point in the network. Then it replays them in the network from final point. Packets sent by tunneling packets forwarded by multi-hop routes and give the attacker node to advantage for next attacks. So, broadcasts of packets in the tunnel as similar as in normal packet sending schemes. It is generally difficult to detect a wormhole attacker by the software only approach such as IDS. Figure 2 shows the wormhole attacks in MANETs.

Figure 2. Wormhole attacks

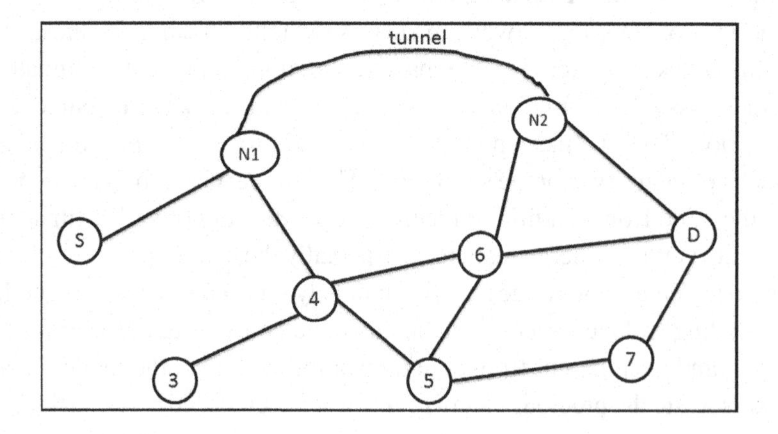

Figure 3. Blackhole attacks

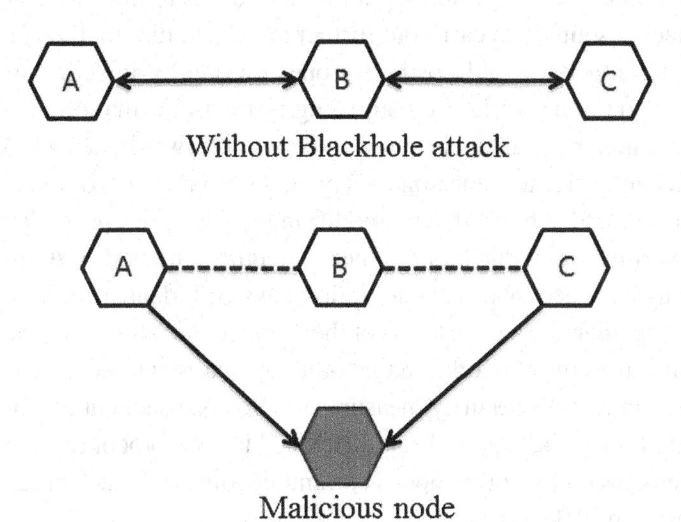

6. **Blackhole Attack:** In this approach (Mahdi, Othman, Ibrahim, Desa, & Sulaiman, 2013), a malicious node forwards the routing information. In flooding based protocols, the malicious node replies the requesting node before the actual node. So it creates an unauthorized route in the network. This malicious node decides whether to drop packets to perform DOS in the middle of the route. Black nodes co-ordinate to launch the attacks. There may be blackhole attacks with each other increasing the security of the attacks. Figure 3 shows how a malicious node affects the routing information in MANET.
7. **Rushing Attack:** In most of the routing protocol, an intermediate node admits and forwards only the first route request. After that it rejecting others request. The rushing attacker exploits this feature to insert the attack. In rushing attack, the attacker hurriedly disseminates the route request packet, faster than any active node can do. A legitimate node asserts a time difference between it received route request packet and forwarded route request packet to avoid collision. However, a rushing attacker may rush a route request instantly without this time difference. As a result, the request from the attacker reaches at other node faster. As a result the receiver keeps route request forwarded from the attacker and discards other. An intermediate node collects route requests and, and forwards that faulty information in RREQ packets.
8. **Spoofing:** In spoofing (Mahdi, Othman, Ibrahim, Desa, & Sulaiman, 2013), an attacker accepts the identity of legitimate nodes in the network. This is done by misrepresenting an IP or medium access control (MAC) address. Usually

this would be one of the first steps for an attacker to intrude into a network. The attacker's aim is to carry out further attacks to disrupt the normal operations. The attacker could obstruct proper routing by injecting false routing packets into the network or by adjusting routing information. Sometimes an attacker finds it advantageous and selectively forward packets. An intruder with this goal tries to impersonate a node within the path of the data flow of interest. It could achieve this by modifying routing data or implying itself as a trustworthy communication partner to neighboring nodes. A compromised node may have access to the encryption keys and identification information. Depending upon the access levels of the impersonated node, the intruders able to reconfigures the network. So that other attackers can more easily join. The attacker can remove security measures to allow subsequent attempts of invasion. Exploiting the loop holes in the MAC layer protocol an attacker places its node between two other nodes communicating with each other (similar to man-in-the-middle attack).

3. SECURED ROUTING PROTOCOLS

In this section some routing protocols are discussed. These protocols include different security parameters. Based on the operation and security features, we classify them as: (i) Secured single path routing and, (ii) Secured multipath routing. They are discussed below.

3.1 Secured Single Path Routing Protocols

Routing protocols can be categorized as on-demand, table-driven or combination of both. The most common protocols in MANETs are: Ad-hoc on-demand distance vector (AODV), Dynamic source routing (DSR), Destination sequenced distance-vector (DSDV), etc. These three protocols do not include any security parameters. Route discovery in AODV and DSR packets includes simple structure. Thus, a malicious node can discover the network structure simply by analyzing this kind of packets. Attackers may be able to determine the role of each node in the network. With all this information more serious attack can be established in order to disrupt network operations. So security is needed for avoiding these disruptions in the network setup. In this section, a few secure single path routing protocols are discussed. These protocols include some security parameters, but at the same time they have suffered with some difficulties to defend every attack.

3.1.1 ARAN: Authenticated Routing for Ad Hoc Networks

The ARAN (Sanzgiri, LaFlamme, Dahill, Levine, Shields, & Royer, 2005) is an on-demand routing protocol. It detects and defends against malicious actions. It introduces authentication, integrity and non-repudiation as part of minimum security policy for the ad hoc surroundings. It consists of a preliminary certification process, a mandatory end-to-end authentication stage. It has an optional stage that provides secure shortest paths. It needs a trusted certificate server (T), before getting into the ad hoc network. Each node in the network requests for a certificate signed by T. The certificate contains the IP address of the node, its public key, a timestamp (the time at which the certificate was created). Also, it has a time at which the certificate expires along with the signature by T.

In ARAN a certificate needs to be revoked, the trusted certificate server T sends a broadcast message to the ad hoc group that declares the revocation. Any node getting this message rebroadcasts it to its neighbors. Revocation notices need to be stored until the revoked certificate would have expired normally. Any neighbors of the node with the revoked certificate needs to reform routing as necessary to avoid transmission through the now un-trusted node. This method is not failsafe. In some cases, the un-trusted node that is having its certificate annulled may be the exclusive connection between two parts of the network. The time at which revoked certificate expires, the un-trusted node is unable to renew the certificate. Additionally, to detect this situation and to hurry the propagation of revocation notices, a node when meets a new neighbor, it exchanges the summary of its revocation notices. If the summaries do not match, the actual signed notices can be forwarded and re-broadcasted to restart propagation of the notice.

Important points:

1. The ARAN protocol defends against exploits using a modification, fabrication and impersonation.
2. The protocol uses of asymmetric cryptography.
3. ARAN is not resistant to the wormhole attack.

3.1.2 A Secure On-Demand Routing Protocol for Ad Hoc Networks (ARIADNE)

ARIADNE (Hu, Johnson, & Perrig, 2005) is an on-demand secure ad hoc routing protocol based on DSR. It relies on highly efficient symmetric cryptography. It guarantees that the target node of a route discovery process can authenticate the initiator. The initiator can authenticate each intermediate node on the path to the destination. Intermediate node cannot remove a previous node in the node list in the RREQ or RREP messages. ARIADNE requires some mechanism to *bootstrap authentic keys*. In particular, each node needs a shared secret key (Ks, D) between a source S and a destination D. Each node communicates with other nodes at higher layer. An authentic TESLA key for each node is required to forward RREQ messages.

ARIADNE provides point-to-point authentication of a routing message using a message authentication code (MAC) and a shared key between the two parties. For authentication of a broadcast packet such as RREQ, ARIADNE uses the TESLA broadcast authentication protocol. Selfish nodes are not taken into account. It copes with attacks performed by malicious nodes that modify and manufacture routing information. It is also immune to the wormhole attack only in its advanced version: using an extension called TIK (TESLA with Instant Key disclosure) that requires tight clock synchronization between the nodes. It is possible to detect anomalies caused by a wormhole based on timing discrepancies.

3.1.3 Secure, Efficient Ad-Hoc Distance Vector (SEAD)

SEAD (Hu, Johnson, & Perrig, 2003) is a proactive secure routing protocol based on the DSDV. In proactive routing protocols node periodically exchange their routing information with other nodes. Each node always knows a current route to all destinations. SEAD authenticates the sequence number and metric of a routing table update message using hash chain elements. In addition, the receiver of SEAD routing information also authenticates the sender. It ensures that the routing information originates from the correct node. The source of each routing update message in SEAD must also be authenticated; otherwise an attacker may create routing loops through the impersonation attacks.

SEAD deals with attackers that modify routing information broadcasted during the update phase. In particular, routing can be disrupted if the attacker modifies the sequence number and the metric field of a routing table update message. It makes use of efficient one-way hash chains rather than relying on expensive asymmetric cryptography operations. It assumes some mechanism for a node to distribute an authentic element of the hash chain that can be used to authenticate all the other elements of the chain. However, SEAD does not handle wormhole attacks.

3.1.4 Security-Aware Ad-Hoc Routing (SAR)

Security-Aware Ad-Hoc Routing (SAR) (Seung, Prasad, & Robin, 2001) is a comprehensive framework for any on demand ad-hoc routing protocol. It needs that nodes having same trust level to be necessarily share a secret key. SAR arguments the routing process using hash digests and symmetric encryption operations. The signed hash digests provides message integrity while the encryption of packets ensures their confidentiality. It also implemented in the AODV protocol and it adds two additional fields of the RREQ packet and one additional to the RREP packet. The first field added to the RREQ packet is the security requirement field and is set by the sender. It indicates the preferred level of trust for the path to the destination. The Second field added to be the security assurance that signifies the maximum level of security provided by the discovered paths. If the security required field has an integer representation, then the security guarantee field will be minimum of all security levels of the participating nodes in the path.

If the security requirement field is represented by vectors, then the security guarantee field value is computed from the security requirement values of the participating nodes in the path. The value thus computed is copied into the additional security guarantee field of the RREP packet. It then sent back to the sender. This value is also copied into the routing table of nodes in the reverse path, to preserve the security information with reference to cashed paths. The highlighted points are:

1. SAR uses security information to dynamically control the choice of routes installed in the routing table.
2. It enables applications to selectively implement a subset of security services based on the cost-benefit analysis.
3. The routes discovered by SAR may not always be the shortest between any two communicating entities in terms of hop count.
4. SAR finds the optimal route if all the nodes on the shortest path satisfy the security requirements.
5. SAR may fail to find the route if the ad hoc network does not have a path on which all nodes on the path satisfy the security requirements in spite of being connected.

A comparison of secured single path routing protocols discussed above is summarized in Table 1.

Table 1. Analysis of secured single path routing protocols

Protocols	Attacks	Security Requirements	Characteristics
ARAN	Interception attack	Satisfied	End-to-end authentication and Link-to-link authentication
ARIDEN	Impersonation attack	Satisfied	End-to-end authentication
SEAD	Wormhole attack	Satisfied	End-to-end authentication
SAR	Wormhole attack	All the shortest path satisfied	Selectively implementation

3.2 Secured Multipath Routing Protocols

In recent times, multipath routing protocols is an option for MANET to overcome the limitations of single path (Tarique, Tepee, Adibi, & Erfani, 2009, Lee, & Gerla, 2001). It uses different paths to discover route between a source and destination node. It gives some benefits such as; reducing end-to-end delay, increased bandwidth utilization, and increased fault-tolerance ability, etc. It also reduces the congestion and balance the network traffic by distributing among multiple paths. In multipath routing mechanisms, paths can be disjoint by two ways: node-disjoint and link-disjoint (Tarique, Tepee, Adibi, & Erfani, 2009). In node-disjoint paths, no nodes are in common except the source and destination. In link-disjoint path, they do not have any common links, but may have one or more common nodes. The advantage of non-disjoint routes is that they can be more easily discovered, since there are no restrictions that require the routes to be a node or link disjoint. The major concerns of multipath routing protocols are: multipath selection criteria, load sharing criteria and path maintenance criteria etc. A multipath scenario is shown in Figure 4.

Most of the multipath routing protocols are based on demand approach. AODV multipath (AODVM) (Marina, & Das, 2006) is a based on AODV. It considers both node-disjoint and link-disjoint paths. It proposes a routing framework to provide robustness to route breaks. The protocol computes node-disjoint paths between source and destination. The main ideas in AOMDV protocol is finding loop-free and disjoint paths based on flood-based route discovery. Split multipath routing (SMR) (Lee, & Gerla, 2001) another AODV variants. It builds multiple routes using route request RREQ/RREP. The objective of SMR is to reduce the frequency of route discovery and to reduce control packets overhead in the network. The data traffic is split into a multiple routes to prevent congestion. The destination node chooses two maximally disjoint paths. It chooses a route from one of the alternative paths, which are maximally available disjoint paths. Some issues in secured multipath are discussed below.

Figure 4. Example of multipath routing

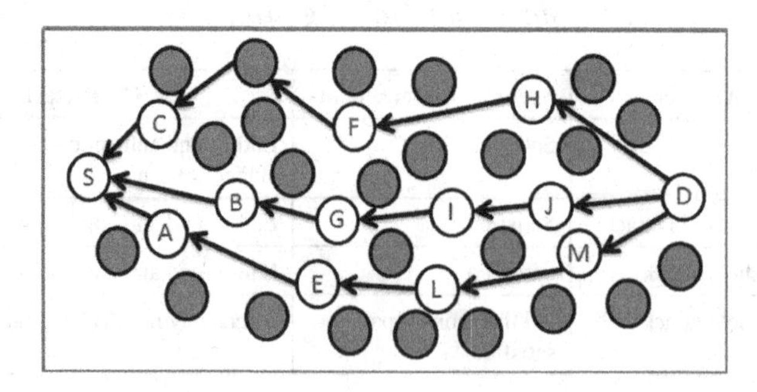

3.2.1 Vulnerabilities in Multipath Routing

In multipath routing, each intermediate node processes the RREQ packets only in the first time it receives. Subsequent RREQs are dropped. This is done in order to minimize the control packet overheads. A succeeding instance of an already processed RREQ query may have propagated through a different path. Whether such protocols discover the complete set of node-disjoint paths depends on the racing conditions of the route request propagation through different paths. If an intermediate node receives first a route request query that prevents discovery of another routing path that belongs to the set of node-disjoint paths (Tarique, Tepee, Adibi, & Erfani, 2009). Then the route request will end up without discovering all existing disjoint paths. Note that a node that experiences the racing phenomenon (Mavropodi, & Douligeris, 2006) will behave as if it was under the Rushing attack, even though no malicious acts take place.

- **Impersonation and Lack of Authentication:** A protocol if requires end-to-end authentication and an intermediate nodes participating in a routing path are not authenticated, then the protocol is subject to impersonation Sybil attacks (Mavropodi, & Douligeris, 2006). A malicious node may present multiple identities. This way a malicious node may participate in more than one, seemingly node-disjoint, routing paths, by presenting a different identity in each path. The adversary may then compromise a small fraction of nodes in selected areas of the network, in order to control the routing paths. The effects of this attack can be maximized if it is combined with the "black-hole" attack, where the attacker responds to all route requests, with fake (nonexistent) short path links. Then, the adversary may manipulate the communication, for example by dropping all the routing paths in critical time periods.
- **Invisible Node:** The broadcast nature of MANET makes multipath routing protocols subject to a special case of Man-In-the-Middle (MIM) attack, and the invisible node (Mavropodi, & Douligeris, 2006) attack. In this situation, a malicious node does not reveal its presence in the routing path. Instead, the invisible node silently repeats the communication between two-hop away nodes, which assume that they communicate directly as one-hop (direct) neighbors. In this way a node may participate in many routing paths. Even if the protocols require authentication of intermediate nodes. Indeed, a node may legitimately participate in one routing path and may also participate "invisibly" in other routing paths. Authentication cannot help, since the invisible node just relays the authenticators.

3.2.2 Secured Multipath Schemes

In this section some secured routing schemes in MANETs are discussed below briefly.

3.2.2.1 Secure Multipath Routing (SecMR)

It is a secure multipath routing mechanism that exhibits authentication in end-to-end and in link-to-link levels (Bagga, & Adhikary, 2014). It protects the integrity of the routing paths. Its mechanism is based on two phases.

The first phase is the neighborhood authentication phase. This phase contains the asynchronous mutual authentication of neighboring nodes. The repeated at periodic time intervals confirms the link-to-link authentication. The use of Certificate Authorities and PKI in a mobile ad hoc environment is a challenging task. Each node broadcasts to its one-hop neighbors a signed message, including the current time and its unique identifier that is included in its certificate. Thus, each node will produce one signature and will validate that signature for its neighbors. The average cost for this phase is one signature generation per node. The interval of the time period of the neighborhood authentication phase is a system parameter and calculates the volatility of the environment. According to the frequency of variations in the connectivity, the time period should be longer as possible. After the verification phase, each node will generate a list of its neighborhood for the current time. The nodes do not re-broadcast the received authentication messages. Since authentication is performed in local neighborhoods and it is not implemented simultaneously from all the nodes, the cost is acceptable and it does not lead to the huge problem. SecMR is a complete set of the existing non-cyclic, node-disjoint paths between a source and a destination node, for a certain maximum hop distance.

The second phase consists of the establishment and maintenance of active routes. The source produces a signed request, which grants the system with end-to-end authentication. Each intermediate node processes all the receiving requests, ensuring that all possible node-disjoint paths will be discovered by the destination. An intermediate node when receives a request through a node that belongs to its authenticated list of neighbors, it will first add itself in the routing path. Secondly, it will create the neighborhood information and the exclude-nodes information that are also appended to the message. The neighborhood information will contain all its authenticated neighbors that have not yet received the request. The exclude information will contain all the nodes that have received the message sometime in the past. The destination when obtains the request it checks its authenticity by checking its signature. Then it will construct the node-disjoint paths and produce a signed reply message, thus protecting the integrity of the used path.

3.2.2.2 Security Protocol for Reliable Data Delivery (SPREAD)

SPREAD (Lou, Liu, Zhang, & Fang, 2009) develops the confidentiality and availability by using multipath routing. The core idea of the SPREAD scheme arises from the following:

A messenger transmits the message from one place to another. The messenger across a hostile ground may break the message easily, if he/she captured. The message if may not fully recovered by challengers, then multiple messengers are deployed. In SPREAD mechanism a source node wishes to send a message to a destination node securely. The source can use a multipath routing algorithm to find multiple paths from the source to the destination with certain properties (e.g., disjoint paths). Then the source finds a secret sharing scheme, depending on the message security level and the availability of multiple paths. It transforms the message into multiple shares. Later the messages are routed to the destination through multipath routing protocol. Destination node then reforms the original message upon receiving a certain number of shares.

SPREAD improves security by covering with the compromised nodes and eavesdropping problem. The hop-by-hop link encryption is used to make it secure. It assumes that if one node is compromised, all the shares traveling through that node are negotiated. A compromised shares means the adversary has a means to decrypt it and it could be used to recover the original message. It also investigates the message eavesdropping problem. Assume that anyone sitting within the transmission range of a transmitting node is able to eavesdrop the transmission of that node. However, it pointed out that an eavesdropped message share does not let drop any useful information before it is decoded. SPREAD manages the issues such as;

1. How to transform the message into multiple shares;
2. How to allocate the shares onto each path; and
3. How to discover the desired multiple paths.

At the source, the message is separated into a number of pieces named shares using a *Threshold Secret Sharing algorithm*. The threshold Secret Sharing algorithm divides the message to N shares with redundancy. The original messages can be recreated in the destination. Each share decodes with different key and transmits on different paths. Even the one attacker or more interrupt one path or more, it is required T shares to reconstruct the original message.

Moreover, the decryption process is hard to recover since each share is encrypted with different keys. The SPREAD is capable of enhancing the confidentiality and availability by encryption method and the threshold secret sharing algorithm. The second issue is how to assign the shares onto each selected path. So that the challenger has least possibility to compromises the message. Considering the case that message is compromised due to the compromised nodes. Let assume that if a node is compromised, all the authorizations of that node are compromised. So the message shares traveling through that node are intercepted and compromised. In an ad hoc network, the topology changes frequently in wireless medium. So packets are dropped during the transmission. To minimize this problem, it is necessary to introduce some redundancy in the SPREAD scheme to improve the reliability. The maximum redundancy can be added to the SPREAD scheme without losing the security is identified. The final issue is the multipath routing means how to find the desired multiple paths in a MANET and how to deliver the shares to the destination using these paths. For SPREAD scheme, it needs independent paths, more definitely, node-disjoint paths, as many as possible, because it deals with the node compromise problem.

3.2.2.3 Securing Data Based Multipath Routing in Ad Hoc Networks (SDMP)

The intent behind this protocol is to separate the initial message into parts. Then the separate message parts combine and encrypt by pairs. After the pairing, the characteristics of multiple paths exploit to increase the robustness of confidentiality. SDMP (Sangulagi, & Naveen, 2014) uses WEP (Wired Equivalent Privacy) link to encoding among corresponding nodes. This also gives a link layer authentication and confidentiality. Confidentiality gained by sending an encrypted combination of the different living paths between the sender and receiver. SDMP uses the mechanism of existing multipath routing and do not provide any assumption about the node-disjoint of the added set-of-paths. In this process there have to be at least three paths in between the source and destination. Among the three paths, SDMP uses one path for signaling. The key message is split into pieces and supplied with a unique identifier. Pieces of message are XOR-ed and each piece is making a pair, which pair are transmitted along a different path. Split of message required a non-redundant version of diversity coding. Redundancy is providing data availability.

Message pair which contains information is sent on the signaling path to give permission for message reconstruction at the destination. Each path has assigned data and allowing path cost function to minimize the time spent at the receiver side. The time spent is also reconstructed the original message. Unless the enemy can give entry to the all communicated piece of message, then the chances of message reconstruction is low. This intends, to negotiate the confidentiality of the original secret message. The attacker needs to get within eavesdropping of the destination or source. At the same time listen to all the paths used and decrypt the WEP encoding of a transmitting message. But, it calls for only a few pieces of transmitting message at the destination. These messages are sufficient to reconstruct the original message. Particularly since one piece of the original message is always sent in its original form on one of the selected paths. The time to send a large message especially increases as number secure paths used are more. However, applying more secure paths in the whole process leads the confidentiality. Hence, there is an exchange between the security and delay of a given message.

3.2.2.4 Secure Message Transmission (SMT)

The secure message transmission (SMT) (Sharma, & Sahni, 2014) schemes covered data integrity, confidentiality and availability. It has no link encoding mechanism. It works on end-to-end basis. It processes only Security Association (SA) between the source and destination nodes. This SA is used to supply data integrity and authentication. It also provided easier end-to-end message encryption. SMT uses multipath routing to increase the data integrity and confidentiality of shared message between source and destination nodes. SMT generally gives more importance on the reliability of data transmission, whereas SPREAD mechanism designed with the confidentiality of data transmission.

In case of node failure SMT provides end-to-end secure and strong feedback process that permission for reconfiguration of the set of paths. The number of successful and unsuccessful transmissions provides each path to continually put with the reliability rating. SMT uses these ratings with a multipath routing algorithm. It then maintains and chooses a maximally secure set of paths. After choosing paths, it adjusts its parameter to remain effective and efficient. It uses a strategy known as IDA (Information Dispersal Algorithm) (Djenouri, Khelladi, & Badache, 2005). IDA separates message to multiple pieces with finite redundancy. Pieces of the message carried on a different node-disjoint path. Message Authentication Code (MAC) is carrying with each piece to give data integrity and original authentication. Redundancy is the ratio between N pieces of transmitting messages and M number of the original message. The SMT scheme guaranteed for reconstruction of the transmitted message at the destination. It also ensured reconstruction, when some piece of messages is lost in the network. It allowed data redundancy paired with multipath routing. During this process some packets are lost which are omitted at the destination. This allows SMT to support real time traffic with Quality of Service requirements. Figure 5 shows the basic operations of SMT protocol.

4. CONCLUSION

The security is very difficult issue due to its behavioral structure of the network and lack of central management in MANET. In this chapter, we made an attempt to discuss some security aspects in MANETs. Different challenges and security threats in mobile ad hoc networks are discussed. We focus basically on secured multipath routing protocols. The existing proposals are typically attacked-oriented. It is found that main focused on routing attacks are classified into eavesdropping attacks, traf-

Figure 5. Secure message transmission protocol

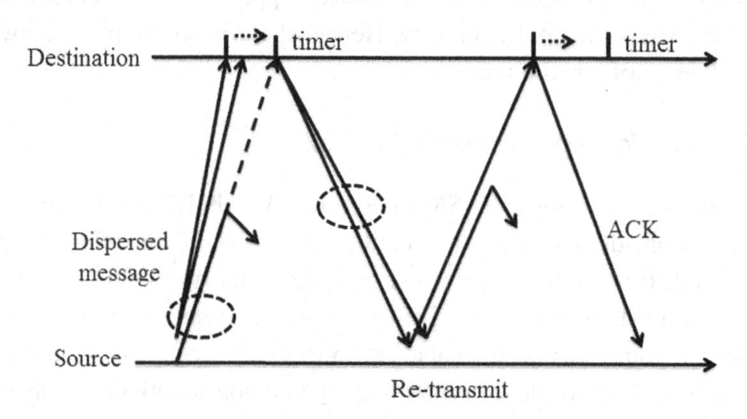

fic analysis, and modification, impersonation, fabrication attacks, wormhole and blackhole. It is also found that there are also many vulnerabilities like phenomenon racing, impersonation and lack of authentication and invisible node which are not properly considered in the literature. The research on security aspect is going on and we found that research on MANET security is still in its early stage.

REFERENCES

Bagga, S., & Adhikary, K. (2014). A Review on Various Protocols and Security Issues in MANET. *International Journal of Advanced Research in Computer and Communication Engineering*, *3*(7), 7478–7482.

Djenojuri, D., Khelladi, L., & Badache, N. (2005). A Survey of Security Issues in Mobile Ad hoc Networks. *IEEE Communications Surveys*, *7*(4), 2–28. doi:10.1109/COMST.2005.1593277

Hu, Y., Johnson, D. B., & Perrig, A. (2003). SEAD: Secure Efficient Distance Vector Routing for Mobile Wireless Ad hoc Networks. *Ad Hoc Networks*, *1*(1), 175–192. doi:10.1016/S1570-8705(03)00019-2

Hu, Y., Johnson, D. B., & Perrig, A. (2005). Ariadne: A Secure On-demand Routing Protocol for Ad hoc Networks. *Wireless Communications and Mobile Computing*, *11*(1-2), 21–28.

Lakhtaria, K. I. (2009). Enhancing QOS and QOE in IMS enabled next generation networks. In *Networks and Communications, 2009. NETCOM'09. First International Conference on* (pp. 184-189). IEEE. doi:10.1109/NetCoM.2009.29

Lakhtaria, K. I. (2012). *Technological Advancements and Applications in Mobile Ad-hoc Networks: Research Trends*. Information Science Reference.

Lakhtaria, K. I., & Nagamalai, D. (2010). Analyzing Web 2.0 Integration with Next Generation Networks for Services Rendering. In Recent Trends in Networks and Communications (pp. 581-591). Springer Berlin Heidelberg.

Lee, S., & Gerla, M. (2001). Split Multipath Routing with Maximally Disjoint Paths in Ad hoc Networks. In *Proceedings of IEEE International Conference on Communications (ICC 2001)*. Helsinki, Finland: IEEE.

Lou, W., Liu, W., Zhang, Y., & Fang, Y. (2009). SPREAD: Improving Network Security by Multipath Routing in Mobile Ad hoc Networks. *Wireless Networks*, *15*(3), 279–294. doi:10.1007/s11276-007-0039-4

Mahdi, S. A., Othman, M., Ibrahim, H., & Desa, J. (2013). Protocols for Secure Routing and Transmission in Mobile Ad Hoc Network: A Review. *Journal of Computer Science*, *9*(5), 607–619. doi:10.3844/jcssp.2013.607.619

Marina, M. K., & Das, S. R. (2006). Ad hoc On-demand Multipath Distance Vector Routing. *Wireless Communications and Mobile Computing*, *6*(7), 969–988. doi:10.1002/wcm.432

Mavropodi, R., & Douligeris, C. (2006). A Multipath Routing Protocols for Mobile Ad Hoc Networks: Security Issues and Performance Evaluation. In I. Stavrakakis & M. Smirnov (Eds.), *Autonomic Communication* (pp. 165–176). Springer Berlin Heidelberg. doi:10.1007/11687818_13

Mohapatra, P., & Krishnamurthy, S. (2005). *Ad Hoc Networks: Technologies and Protocols*. Springer Science.

Mueller, S., Tsang, R., & Ghosal, D. (2004). Multipath Routing in Mobile Ad hoc Networks: Issues and Challenges. In Performance Tools and Applications to Networked Systems (pp. 209-234). Springer Berlin Heidelberg. doi:3_10 doi:10.1007/978-3-540-24663

Raghavendran, C. V., Satish, G. N., & Varma, P. S. (2013). A Study on Contributory Group Key Agreements for Mobile Ad Hoc Networks. *International Journal of Computer Network and Information Security*, *5*(4), 48–56. doi:10.5815/ijcnis.2013.04.07

Sangulagi, P., & Naveen, A. S. (2014). Efficient Security Approaches in Mobile Ad-Hoc Networks: A Survey. *International Journal of Research in Engineering and Technology*, *3*(3), 14–19.

Sanzgiri, K., LaFlamme, D., Dahill, B., Levine, B. N., Shields, C., & Royer, E. M. (2005). Authenticated Routing for Ad hoc Networks. *IEEE Journal on Selected Areas in Communications*, *23*(3), 598–610. doi:10.1109/JSAC.2004.842547

Seung, Y., Prasad, N., & Robin, K. (2001). Security-aware Ad hoc Routing for Wireless Networks. In *Proceedings of the 2nd ACM international symposium on Mobile ad hoc networking & computing*, (pp. 299-302). ACM.

Sevil, S., Clark, J. A., & Tapiador, J. E. (2010). Security Threats in Mobile Ad Hoc Networks. In Security of Self-Organizing Networks: MANET, WSN, WMN, VANET (pp. 127-146). CRC Press.

Sharma, N., & Sahni, P. (2014). Secure Routing & Data Transmission in Mobile Ad Hoc Networks. *International Journal of Innovative Research in Technology*, *1*(2), 407–412.

Singh, T. P., Kaur, S., & Das, V. (2012). Security Threats in Mobile Ad hoc Network: A Review. *International Journal of Computer Networks and Wireless Communications*, *32*(6), 27–34.

Tarique, M., Tepee, K. E., Adibi, S., & Erfani, S. (2009). Survey of Multipath Routing Protocols for Mobile Ad hoc Networks. *Journal of Network and Computer Applications*, *32*(6), 1125–1143. doi:10.1016/j.jnca.2009.07.002

Yadav, S., Jain, R., & Faisal, M. (2012). Attacks in MANET. *International Journal of Latest Trends in Engineering and Technology*, *1*(3), 123–126.

Yang, H., Haiyun, L., Fan, Y., Songwu, L., & Zhang, L. (2004). Security in Mobile Ad hoc Networks: Challenges and Solutions. *IEEE Wireless Communications*, *11*(1), 38–47. doi:10.1109/MWC.2004.1269716

KEY TERMS AND DEFINITIONS

Active Attacks: Attacker attempts to modify user data and disrupts normal functionality of the network.

Blackhole Attack: Malicious node that falsely replies route information.

Denial of Service (DOS): In this attack a malicious node makes an attempt to prevent legitimate nodes from accessing information or services.

Flooding Attack: The attack that is designed to bring a network or service down by flooding it with large amounts of traffic.

MANETs: A new paradigm of wireless network, offering unrestricted mobility without any underlying infrastructure.

Multipath Routing: It is a routing technique which uses multiple alternative paths.

Passive Attacks: Attacker snoops data exchange without altering it.

Chapter 5
An Infrastructure for Wireless Sensor–Cloud Architecture via Virtualization

S. P. Anandaraj
SR Engineering College, India

S. Poornima
SR Engineering College, India

ABSTRACT

A typical WSN contains spatially distributed sensors that can cooperatively monitor the environment conditions, like second, temperature, pressure, motion, vibration, pollution and so forth. WSN applications have been used in several important areas, such as health care, military, critical infrastructure monitoring, environment monitoring, and manufacturing. At the same time. WSN Have some issues like memory, energy, computation, communication, and scalability, efficient management. So, there is a need for a powerful and scalable high-performance computing and massive storage infrastructure for real-time processing and storing the WSN data as well as analysis (online and offline) of the processed information to extract events of interest. In this scenario, cloud computing is becoming a promising technology to provide a flexible stack of massive computing, storage, and software services in ascalable and virtualized manner at low cost. Therefore, sensor-cloud (i.e. an integrated version of WSN & cloud computing) infrastructure is becoming popular nowadays that can provide an open flexible, and reconfigurable platform for several monitoring and controlling applications.

DOI: 10.4018/978-1-4666-8687-8.ch005

INTRODUCTION

Motivation

A typical sensor network may consists of a number of sensor nodes acting upon together to monitor a region and fetch data about the surroundings. A typical WSN contains self-regulated sensors that can cooperatively monitor the environmental conditions, like sound, temperature, pressure, motion, vibration, pollution, fire like, and other application dependent events. Each node in a sensor is loaded with a raio transceiver or some other wireless communication device, a small microcontroller, and an energy source most often cells/battery. WSNs have some of the limitations, like in terms of memory, energy, computation, communication and scalability, efficient management of the large number of WSNs data.

Cloud computing allows the systems and users to use Platform as a Service (PaaS), for example, Operating System (Oss), Infrastructure as aService (IaaS), for example, storages and servers and Software as a Service(SaaS), for example, application level programs, and so forth at a very low cost which are being provided by several cloud providers (e.g., Amazon, Google, and Microsoft) on the basis of pay per use services (Atif Alanri, 2013). Cloud Computing platform dynamically available, configures, and updates the servers as and when needed by end uses. The limitations of WSNs are the pluspoints in the Cloud Computing.

This is the reason why the integrations of cloud computing & WSNs will lead to greater benefits & efficiency.

What Is Sensor-Cloud Infrastructure?

Sensor-Clod infrastructure i.e. integrated version of Wireless Sensor networks and Cloud Computing is powerful and scalable high-performance computing and massive storage infrastructure for real-time processing and storing of the WSN data (online as well as previously collected ofline) as well as analysis of the processed information to extract events of interest.

Some of the Definitions of Sensor Cloud Architecture

AN Infrastructure that allows truly pervasive computation using sensors as an interface between physical and cyber worlds, the data-compute clusters as the cyber backbone and the internet as the communication medium (Sajjad Hussain Shah, 2013).

It is a unique sensor data storage, visualization and remote management platform that leverage [sic] powerful cloud computing technologies to provide excellent data scalability, rapid visualization, and user programmable analysis. It is originally designed to support long-term deployments of micro strain wireless sensors, Sensors-Cloud now supports any web-connected third party device, sensor, or sensor network through a simple open data API (J.Yick, 2008)

When WSN is integrated with cloud computing environment, several shortfalls of WSN like storage capacity of the data collected on sensor nodes and processing of these data together would become much easier. Since cloud Computing provides a vat storage capacity and processing capabilities, it enables collecting the huge amount of sensor data by linking the WSN and cloud through the gateways on both sides, that is, sensor gateway and cloud gateway. Sensor gateways collects information from the sensor nodes of WSN compresses it, and transmits it back to the cloud gateway which in turn decompresses it and stores it in the cloud storage server, which is sufficiently large (S.K. Dash, 2012). Sensor- Cloud is a new paradigm for cloud computing that uses the physical sensors to accumulate its data and transmittal sensor data into a cloud computing infrastructure (L.P.D. Kumar, 2012). Sensor- Cloud infrastructure is used that enables the sensors to be utilized on a digital infrastructure by virtualizing the physical sensor on a cloud computing platform according to the need of the user and application. These virtualized sensors on a cloud computing platform are dynamic in nature and hence facilitate automatic provisioning of its services as and when required by users.

THEORETICAL BACKGROUND AND LITREATURE SURVEY

Wireless Sensor Networks

A Wireless Sensor networks (WSN) consists of spatially distributed autonomous sensors to monitor physical or environmental conditions, such as temperature, sound, pressure, etc., and to cooperatively pass their data through the network to a main location (H.T. Dinh, 2011). Wireless Sensor Network applications have been used in important areas, like healthcare, military, critical infrastructure monitoring, and manufacturing.

WSN applications are listed in (Lan, K.T, 2010) shows the usefulness in respective areas.

- Military:
  - Military situation awareness.
  - Logistics in urban warfare.
  - Battlefield surveillance.
  - Computing intelligence, surveillance, reconnaissance, and targeting systems.

- Mobile wireless low-rate networks for precision location:
  - Including industrial, retail, hospital, residential, and office environments, while maintaining low-rate data communications for monitoring, messaging and control.
- Airports:
  - Smart badges and tags.
  - Wireless luggage tags.
  - Passive mobility (e.g., attached to a moving object not under the control of the sensor node).
- Medical/Health:
  - Monitoring Peoples locations and health conditions.
  - Sensors for: blood flow, respiratory rate, ECG Electrocardiogram.
  - Monitor patients and assist disabled patients.
- Ocean/Weather Monitoring:
  - Monitoring Fish, Water Displacements/ NOAA data.

Even if WSN have the vast area of application but it also have limitation which restricts its application and/or area. WSNs have to face many limitations due to its architecture regarding their communications (like power considerations, storage capacity, reliability, mobility, etc.) and resources (like power considerations, storage capacity, processing capabilities, bandwidth availability, etc.). Besides, WSN has its own resource and design constraints. Design constraints are application specific and dependent on monitored environment. Based on the monitored environment, network size in WSN varies. For monitoring large environment, there is limited communication between nodes due to obstructions into the environment, which in turn affects the overall network topology (or connectivity) (Atif Alanri, 2013).

The Sensor-Cloud Architecture i.e. Integration of WSN with Cloud Computing focuses on overcoming the limitations of the WSN by using the advantages of Cloud Services.

CLOUD COMPUTING

Cloud computing can be defined by many definitions/ways, some of the definition is =cloud computing is a model for enabling convenient, on demand network access to a shared pool of configurable computing resources (e.g., networks, servers, storage, applications, and services) that can be rapidly provisioned and released with minimal management effort or service provider interaction (L.P.D. Kumar, 2012).

Cloud is a new consumption and deliver model for many IT-based services, in which the user sees only the service,, and has no need to know anything about the technology or implementation" (Lan, K.T, 2010).

Cloud computing is a model for enabling environment, on demand network access to shared pool of configurable computing resources (Lan, K.T, 2010). Cloud computing is the delivery of computing services over the Internet. Cloud services allow individuals and businesses to use software and hardware that are managed by third parties at remote locations. Examples of cloud services include online file storage, social networking sites, webmail, and online business applications. The cloud computing model allows access to information and computer resources from anywhere that a network connection is available. Cloud computing provides a shared pool of resources, including data storage space, networks, computer processing power, and specialized corporate and user applications.

Cloud computing provides services as –IaaS, PaaS, SaaS, SaaS provides board market solutions where the vendor provides access to hardware and software products through portal interface (Lan, K.T, 2010).. PaaS allows the consumer to run the specified application on the platform. In these types of services, consumer have no control over the infrastructure as well as on the installed applications (S. Guo, 2009). IaaS provides consumers with the benefit to consume the infrastructure that includes processing power, data storage, and networks, etc. The consumer can run multiple applications without worrying about maintenance of underlying infrastructure (Lan, K.T, 2010).

Cloud computing platform dynamically provisions, configures, and reconfigures the servers as and when needed by end users. These servers can be in the form of virtual machines or physical machines in the cloud. Cloud Computing renders the two major trends in IT:

1. Efficiency, which is achieved through the highly scalable hardware and software resources, and
2. Agility, which is achieved through parallel batch processing, using computer intensive business analytics and real-time mobile interactive applications that respond to user requirements.

Cloud architecture have benefits that the end users need not to worry about the exact location of servers and switch to their application by connecting to the server on cloud and start working without any hassle (Sajjad Hussain Shah, 2013) .

Characteristics

The characteristics of cloud computing include on-demand self-service, broad network access, resource pooling, rapid elasticity and measured service. On-demand self-service means that customers (usually organizations) can request and manage their own computing resources. Broad network access allows services to be offered over the Internet or private networks. Pooled resources means that customers draw from a pool of computing resources, usually in remote data centres. Services can be scaled larger or smaller; and use of a service is measured and customers are billed accordingly.

Service Models

The cloud computing service models are Software as a Service (SaaS), Platform as a Service (PaaS) and Infrastructure as a Service (IaaS). In a Software as a Service model, a pre-made application, along with any required software, operating system, hardware, and network are provided. In PaaS, an operating system, hardware, and network are provided, and the customer installs or develops its own software and applications. The IaaS model provides just the hardware and network; the customer installs or develops its own operating systems, software and applications. Deployment of cloud services: Cloud services are typically made available via a private cloud, community cloud, public cloud or hybrid cloud. Generally speaking, services provided by a public cloud are offered over the Internet and are owned and operated by a cloud provider. Some examples include services aimed at the general public, such as online photo storage services, e-mail services, or social networking sites. However, services for enterprises can also be offered in a public cloud.

- In a *private cloud*, the cloud infrastructure is operated solely for a specific organization, and is managed by the organization or a third party.
- In a *community cloud*, the service is shared by several organizations and made available only to those groups. The infrastructure may be owned and operated by the organizations or by a cloud service provider.
- A *hybrid cloud* is a combination of different methods of resource pooling (for example, combining public and community clouds).

The limitations of Wireless sensor network are the major plus points of cloud Computing. The sensor Cloud Architecture takes the advantages of this points and provide better system.

Sensor Cloud Architecture

General Architecture

When a user requests, the service instances (e.g., virtual sensors) generated by cloud computing services are automatically provisioned to them (Kushida, 2012). Some previous studies on physical sensors focused on routing. Clock synchronization, data processing, power management, OS, localization, and programming (Atif Alanri, 2013). There are very less work have been done which concentrate on physical sensor management because these physical sensors are bound closely to their specific application as well as to its tangible users directly. However, users, others than their relevant sensors should be supervised by some special sensor-management schemes. The Sensor-Cloud infrastructure would subsidize the sensor system management, which ensures that the data-management usability of sensor resources would be fairly improved.

There exists no application that can make use of every kind of physical sensors at all times; instead, each application required pertinent physical sensors for its fulfilment. To realize the concept, publish/subscription mechanism is being employed for choosing the appropriate physical sensor (Atif Alanri, 2013). Server- Cloud infrastructure provides the facility to user to create the template/virtual special group of sensor nodes whose data will be collaborated for the specific applications. These service templates/virtual groups are reconfigurable according to the user needs. Once service instances become useless, they can then be deleted quickly by users to avoid the utilization charges for these resources. Every sensor node, application program senses the application and sends the sensor data back to the gateway in the cloud directly through of the base station. Sensor- Cloud infrastructure provides service instances (virtual sensors) automatically to the end users as and when requested, in such a way that these virtual sensors are part of their IT resources (like disk storage, CPU, Memory, etc.) (R. S. Ponmagal, 2011). These template services/virtual groups instances and their associated appropriate sensor data can be used by the end users via a user interface through the web crawlers as described in Figures 3-1.

The physical sensors are ranked on a basis of their sensor readings as well as on their actual distance from an event. S. Guo and Z. Zhong (2009) proposed a technique (FIND) to locate physical sensors having data faults by assuming a mismatch between the distance rank and sensor data rank. However, the study led by FIND aims at the assessment of physical sensors faults, and there is a close relation between the virtual and physical sensors and hence a virtual sensor will provide incorrect results if their relevant physical sensors are faulty.

Current Approaches

In just few years Sensor-Cloud Infrastructure gaining the popularity & some of the architecture are introduced. Some of them are as follows,

- **A New Model of Accelerating Service Innovation with Sensor-Cloud Infrastructure:** A novel approach to integrate WSN with cloud computing for effective applications and Service Provisioning in WSN. Introduced the third party application by inter-cloud connectivity. Authors also proposed Research Agenda (Madoka Yuriyama, 2010).

- **Sensor-Cloud Infrastructure-Physical; Sensor Management with Virtualized Sensors on Cloud Computing:** The most basic and latest when proposed, architecture proposed by authors who proposed the making of virtual group AKA Service Template by aggregating data from the different types of physical sensors. Then those ST/Virtual groups can be re-used, configurable and sharable (V.Rajesj, 2010).
- **Integration of Wireless Sensor-Network with Cloud:** Introduced the framework which focuses on reliability, availability and extensibility. Combining SOA Roles in proposed Architecture- in which sensor nodes are data/service providers & sinks are consumes & clients make request to *integration Controller* by SOAP Clients (Ishi, Y, 2012).
- **Integrating Wireless Sensor Networks with Cloud Computing:** Authors proposed the frame work in which WSN transfers data to the Cloud system, which will be available to blogs, virtual communities, and social networking applications and many more (Liu, J, 2012).
- **A Novel Architecture Based on Cloud Computing for Wireless Sensor Network:** The frame work proposed represents cloud as virtual sink node for controlling the data from the WSN and simulation of the frame work prove the improvement in performance (Peng Zhang, 2013).
- **New Framework to Integrate Wireless Sensor Networks with Cloud Computing:** Proposed framework uses the *Integration Controller* for integrating WSN with Cloud Computing. Integration Controller collects data from WSNs & passes to Cloud Application which provides the service to the users depending upon the need of individual (Lan, K.T, 2010).

These all frameworks available approaches follows one common base architecture of making user to subscribe for different/same types of physical sensor and analyse the data collected by the same. For that it creates the ST or virtual group.

These all frameworks are presented with respect to some specific application. It may possible that one framework is not a best for the different situation/application. So one can choose its application based on application area.

Cloud Computing vs. Sensor Cloud Computing

Cloud computing is a computing paradigm, where a large pool of systems are connected in private or public networks, to provide dynamically scalable infrastructure for application, data and file storage. With the advent of this technology, the cost of computation, application hosting, content storage and delivery is reduced significantly. Cloud computing refers to both the applications delivered as services over the Internet and the hardware and systems software in the datacenters that provide those services (L.P.D. Kumar, 2012). A cluster of computer hardware and software that offer the services to the general public (probably for a price) makes up a 'public cloud'. Computing is therefore offered as a utility much like electricity, water, gas etc. where you only pay per use. For example, Amazon's Elastic cloud, Microsoft's Azure platform, Google's App Engine and Salesforce are some public clouds that are available today. However, cloud computing does not include private clouds which refer to data centers internal to an organization. Therefore, cloud computing can be defined as the aggregation of computing as a utility and software as a service. Virtualization of resources is a key requirement for a cloud provider for it is needed by statistical multiplexing that is required for scalability of the cloud, and also to create the illusion of infinite resources to the cloud user holds the view that different utility computing offerings will be distinguished based on the level of abstraction presented to the programmer and the level of management of the resources (Atif Alanri, 2013).

Sensor cloud collects and processes information from several sensor networks enables information sharing on big-scale and collaborate the applications on cloud among users. It integrates several networks with number of sensing applications and cloud computing platform by allowing applications to be cross-disciplinary that may spanned over organizational ranges (Sajjad Hussain Shah, 2013). Sensor cloud enables users to easily gather, access, processing, visualizing and analysing, storing, sharing and searching large number of sensor data from several types of applications. These vast amount of data are stored, processed, analysed and then visualized by using the computational IT and storage resources of the cloud.

SENSOR-CLOUD SERVICE LIFE-CYCLE MODEL AND ITS LAYERED STRUCTURE

Sensor-Cloud is useful for many applications, mainly where data from large sensor networks needs to be collected, monitored and viewed remotely. The case for WSN enables novel and attractive solutions for information gathering across the spectrum of endeavour including business, transportation, health-care, environmental monitoring, and industrial automation. Despite these advances, the exponentially increasing data extracted from WSN is not getting enough use due to the lack of skill, time and money with which the data might be better explored and stored for future need. The next generation of WSN will benefit when sensor data is added to virtual communities, blogs and social network applications. This transformation of data derived from sensor networks into a valuable resource for information hungry applications will benefit from techniques being developed for the emerging Cloud Computing. Traditional High Performance Computing approaches may be replaced or find a place in data manipulation prior to the data being moved into the Cloud. In this paper, a novel framework is proposed to integrate the Cloud Computing model with WSN. Deployed WSN will be connected to the proposed infrastructure. Users request will be served via three service layers (IaaS, PaaS, SaaS) either from the archive, archive is made by collecting data periodically from WSN to Data Centres (DC), or by generating live query to corresponding sensor network (S.K. Dash, 2012).The service life-cycle model and the layered structure of Sensor-Cloud infrastructure are illustrated as follows-

Service Life Cycle Model of Sensor-Cloud

Before creating the service instances within Sensor-Cloud infrastructure, preparation phase is needed, and this includes the following –

- Preparing the IT Resources (processors, storage, disk, memory, etc).
- Preparing the physical sensor devices.
- Preparing the service templates.

Figure 1 depicts the Sensor-Cloud Service Life Cycle. The users of the Sensors can select the appropriate service template and request the required service instances. These service instances are provided automatically and freely to the users, which can then be deleted quickly when they become useless. From a single service template, multiple numbers of service instances can be created. Service provider regulates the service templates and can add new service templates as and when required by different number of users.

Layered Structure of Sensor-Cloud

Figure 1 depicts the layered architecture of the Sensor-Cloud platform, which is divided mainly into three layers:

1. User and application Layers.
2. Sensor-Cloud and Virtualization Layers.
3. Template Creations and tangible sensor layers.

Layer 1: This layer deals with the users and their relevant applications. Several users want to access the valuable sensor data from independent from the platform for a variety of applications. This structure allows users of

Figure 1. Sensor-cloud infrastructure life cycle

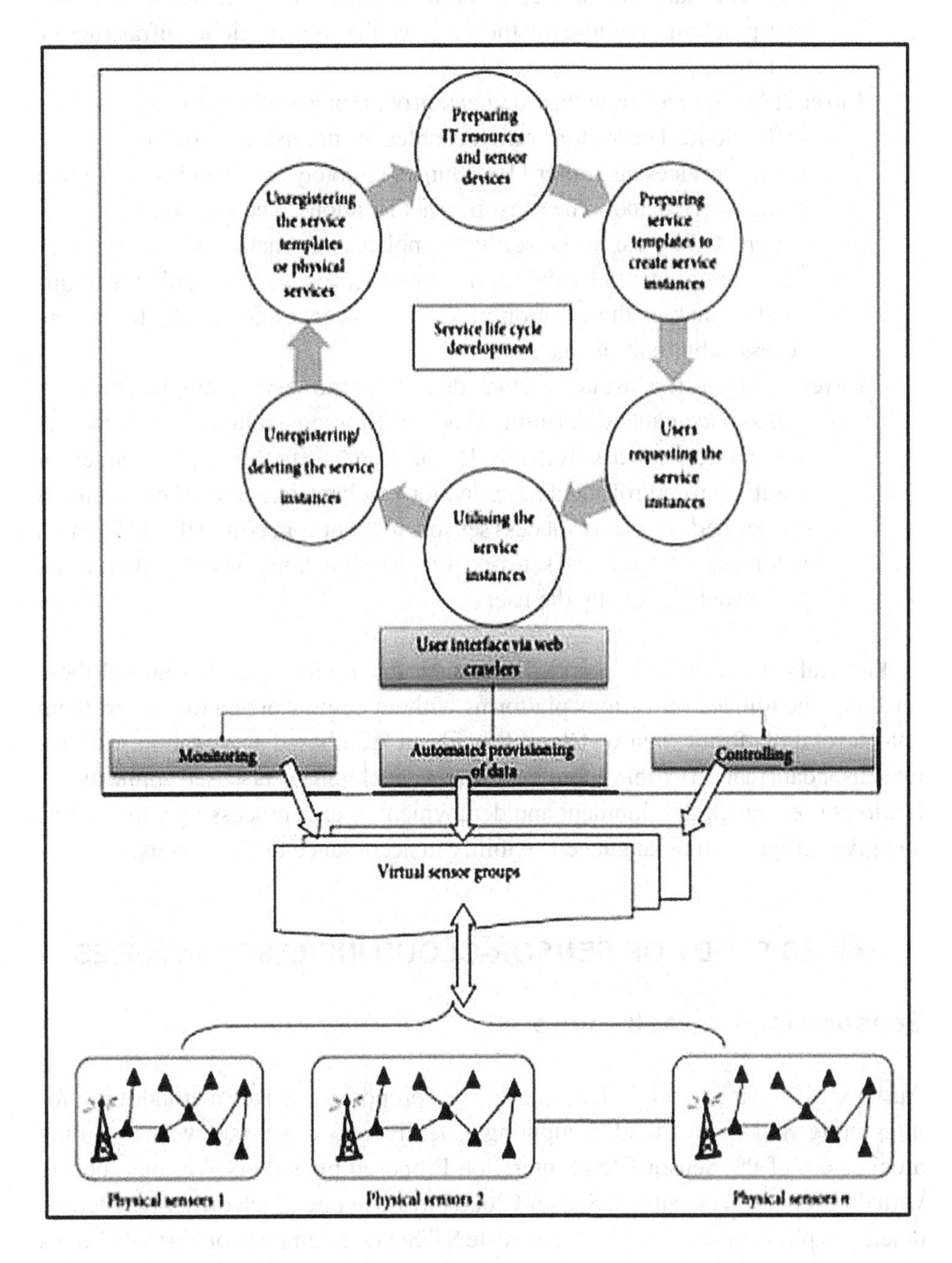

different platforms to access and utilize the sensor data without facing any problems because of the high availability of cloud infrastructure and storage.

Layer 2: This layer deals with virtualization of the physical sensors and resources in the cloud. The Virtualization enables the provisioning of Cloud-based sensor services and other IT resources remotely to the end-user without being worried about the sensors exact locations. The virtualized sensors are created by using the service templates automatically. Service templates are prepared by the service providers as service catalog, and this catalog enables the creation of service instances automatically hat are accessed by multiple users.

Layer 3: This is the last layer which deals with the service template creation and service catalog definition layer in forming catalog menu. Physical sensors are located and retrieved from his layer. Since each physical sensor has its own control and data collection mechanism, standard mechanisms are defined and used to access sensors without concerning the differences among various physical sensors. Standard functions are defined to access the virtual sensors by the users.

Physical sensors are XML encoded that enable the services provided through these sensors to be utilized on various platforms without being worried to convert them onto several platforms. Sensor-Cloud Provides a Web based aggregation platform for sensor data that is flexible enough to help in developing user-based applications. It allows users quick development and deployment of data processing applications and gives programming language flexibility in accordance to their needs.

DETAILED STUDY OF SENSOR-CLOUD INFRASTRUCTURES

Sensor-Cloud Infrastructure

Authors M.Yuiyama and T. Kushida (2012) proposed a novel methodology for integrating WSN with Cloud Computing. The Figure 3 shows the overview of the architecture of the Sensor Cloud interation Proposed by authors. Various sensors with different owners can join Sensor Cloud infrastructure. Each owner registers or deletes its physical sensors. User can create ST/Service Template or Virtual Groups y subscribing data or particular physical sensors. User can easily add, remove or share the configuration o template as sensors as per the need.

Figure 2. Sensor-cloud layered model

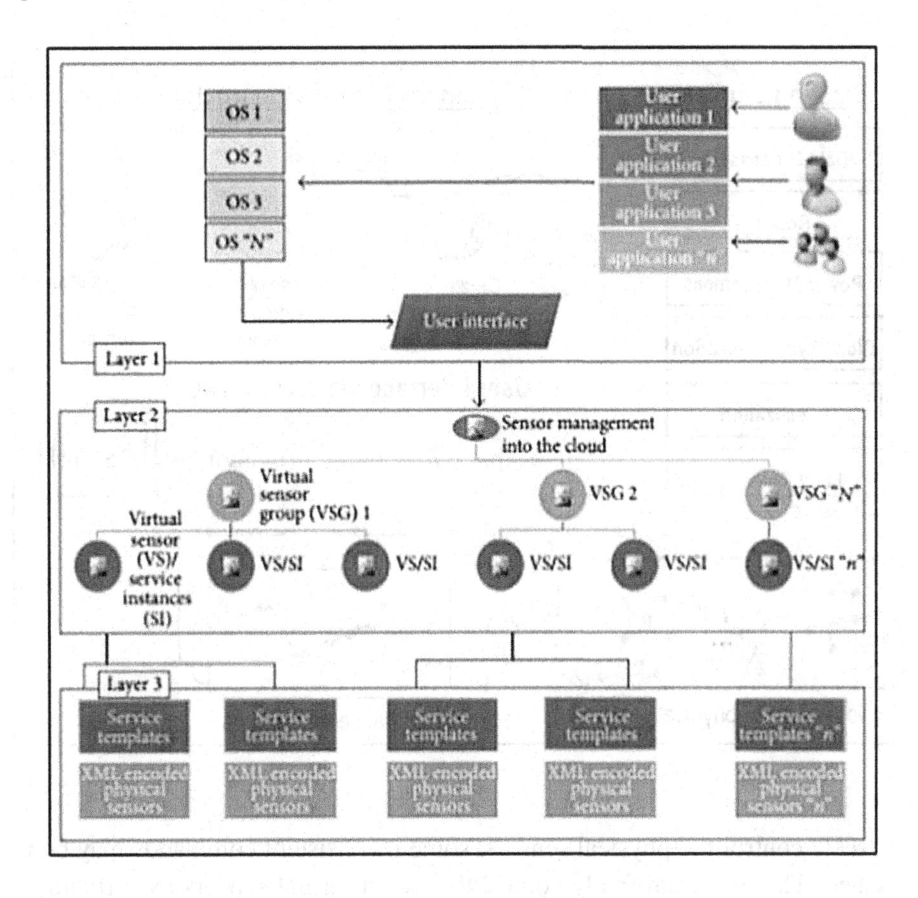

Design Considerations

- **Virtualization:** In the real scenario there are different kind of sensors are scattered over the spatial area. We propose virtual sensor and virtual sensor group in order for the users to be able to use sensors without worrying about the locations and the specification of physical sensors. Figure 4 describes the relationship among virtual sensor groups, virtual sensors, and physical sensors. Each virtual sensor is created from one or more physical sensors which is dependent on the user application area. A virtual sensor group is created from one or more virtual sensors. Users can create virtual sensor groups and freely use the virtual sensors included the groups as if they owned sensors. For example, they can activate or inactivate their virtual sensors, check their status, and set the frequency of data collection from them. If multiple user

Figure 3. Sensor cloud infrastructure-overview

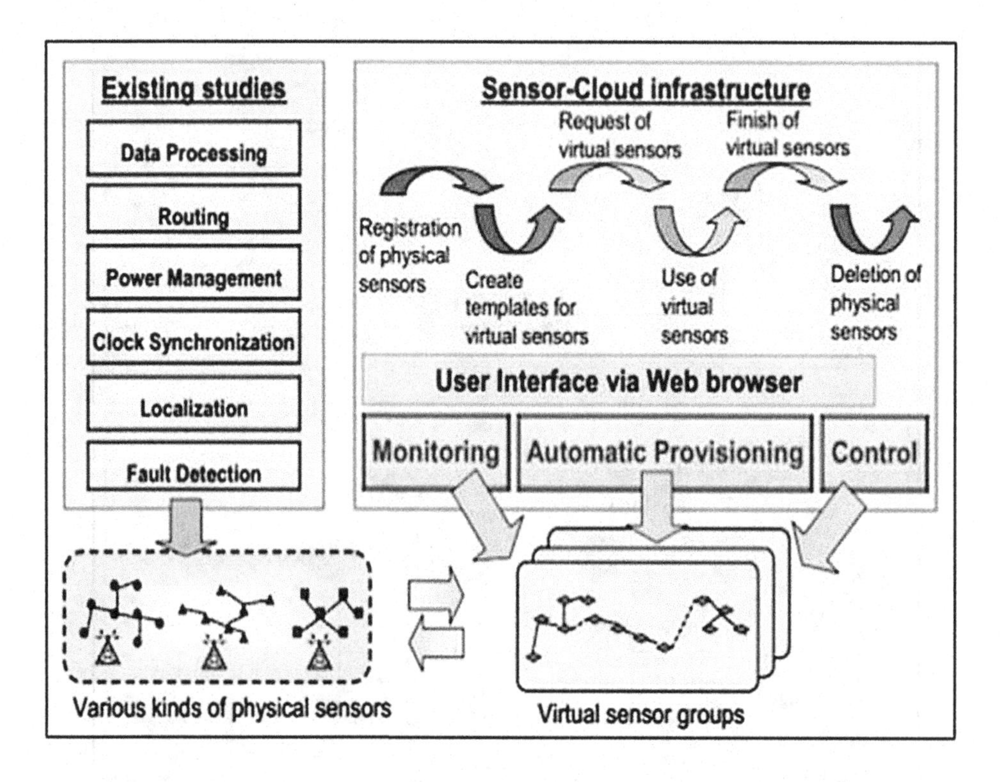

freely control the physical sensors, some inconsistent commands may be issued. The users can freely control their own virtual sensors by virtualizing the physical sensors as virtual sensors.

- **Standardization and Automation:** Different kinds of physical sensors have different functions in terms of sensing the environment. Each physical sensor provides its own functions for control and data collection. Standard like Sensor Markup Language/SML mechanism enables users to access sensors without concern for the differences among the physical sensors. Sensor-Cloud infrastructure translates the standard functions for the virtual sensors into specific functions for the different kinds of physical sensors. Automation (in terms of response of data), improves the service delivery time and reduces the cost. Sensor-Cloud infrastructure prepares templates for the specifications of various types of physical sensors. When users select the template of a virtual sensor or virtual sensor group. Sensor-Cloud infrastructure dynamically and automatically provisions the virtual sensors in that virtual sensor group from the templates. Sensor-Cloud infrastructure is as on demand service delivery and supports the full lifecycle of service delivery from the registration of physical sensors through creating templates, requesting of virtual sensors, provisioning, starting and finishing to use virtual sensors, and deleting the physical sensors.

- **Monitoring:** Because the application has troubles, if it cannot use the sensor data from the virtual sensors, the application owner should check whether or not the virtual sensors are available and monitor their status for sustaining the quality of service. The users can check the status and the availability of the virtual sensors by the monitoring mechanism of Sensor-Cloud infrastructure.
- **Grouping:** Although there are many kinds of physical sensors, each application does not have to use all of them. Each application uses some types of sensors or when the sensors which match certain constraints such as a location. Sensor-Cloud infrastructure can provide virtual sensor groups. For example, a user can see the access control and the frequency of data collection of virtual sensor groups. Sensor-Cloud infrastructure prepares typical sensor groups and users can create new virtual sensor groups by selecting virtual sensors.

Figure 4. Proposed system architecture

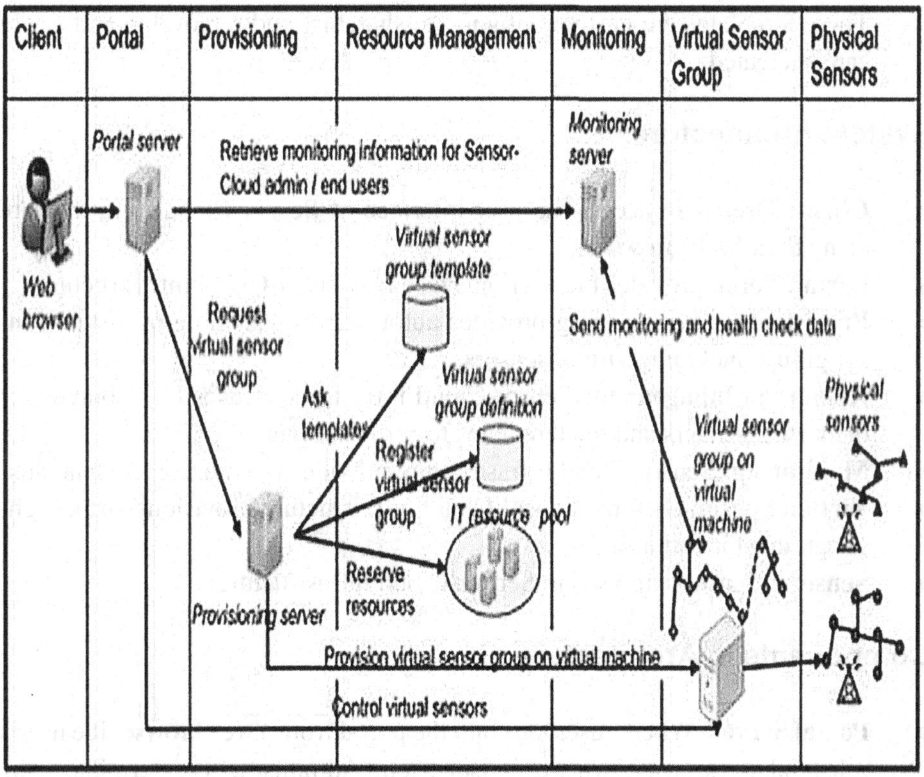

Entities Involved

- **Sensor Owner:** A Sensor owner is a person who owns has physical sensors which are deployed over the area of interest. One of the possible advantage for sensor owner could be rental fees for using the physical sensors. The fees reflects the actual usage of the physical sensors. A sensor owner registers the physical sensors with their properties to Sensor-Cloud infrastructure. The owner deletes the registration of them when owner quits sharing them.
- **Sensor-Cloud Administrator:** The Sensor-Cloud Infrastructure services are continuously monitored and managed by an actor called as Sensor-Cloud Administrator. The Information Technology resources for the virtual sensors, integration of user interfaces and management is handled by Administrator. The sensor-cloud Administrator also takes care of template generation for the virtual sensors and also designing of distinctive virtual sensor patterns. The delivery of Sensor-Cloud infrastructural services are billed and charged appropriately by the Administrator.
- **End User:** An Actor owning various tools or applications, which depends on the sensor data is called End User. An End User request the use of virtual sensors or virtual sensor groups that satisfy the requirements from the templates. These templates are easily configurable sharable, and removable and easily can be created.

System Architecture

- **Client:** Users can access the user interface of Sensor-Cloud infrastructure using their Web Browsers.
- **Portal:** Portal provides the user interface for Sensor-Cloud Infrastructure.
- **Provisioning:** Provisioning provides automatic provisioning of virtual sensor groups including virtual sensors.
- **Resource Management:** Sensor-Cloud infrastructure uses IT resources for the virtual sensors and the templates for provisioning.
- **Monitoring:** Sensor-Cloud Infrastructure provides monitoring mechanisms.
- **Virtual Sensor Groups:** Sensor Cloud infrastructure provisions virtual sensor groups for end users.
- **Sensors:** Sensors are used in Sensor-Cloud infrastructure.

Components in Architecture

- **Portal Server:** When a user logs into the portal from a web browser the user's role (end user, sensor owner or sensor-Cloud Administrator) determinates the available operations. The portal server shows the end users the menus for logging in, logging out, requesting for provisioning or destroying virtual sensor groups, monitoring their virtual sensors, controlling them, creating templates of virtual sensor groups and checking their usage-related charges. The portal server gives sensor owners the menus for logging in, logging out, registering or deleting physical sensors, and checking the usage-related rental fees. One of the menus or Sensor-Cloud administrators is for creating, modifying, and deleting the templates for virtual sensors or virtual sensor groups.

- **Provisioning Server:** The provisioning server provisions the virtual sensor groups for the requests from the portal server. It contains a workflow engine and predefined workflows. It executes the workflows in the proper order. First, it checks and reserves the IT resource pool when it receives a request for provisioning. It retrieves the templates of virtual sensors on the existing or a new virtual server. After provisioning, the provisioning server updates the definitions of the virtual sensor groups. The virtual servers are provisioned with the agents for monitoring.
- **Virtual Sensor Group:** A virtual sensor groups is automatically provisioned on a virtual server by the provisioning server. Each virtual sensor group is owned by end user and has one or more virtual sensors. The end user can control the virtual sensors. For example, they can activate or inactivate their virtual sensors, set the frequency of data collection from them, and check their status. The virtual sensor groups are controlled directly or from a web browser.
- **Monitoring Server:** The monitoring server receives the data about virtual sensors from the agents in the virtual servers and the servers. It stores the received data in a database. The monitoring information for the virtual sensors is available sing a Web browser. The Sensor-Cloud administrators are also able to monitor the status of the servers.

Data Centric-Sensor-Cloud Infrastructure Framework

The authors proposed a data centric Sensor-Cloud Infrastructure in which the data always will be available for users from Cloud. The proposed architecture is integration of Cloud Computing with Wireless Sensor Networks through Internet & Service Oriented Architecture (SOA). The major components of the framework are Data Processing unit (DPU), Publishes/Subscriber Broker, Request Subscriber, Identity and Access Management Unit (AMU) and Data Repository (DR). The data gathered from WSN is passed through gateway to DPU, which process data and add it to DR. In order to access the stored data from cloud services, user connects through secured IAMU, on successful connection establishment user will be given the access according to the account policies. User data request is forwarded to RS which creates a request subscription and forward he subscription to the Pub/Sub broker, when DPU receives the data from gateway, it forwards the data to Pub/Sub Broker.

The user can access the data from any location in the world. The sensor networks are ideally considered to be energy efficient and it's the major criticality of the network that must be answered. In addition to that, short range hop communications is preferred in order to communicate with a long range destination. Therefore, the information from source is distributed across intermediate nodes in the path towards destination node.

Figure 5. Sensor-cloud proposed architecture

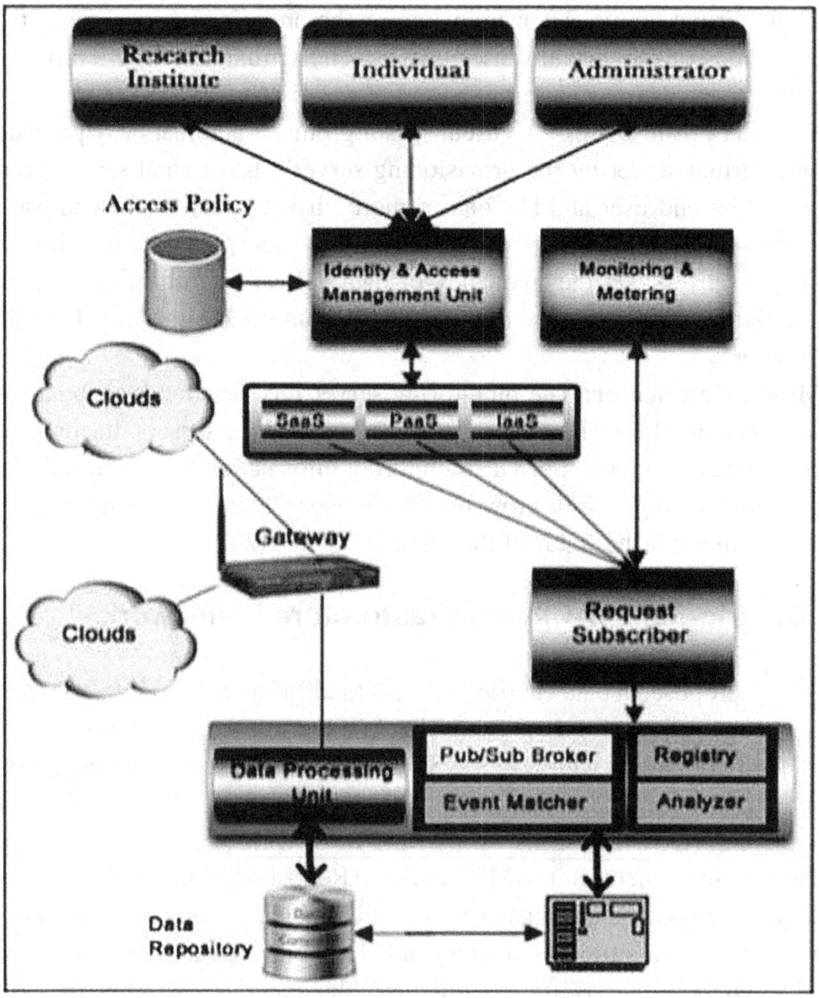

Components for Message Exchange in Architecture

- **Access Control Enforcement Unit:** ACEU is used to authenticate the user and it is consists of EN and three servers i.e. Authenticate Server (AS), Ticket Granting Server (TGS) and SS. The request received by EN I sent to AS. EN implements Kerberos in order to authenticate the user with AS.
- **Access Control Decision Unit:** ACDU is used to enforce the policy rules. It consists of RBAC protocol and policy storage. It communicates with ACEU through SS. After successful authentication, user is given the access to the resources as constrained by the access policies.

Communication flow between User and IAMU: The following Diagram shows the overall communication in the system

SERVICE ORIENTED-SENSOR-CLOUD INFRASTRUCTURE FRAMEWORK

Madoka Yuriyama Takayuki Kushida (2010), proposed the extended version, which focuses on the service availability or end users with the help of Cloud Services. Proposed System model supports the automation like most other frameworks of Sensor-Cloud Infrastructure by creating Service Instances automatically whenever requested. The service providers have provided services with the different configurations by each service requester's requirements before cloud computing service. The service providers prepare service templates as service catalog and request a service instance. This cloud computing service model is giving big impacts to the service because service requester can use necessary service instance quickly when service requester would like a start new service or to need additional resources for extending existing service.

Existing Sensor services use sensors directly or sensor data provided by other sensor services. Sensor services know the details of sensors such as specification, configuration and location. Cloud computing service's advantage is the cloud computing service provider optimizes everything below the service boundary i.e. all code/implementation complexity is hidden from the user & users just have to use the service templates which is ultimately the results or analysis of data.

Figure 6. Communication flow in proposed architecture

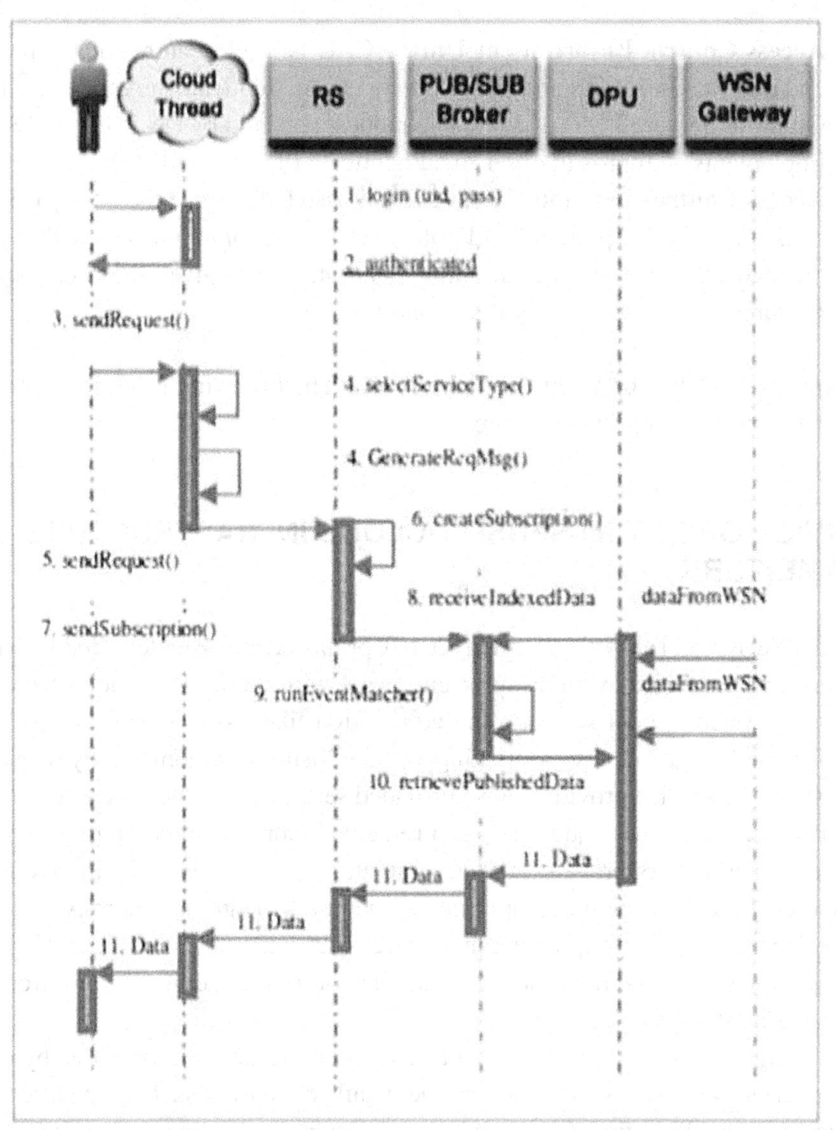

Service Lifecycle

- **Service Catalog Definition Phase:** The service providers have created the service instances with the different configurations by each service requester's requirements before cloud computing service. Service providers prepare service templates as service catalog. The service catalog includes the menus

Figure 7. Service module

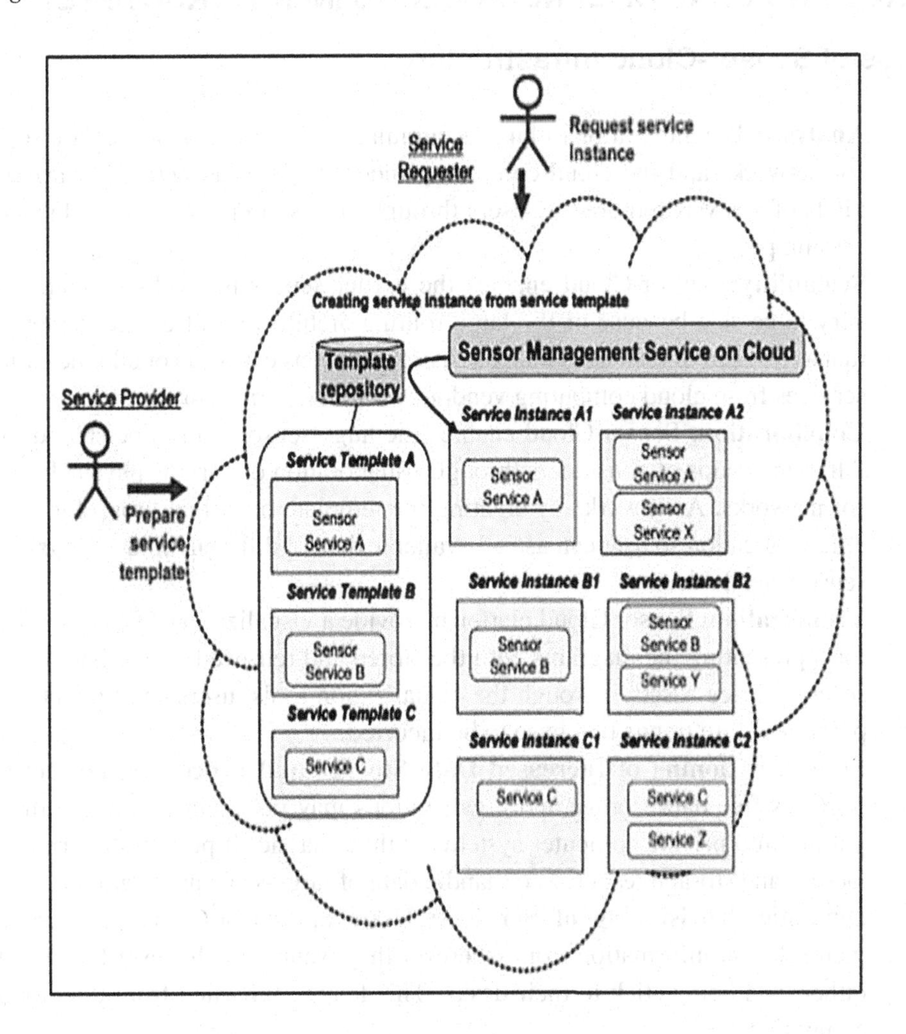

describing service's specifications including grouped sensors. For example, a menu describes Linux (OS), database software (Middleware), traffic analysis service (Service), the sensors in Tokyo (grouped sensors). The service providers prepare the service templates and define them as the service catalog menus. The service instances are created by using service templates automatically. Preparing service templates and defining service catalog enable to create service instances automatically and to duplicate the same specification's service instances for various users.

PROS AND CONS OF SENSOR-CLOUD INFRASTRUCTURE

Pros of Sensor-Cloud Infrastructure

- **Analysis:** The integration of huge accumulated sensor data from several sensor networks and the cloud computing model make it attractive or various kinds of analysis required by users through provisioning o the scalable processing power.
- **Scalability:** Sensor-Cloud enables the earlier sensor networks to scale on very large size because of the large routing architecture of cloud. It means that as he need or resources increases, organizations can scale or add the extra services from cloud computing vendors without having actually own it.
- **Collaboration:** Sensor-Cloud enables the huge sensor data to be shared by different groups of consumers through collaboration of various physical sensor networks. As it works by creating the template or virtual group for specific application so user can use all available data of different kind of physical sensors available.
- **Visualization:** Sensor Cloud platform provide a visualization API to be used for representing the diagrams with the stored and retrieved sensor data from several device assets. Through the visualization tools, users can predict the possible future trends that have to be incurred.
- **Free Provisioning of Increased Data Storage and Processing Power:** It provides free data storage and organizations may put their data rather than putting into private computer systems without hassle. It provides enormous facility and storage resources to handle data of large-scale applications.
- **Dynamic Provisioning of Services:** Users of Sensor-Cloud can access their relevant information from wherever they want and whenever they need rather than being stick to their desks. This feature inherited from the cloud computing.
- **Multitenancy:** The number of services from service providers can be integrated easily through cloud and internet for numerous service innovations to meet user's demand. Sensor-Cloud allows the accessibility to several numbers of data centres placed anywhere on the networks world (again the feature inherited from the cloud computing).
- **Automation:** Automation played a vital role in provisioning of sensor-cloud computing services. Automation of services improved the delivery time to a great extent.

Figure 8. Service providing phase

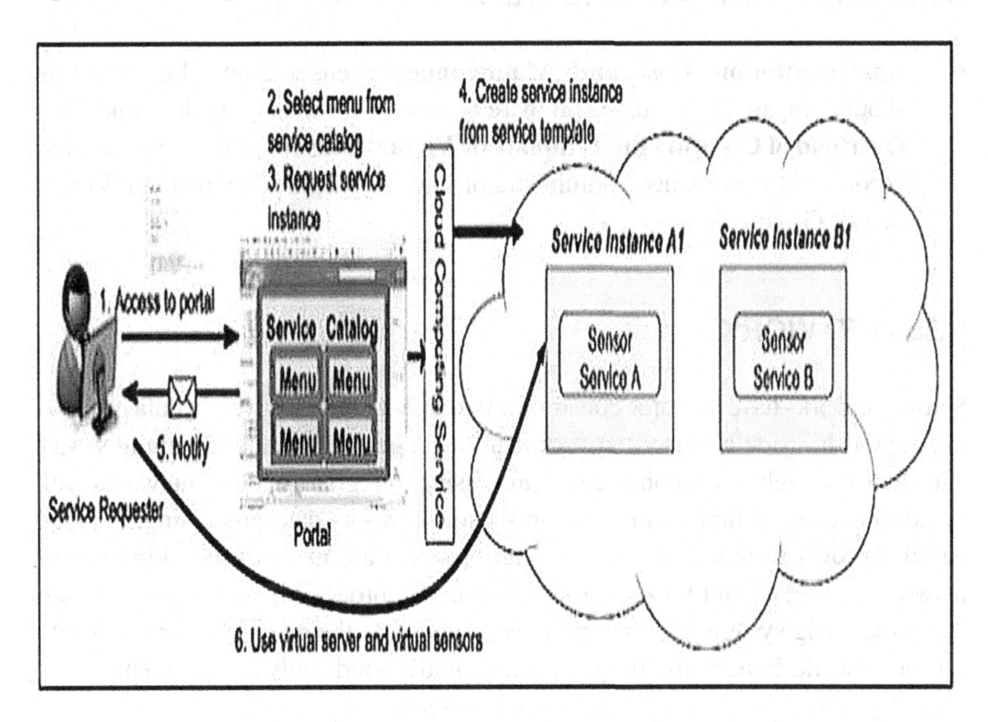

- **Flexibility:** Sensor-Cloud provides more flexibility to its users than the past computing methods. It provides flexibility to use random applications in any number of times and allows sharing of sensor resources under flexible usage environment.
- **Quick Response Rime:** The integration of WSN's with loud provides a very quick response to the user, that is, in real-time due to the large routing architecture of Cloud Computing. The quick response time of data feeds from several sensor networks or devices allows users to make critical decision in near real time.
- **Resource Optimizations:** Sensor-Cloud infrastructure enables the resource optimizations by allowing the sharing of resources for several number of applications. The integration of sensors with cloud enables gradual reduction cost and achieves higher gains of services. With Sensor-Cloud, both the small and midsized organizations can benefit from an enormous resource infrastructure without having to involve and administer it directly by cloud computing.

Cons of Sensor-Cloud Infrastructure:

- **Implementation Cost and Maintenance:** Sensor-Cloud Infrastructure should prepare IT resource and there is the cost for managing the same.
- **Overload of Creating the Template or Virtual Sensor Groups:** Sometimes, according to user's need administrator have to prepare Template or Virtual Sensor Group.

RELATED WORK

Sensor networks have resource constraint. We cannot use heavyweight data processing approach inside a sensor network. But huge, heterogeneous streaming sensor data demands high level sophisticated processing. Integrating sensor networks with cloud can solve challenges involved in sensor networks data processing. A cloud based sensor data processing system can process data in two ways- First sensor network sense data and forward it to the cloud for processing and secondly cloud data processing system first process query, subdivide the query and forwards into sensor network. In response of query sensor nodes sends only required data.

Existing Cloud Based Platform for Sensor Data Processing

Different authors use different cloud framework for data processing in sensor network. CLOUDVIEW is a framework for storage, processing and analysis of massive machine maintenance data collected from large number of sensors embedded in industrial machines in cloud computing environment. This hybrid (offline and online) system uses distributed Hadoop and Map-Reduce for storage and parallel processing of streaming time series sensor data. System includes data collection (data aggregation, pre-processing, filtering, storing), case base creation, updating and maintenance, feature reduction, feature extraction and fault prediction using CBR (Case Base Reasoning) and rough set theory. But the problem is Hadoop supports batch processing style. So problem may arise for interactive real time processing. In authors concentrate on maintaining the isolation and consistency property of sensor transaction. Some authors use Apache Hadoop and Map-Reduce to process semi structured and unstructured distributed large scale sensor data. In the authors propose a multilevel and flexible data processing model for sensor network. In this system, coordinator and gateway are used in each sensor network to collect and store sensor data. They provide local storage facility. Different sensor networks are connected to a local server which provides local storage backup and computing functionality. Local server provides uniform data access interface whose specification are stored at cloud and used by clients. This system uses different databases (SQLite, TinyDB, MySQL and MongoDB). To integrate ERP and sensor network to minimize resource usage of senor networks. They use processing and prediction model of cloud based ERP system to process data collected from sensor network, generate some decision and forwards this decision/control information back to sensor network to turn the sensors on/off and adjust their transmission frequency to reconfigure the sensor network. But the problem is sensor networks transmit all the data which increases power and energy consumption.

Sensor Network Query Processing Framework Consideration:

To design a cloud framework for sensor data processing, there is a need to consider the following points.

- The system has support for stored or historical sensor data processing as well as real time series streaming data processing. In this case we can consider relational database or Google Big-table approach for stored structured data and Apache/Google Hadoop for unstructured/semi structured streaming data processing.
- System must have support for one time, continuous, periodic query processing. As wsn is changed dynamically query optimization plan must be adaptive instead of static as in traditional database.
- Support system defined and user defined grouped aggregation which reduces data transmission.
- Explicit support of data pre-processing (especially uncertainty or discrepancy management).
- Approximate query processing with certain level of error tolerance.
- System must support acquisitional query processing, on demand query processing and query on collected stored data
- System can support WSN resource management using processed data.
- Explicit fault management, data consistency management and loads balancing.
- Framework also includes power consumption monitoring and network legibility issues.

CONCLUSION

Typical WSN have very large domain of applications but it still legging because of its limitation like limited energy, storage, processing power, bandwidth, range etc. There is the way to overcome its limitation by cloud computing architecture which is the best fit to overcome all limitations.

So, the integrated version of Cloud Computing of WSN, i.e Sensor-Cloud Infrastructure will provide the good way to enhance the area of applications of WSN.

REFERENCES

Alanri, A., Ansari, W. S., Hassan, M. M., Hossain, M. S., Alelaiwi, A. H., & Hossain, M. A. (2013). A Survey on Sensor-Cloud Architecture, Applications, and Approaches. *International Journal of Distributed Sensor Networks.*

Dash, S. K., Sahoo, J. P., Mohapatra, S., & Pai, S. P. (2012). Sensor-Cloud assimilation of Wireless sensor network and the cloud. *Advances in Computer Science and Information Technology Networks and Communications, Springer*, *84*, 455–464. doi:10.1007/978-3-642-27299-8_48

Dash, S. K., Sahoo, J. P., Mohapatra, S., & Pati, S. P. (2012). Sensor-cloud: assimilation of wireless sensor network and the cloud. *Advances in Computer Science and Information Technology, 84*, 455-464. doi:10.1007/978-3-642-27299-8_48

Ding, Z., Guo, L., & Yang, Q. (2014). RDB-KV: A Cloud Database Framework for Managing Massive Heterogeneous Sensor Stream Data. In *Proceedings of IEEE Second International Conference on Intelligent System Design and Engineering Application* (ISDEA), (pp. 653-656). Retrieved from http://en.wikipedia.org/wiki/BigTable

Dinh, H. T., Loe, C., Niyato, D., & Wang, P. (2011). *A Survey of Mobile Cloud Computing: Architecture, Applications and Approaches*. Wireless Communications and Mobile Computing-Wiley Online Library.

Fok, C., Roman, G., & Lu, C. (2007). Towards A Flexible Global Sensing Infrastructure. *ACM SIGBED Review, 4*(3), 1-6.

Grgen, L., Roncancio, C., Labb, C., & Olive, V. (2006). Transactional issues in sensor data management. In *Proceeding DMSN '06 Proceedings of the 3rd workshop on Data management for sensor networks: in conjunction.* doi:10.1145/1315903.1315910

Guo, S., Zhong, Z., & He, T. (2009). FIND: Faulty node detection for wireless sensor networks. In *Proceedings of the 7th ACM Confeence on Embedded Networked Sensor Systems (Sensys'09)*, (pp. 252-266). ACM.

Intelligent Systems Centre Nanyang Technological University. (n.d.). Available: http://www.ntu.edu.sg/intellisys

Ishi, Y., Kawakami, T., Yoshihisa, T., Teranishi, Y., et al. (2012). Design and Implementation of Sensor Data Sharing Platform for Virtualized Wide Area Sensor Networks. In *Proceedings of Seventh International Conference on P2P, Parallel, Grid, Cloud and Internet Computing* (3PGCIC), (pp. 333-338). Academic Press.

Kamaljit, I. L., & Patel. (n.d.). Comparing Different Gateway Discovery Mechanism for Connectivity of Internet & MANET. *International Journal of Wireless Communication and Simulation, 2*(1), 51-63.

Kumar, L. P. D., Grace, S. S., Krishnan, A., & Manikandan, V. M. (2012). Data Filtering in Wireless Sensor Networks using neural networks for storage in cloud. In *Proceedings of the IEEE International Conference on Recent Trends in Information Technology* (ICRTIT'11). IEEE.

Kushida, M., & Yuriyama, T. (2012). Sensor-Cloud Infrastructure physical sensor management with virtualized sensors on cloud computing. Advance in Computer Science and Information Technology Networks and Communications, 84, 455-464.

Lakhtaria, K. I. (2009). Enhancing QOS and QOE in IMS enabled next generation networks. In *Networks and Communications, 2009. NETCOM'09. First International Conference on* (pp. 184-189). IEEE. doi:10.1109/NetCoM.2009.29

Lakhtaria, K. I. (2010). *Analyzing Zone Routing Protocol in MANET Applying Authentic Parameter*. arXiv preprint arXiv:1012.2510.

Lakhtaria, K. I. (2012). *Efficient detection of malicious nodes by implementing OpenDNS and statistical methods*. Paper presented at 4th International Conferences on IT and Businesses Intelligence.

Lakhtaria, K. I., & Jani, D. N. (2010). *Design and Modeling Billing solution to Next Generation Networks*. arXiv preprint arXiv:1008.1851.

Lakhtaria, K. I., & Nagamalai, D. (2010). Analyzing Web 2.0 Integration with Next Generation Networks for Services Rendering. In Recent Trends in Networks and Communications (pp. 581-591). Springer Berlin Heidelberg.

Lan, K. T. (2010). What's Next? Sensor+Cloud?. In *Proceeding of the 7th International Workshop on Data Management for Sensor Neworks*. ACM Digital Library.

Liu, J., Chen, J., Peng, L., & Cao, X. (2012). An open, flexible and multilevel data storing and processing platform for very large scale sensor network. In *Proceedings of 14th International Conference on Advanced Communication Technology* (ICACT), (pp. 926-930). Academic Press.

Madoka, Y. T. K. (2010). Sensor-Cloud Infrastructure-Physical Sensor Management with Virtualized Sensors on Cloud Computing. *13th International Conference on Network-Based Information Systems*.

Ponmagal & Raja. (2011). An extensible cloud architecture model for heterogenous sensor services. *International Journal of Computer Science and Information Security, 9*.

Rajesj, V., Gnanasekar, J. M., Ponmaga, R. S., & Anbalagan, P. (2010). Integration of Wireless Sensor Network with Cloud. *International Conference on Recent Trends in Information, Telecommunication and Computing*. doi:10.1109/ITC.2010.88

Rolim, C. O., Koch, F. L., Westphall, C. B., Werner, J., Fracalossi, A., & Salvador, G. F. (2010). A cloud computing solution for patient's data collection in health care institutions. In *Proceedings of the 2nd International Conference on Ehealth, Telemedicine, and Social Medicine*, (pp. 95-99).

Sensor-Cloud. (n.d.). Available: hp://www.sensorcloud.com/system-overview

Shah, Khan, Ali, & Khan. (2013). A new Framework to Integrate Wireless Sensor Networks with Cloud Computing. In *Proceedings of Aeropace Conference*. IEEE.

Wireless Sensor Network. (n.d.). In *Wikipedia, the free encyclopedia*. Available: http://en.widipedia.org/wiki/wireless_sensor_network

Yick, J., Mukherjee, B., & Ghosal, D. (2008). *Wireless Sensor Network Survey*. Elsevier.

Yuriyama, M., & Kushida, T. (2010). Sensor-Cloud infrastruture physical sensor management with virtualized sensors on cloud computing. In *Proceedings of the IEEE 13th International Conference on Network-Based Information Systems* (NbiS'10). IEEE.

Yuriyama, M., Kushida, T., & Itakura, M. (2011). A new model of accelerating service innovation with sensor-cloud infrastructure. In *Proceedings of the annual SRII Global Conference (SRII'11)*, (pp. 308-314). doi:10.1109/SRII.2011.42

Zhang, P., Yan, Z., & Sun, H. (2013). A Novel Architecture Based on Cloud Compting for Wireless Sensor Network. In *Proceedings of the 2nd International Conference on Computer Science and Electronics Engineering*.

Chapter 6
A Survey on Wireless Sensor Networks

Homero Toral-Cruz
University of Quintana Roo, Mexico

Romeli Barbosa
University of Quintana Roo, Mexico

Faouzi Hidoussi
University Hadj Lakhdar of Batna, Algeria

Miroslav Voznak
Technical University of Ostrava (VSB), Czech Republic

Djallel Eddine Boubiche
University Hadj Lakhdar of Batna, Algeria

Kamaljit I Lakhtaria
Gujarat University, India

ABSTRACT

Wireless sensor networks (WSN) have become one of the most attractive research areas in many scientific fields for the last years. WSN consists of several sensor nodes that collect data in inaccessible areas and send them to the base station (BS) or sink. At the same time sensor networks have some special characteristics compared to traditional networks, which make it hard to deal with such kind of networks. The architecture of protocol stack used by the base station and sensor nodes, integrates power and routing awareness (i.e., energy-aware routing), integrates data with networking protocols (i.e., data aggregation), communicates power efficiently through the wireless medium, and promotes cooperative efforts of sensor nodes (i.e., task management plane).

INTRODUCTION

Wireless sensor networks (WSN) have become one of the most attractive research areas in many scientific fields for the last years. WSN consists of several sensor

DOI: 10.4018/978-1-4666-8687-8.ch006

nodes that collect data in inaccessible areas and send them to the base station (BS) or sink (Akyildiz, Su, Sankarasubramaniam, & Cayirci, 2002a). At the same time sensor networks have some special characteristics compared to traditional networks, which make it hard to deal with such kind of networks.

The architecture of protocol stack used by the base station and sensor nodes (Akyildiz, Su, Sankarasubramaniam, & Cayirci, 2002b), integrates power and routing awareness (i.e., energy-aware routing), integrates data with networking protocols (i.e., data aggregation), communicates power efficiently through the wireless medium, and promotes cooperative efforts of sensor nodes (i.e., task management plane). The sensor network protocol stack is much like the traditional protocol stack (Maraiya, Kant, & Gupta, 2011), with the following layers: *physical layer, data link layer, network layer, transport layer, application layer, power management plane, mobility management plane,* and *task management plane.*

WSNs have various applications; examples include *military applications, environmental monitoring, medical application, home application, industrial* and *commercial application.*

WSNs are deployed in physical harsh and hostile environments where nodes are always exposed to physical security risks damages (Alrajeh, Khan, & Shams, 2013). In WSNs, one of the most important constraints is the low power consumption requirement. Sensor nodes carry limited, generally irreplaceable, power sources. Therefore, they must have inbuilt trade-off mechanisms that give the end user the option of prolonging network lifetime at the cost of lower throughput or higher transmission delay (Akyildiz, Su, Sankarasubramaniam, & Cayirci, 2002a). In order to acquire energy efficiency, various hierarchical or cluster-based routing methods, originally proposed in wire networks, are well-known techniques with special advantages related to scalability and efficient communication.

In a hierarchical architecture, higher energy nodes can be used to process and send the information, while low-energy nodes can be used to perform the sensing in the proximity of the target. The creation of clusters and assigning special tasks to cluster heads can greatly contribute to overall system scalability, lifetime, and energy efficiency. The main aim of hierarchical routing is to efficiently maintain the energy consumption of sensor nodes by involving them in multi-hop communication. Cluster formation is typically based on the energy reserve of sensors and sensor's proximity to the cluster head. Several cluster-based routing protocols are proposed in the literature such as *LEACH* (Heinzelman, Chandrakasan, & Balakrishnan, 2002), *TEEN* (Lee, Noh, & Kim, 2013), *APTEEN* (Manjeshwar & Agrawal, 2002), *HEED* (Younis & Fahmy, 2004), *PEGASIS* (Lindsey, & Raghavendra, 2002) and *HEEP* (Boubiche, & Bilami, 2011). Where *LEACH* is one of the first hierarchical routing approaches for sensors networks.

Organizing the network nodes in chains clusters avoids the bad energy dissipation in *LEACH* protocol and reduces the routing delay generated by *PEGASIS* protocol (Boubiche, & Bilami, 2011). Based on the chains clustering approach, in each cluster adjacent chains node are formed and the most powerful node is selected to be the cluster head (CH). All nodes will transmit their collected data to their CH using neighboring chains of nodes. Then CHs transmit the received data directly to the base station, or indirectly through the neighboring CHs. Transmitting collected data through the neighboring chains nodes can reduce transmission distances and optimize energy consumption. Data aggregation is applied by each node in a chain, to reduce the amount of exchanged data between nodes and their CH, which preserves energy reserves (Boubiche, & Bilami, 2011).

Researchers have only recently started to study the sensor movement and unique attributes of mobile sensor networks since the sensor networks were originally assumed to consist of only static nodes. It has been suggested that the mobility of sensor nodes improves the sensing coverage. Robotic Fleas project in Berkeley; Robomote and Parasitic Mobility were attempts to enable mobility in sensor networks. In *LEACH* (Heinzelman, Chandrakasan, & Balakrishnan, 2002), mobility is not supported directly. In a round time, if a node moves away from the current cluster-head it has to spend more energy to keep in touch with the current cluster-head. To mitigate this issue, *M-LEACH* is proposed in (Nguyen, Defago, Beuran, & Shinoda, 2008).

Additional to *M-LEACH*, there are also some protocols for mobile sensor network (MSN) like *LEACH-Mobile* (Kim & Chung, 2006), *LEACH-ME* (Kumar, Vinu, & Jacob, 2008), *Energy Efficient Mobile Wireless Sensor Network Routing Protocol* (*E2 MWSNRP*) (Sara, Kalaiarasi, Pari, & Sridharan, 2010) and the *Grid Based Energy Efficient Routing* (*GBEER*) (Kweon, Ghim, Hong, & Yoon, 2009) that was proposed for communication from multiple sources to multiple mobile sinks in wireless sensor network.

Additional to the energy and mobility constraints, most of WSNs are vulnerable to many types of security attacks due to open wireless medium, multi-hop decentralized communication, and deployment in hostile and physically nonprotected areas (Alrajeh, Khan, & Shams, 2013). Based on (Karlof, & Wagner, 2003) research, we can classify routing attacks into six categories: *sink hole attack, black hole attack, selective forwarding, sybil attack, hello flood attack and misdirection.*

In addition to energy efficiency, mobility and security there still exist some other open research issues, such as: *integration of sensor networks and the Internet, WSNs in challenging environments (wireless underground sensor networks (WUSNs) & wireless underwater sensor networks (UWSNs))* and *wireless multimedia sensor networks.*

The intent of this chapter is to present an overview of wireless sensor networks. Firstly, the basic concepts of wireless sensors networks (WSNs) are presented. Secondly, the architecture of protocol stack and the main applications of WSN are explored. Thirdly, the most important cluster-based routing protocols for wireless sensors network and mobile sensor network (MSN) are described. Fourth, a classification of routing attacks is presented. Finally the main open research issues, future research directions and conclusions around this work are presented.

BASIC CONCEPTS OF WSN

Sensor networks represent a significant improvement over traditional sensors, which are deployed in the following two ways (Akyildiz, Su, Sankarasubramaniam, & Cayirci, 2002a): 1) Sensors can be positioned far from the actual phenomenon. Therefore, large sensors that use some complex techniques to distinguish the targets from environmental noise are required. 2) Several sensors that perform only sensing can be deployed. The positions of the sensors and communications topology are carefully engineered.

Early sensor networks involved simple transducers that convert a measured variable into a signal that can be transmitted to a central processing system for analysis. These sensor networks were based on a star topology, with single-hop point-to-point links between the sensor and the central base station. The power requirements of single-hop links limited the range of the network, unless a significant power supply is available at each node (Loo, C. E., Yong, M., Leckie, C., & Palaniswami).

In order that sensor networks can be used for various application areas, wireless ad hoc networking techniques are required. Although many protocols and algorithms have been proposed for traditional wireless ad hoc networks, they are not well suited to the unique features and application requirements of sensor networks. Many researchers are currently engaged in developing schemes that fulfill these requirements. The main differences between sensor networks and ad hoc networks are (Akyildiz, Su, Sankarasubramaniam, & Cayirci, 2002b):

- The number of sensor nodes in a sensor network can be several orders of magnitude higher than the nodes in an ad hoc network.
- Sensor nodes are densely deployed.
- Sensor nodes are prone to failures.
- The topology of a sensor network changes very frequently.
- Sensor nodes mainly use a broadcast communication paradigm, whereas most ad hoc networks are based on point-to-point communications.

- Sensor nodes are limited in power, computational capacities, and memory.
- Sensor nodes may not have global identification (ID) because of the large amount of overhead and large number of sensors.

Wireless sensor networks are composed of several sensor nodes, where the main objective of a sensor node is to collect information from its surrounding environment and transmit it to one or more points of centralized control called base stations (Akyildiz, Su, Sankarasubramaniam, & Cayirci, 2002a; Karlof, & Wagner, 2003; Alrajeh, Khan, & Shams, 2013). A base station is typically many orders of magnitude more powerful than a sensor node, with high bandwidth links for communication amongst themselves. It can be a gateway to another network, a powerful data processing, a storage center, or an access point for human interface and can be used as a nexus to disseminate control information into the network or extract data from it (Karlof, & Wagner, 2003). On the other hand, sensor nodes are constrained to use lower-power, lower-bandwidth, shorter-range radios, also they have the capability of self-healing and self-organizing. They are decentralized and distributed in nature and they form a multi-hop wireless network to allow sensors to communicate to the nearest base station (Karlof, & Wagner, 2003; Alrajeh, Khan, & Shams, 2013).

One of the advantages of WSNs is their ability to operate unattended in harsh environments, where the monitoring schemes are risky, inefficient and sometimes infeasible. Therefore, sensors are expected to be deployed randomly in the area of interest by a relatively uncontrolled means. Given the vast area to be covered, the short lifespan of the battery-operated sensors and the possibility of having damaged nodes during deployment, large population of sensors is expected in most WSNs applications. Designing and operating such large size network would require scalable architectural and management strategies. In addition, sensors in such environments are energy constrained and their batteries cannot be recharged. Therefore, designing energy-aware algorithms becomes an important factor for extending the lifetime of sensors. Grouping sensor nodes into clusters has been widely pursued by the research community in order to achieve the network scalability objective (Abbasi & Younis, 2007).

ARCHITECTURE OF PROTOCOL STACK

The protocol stack is a combination of different layers and consists of the physical layer, data link layer, network layer, transport layer, application layer, power management plane, mobility management plane and task management plane. Each layer has a set of protocols with different operations and integrated with other layers. The protocol stack used by the sink and sensor nodes is shown in Figure 1.

Figure 1. The WSN protocol stack

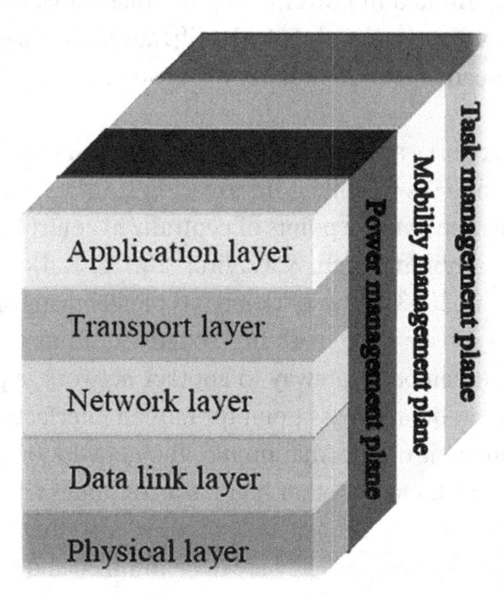

Physical Layer

The physical layer is responsible for frequency selection, carrier frequency generation, signal detection, modulation, and data encryption. Frequency generation and signal detection have more to do with the underlying hardware and transceiver design. The physical layer presents many open research and issues that are largely unexplored. Open research issues include modulation schemes, strategies to overcome signal propagation effects and hardware design (Kifayat, Merabti, Shi, & Llewellyn, 2010).

Data Link Layer

The data link layer is responsible for the multiplexing of data streams, data frame detection, and medium access and error control. It ensures reliable point-to-point and point-to-multipoint connections in a communication network. The data link layer is combination of different protocols includes: medium access control (*MAC*) and *error control*.

- **MAC Protocols:** Organize thousands of sensor nodes and establish the communication link using hop by hop to transfer data and fairly and efficiently share communication resources between sensor nodes.
- **Error Control Schemes:** Are another important function of the data link layer to error control of transmission data. Two important types of error con-

trol in communication networks are the forward error correction (FEC) and automatic repeat request (ARQ) techniques. Research work in all of these areas is continuing to provide improvements.

Network Layer

The network layer is to provide internetworking with external networks like other sensor networks, in one scenario, the sink nodes can be used as a gateway to other networks. The network layer in a WSN must be designed with the following considerations in mind: Power efficiency, WSNs are data-centric networks WSNs have attribute-based addressing and Sensor nodes are location aware. The Link layer handles how two nodes talk to each other, the network layer is responsible for deciding which node to talk to.

Transport Layer

The transport layer comes into play when the system needs to communicate with the outside world. Transmitting data from sink to outside user is a problem because WSNs do not use global identification and attribute based naming is used for sending the data. Very little research has been done at the transport layer.

Application Layer

The Application layer contains the logic required for data acquisition and processing. A simple application might measure quantities such as temperature, humidity or luminosity in regular intervals and forward the data to a sink node. Other applications might also process measured data, serve data requests or send messages in response to external events. Furthermore applications also need to decide which nodes to forward their data to. This can either be specific nodes or a high-level destination such as data sinks.

Power Management Plane

The power management plane manages how a sensor node uses its power and manages its power consumption among the three operations (sensing, computation, and wireless communications). For instance, to avoid getting duplicated messages, a sensor node may turn off it receiver after receiving a message from one of its neighbors. Also, a sensor node broadcasts to its neighbors that it is low in power and can't take part in routing messages. The remaining power is reserved for sensing and detecting tasks.

Mobility Management Plane

The mobility management plane detects and registers the movement of sensor nodes, so a route back to the user is always maintained, and the sensor nodes can keep track of who their neighboring sensor nodes are. By knowing who the neighbors are, sensor nodes can balance their power and task usage.

The Task Management Plane

The task management plane (i.e., cooperative efforts of sensor nodes) balances and schedules the events' sensing and detecting tasks from a specific area. Hence; not all of the sensor nodes in that specific area are required to carry out the sensing tasks at the same time. Depending on their power level, some nodes perform the sensing task more than others.

APPLICATION OF WIRELESS SENSOR NETWORK

Sensor networks have a variety of applications. Examples include military applications, environmental monitoring (which involves monitoring air, soil and water), medical application, home application, industrial and commercial application.

Military Applications

Most of the elemental knowledge of sensor networks is basic on the defense application at the beginning, especially two important programs the Distributed Sensor Networks (DSN) and the Sensor Information Technology form the Defense Advanced Research Project Agency (DARPA), sensor networks are applied very successfully in the military sensing.

In fact, it is very difficult to say for sure whether motes were developed because of military and air defense needs or whether they were invented independently and were subsequently applied to army services. Regarding military applications, the area of interest extents from information collection, generally, Battlefield Surveillance and Tracking and shooter detection system.

- **Battlefield Surveillance and Tracking:** In battlefield critical territory, approach routes, paths and straits can be rapidly covered with sensor networks and closely watched for the activities of the opposing forces. As the operations evolve and new operational plans are prepared, new sensor networks can be deployed anytime for battlefield surveillance.

- **Shooter Detection System:** The Boomerang sniper detection system (Raytheon BBN Technologies, 2013) has been developed for accurate sniper location detection by pinpointing small-arms fire from the shooter. It has been used by the military, law enforcement agencies, and municipalities.

The system operates when the vehicle is stationary or moving, using a single mast-mounted, compact array of microphones. Boomerang detects small arms fire traveling toward the vehicle for bullet trajectories passing within approximately 30 meters of the mast and shooters firing at maximum effective weapons ranges. Incoming round detection is determined in under a second. Significant efforts have been implemented to prevent system false alarms caused by non-ballistic events such as road bumps, door slams, wind noise, tactical radio transmissions, and extraneous noise events (vehicle traffic, firecrackers, and urban activity). The system does not alert when shots are fired from the vehicle.

Boomerang is easily integrated with other Boomerang product options, such as the Situation Awareness System, as well as other third-party systems. Through its intuitive System Integration Kit and simple Ethernet interface, Boomerang output can be used to slew camera devices, feed remote weapons station equipment, or report shooter position to an existing Tactical Operations Center. Whether stand alone or in combination with other systems, Boomerang increases target detection and survivability.

Environmental Applications

There are various environmental applications of sensor networks including tracking the movements of birds, small animals and insects; monitoring environmental conditions that affect crops and livestock; irrigation; macro instruments for large-scale Earth monitoring and planetary exploration; chemical/ biological detection; precision agriculture; biological, Earth and environmental monitoring in marine, soil and atmospheric contexts; forest fire detection; meteorological or geophysical research; flood detection; bio-complexity mapping of the environment; and pollution study (Zhang, Sadler, Lyon, & Martonosi, 2004). We will illustrate below some projects related to the environmental applications.

- **Volcano Monitoring:** WSNs have also been used in extreme environments, where continuous human access is impossible. Volcano monitoring is an example of these extreme applications, where a network of sensors can be easily deployed near active volcanoes to continuously monitor their activities and provide data at a scale and resolution not previously possible with existing tools.

Two WSNs on active volcanoes were deployed by this project (Kifayat, Merabti, Shi, & Llewellyn, 2010). Their initial deployment at Tungurahua volcano, Ecuador, in July 2004 served as a proof-of-concept and consisted of a small array of wireless nodes capturing continuous infrasound data. Their second deployment at Reventador volcano, Ecuador, in July/August 2005 consisted of 16 nodes deployed over a 3 km aperture on the upper flanks of the volcano to measure both seismic and infrasonic signals with a high resolution (24 bits per channel at 100Hz) (Kifayat, Merabti, Shi, & Llewellyn, 2010).

- **ZebraNet:** ZebraNet (Zhang, Sadler, Lyon, & Martonosi, 2004) is a wireless sensor network deployed in Kenya to track two species of zebras. It's design for monitoring and tracking wildlife. ZebraNet uses nodes significantly larger and heavier than motes. The architecture is designed for an always mobile, multi-hop wireless network. In many respects, this design does not fit with monitoring the Leach's Storm Petrel at static positions (burrows). ZebraNet, at the time of this writing, has not yet had a full long-term deployment so there is currently no thorough analysis of the reliability of their sensor network algorithms and design.

Medical Applications

WSNs can also be useful in the health sector through telemonitoring of human physiological data, tracking and monitoring of doctors and patients inside a hospital (Gao, Greenspan, Welsh, Juang, & Alm, 2005) there are several ongoing projects to use WSNs in the medical sector. We describe a few of them here:

- **UbiMon Project:** UbiMon (ubiquitous monitoring environment for wearable and implantable sensors) is the architecture for distributed mobile monitoring, developed at Imperial College London. The aim of this system is to provide continuous management of patients under their natural physiological states so that transient but life threatening abnormalities can be detected and predicted (Ng et al, 2004).
- **Alarm-Net Project:** Alarm-Net is a wireless sensor network for assisted-living and residential monitoring being developed on University of Virginia. The Alarm Net system integrates heterogeneous devices, some wearable on the patient and some placed inside the living space. Together they perform a health mission specified by a healthcare provider. Data are collected, aggregated, pre-processed, stored, and acted upon, according to a set of identified system requirements (Wood et al, 2006).

Home Applications

Along with developing military, environmental and medical application of sensor network it is no so hard to image that home application will step into our normal life in the future. Many concepts are already designed by researcher and architects, like "Smart Environment: Some are even realized. Let's see the concept "the intelligent home": After one day hard work you come back home. At the front door the sensor detects you are opening the door, then it will tell the electric kettle to boil some water and the air condition to be turned on.

- **Smart Environment:** Sensor nodes can be embedded into furniture and appliances, and they can communicate with each other and a room server. The room server can also communicate with other room servers to learn about services they offer like printing, scanning and faxing.

Industrial and Commercial Application

Consequently, many different industrial and commercial applications have been developed based on wireless sensor networks. Specific applications for industrial and commercial spaces include (Adams, 2004):

- Warehouses, fleet management, factories, supermarkets, office complexes.
- Gas, water, and electric meters.
- Smoke, CO, and H_2O detectors.
- Refrigeration cage or appliance.
- Equipment management services and preventive maintenance.
- Security services (including peel-n'-stick security sensors).
- Lighting control.
- Assembly line and workflow and inventory.
- Materials processing systems (heat, gas flow, cooling, chemical).
- Materials processing systems (heat, gas flow, cooling, chemical).
- Remote monitoring from corporate headquarters of assets, billing, and energy Management.

CLUSTER-BASED ROUTING PROTOCOLS FOR WSN

There are different ways by which we can classify the sensor networks' routing protocols. According to network structure, these routing protocols can be classified as flat, hierarchical, and location-based protocols. In order to acquire energy efficiency, various hierarchical or cluster-based routing methods, originally proposed in wire networks, are well-known techniques with special advantages related to scalability and efficient communication. In cluster-based routing, nodes will play different roles or functionalities, aiming at routing techniques clustering the nodes with different roles so that the heads of the cluster can do some data aggregation or confusion in order to save power, in the next subsection we will describe some challenges of clustering.

Challenges of Routing Protocols

Wireless Sensor Networks present vast challenges in terms of implementation. There are several key attributes that designers must carefully consider, which are of particular importance in wireless sensor networks. (Joshi & Priya, 2011) describe these challenges as the following:

- Cost of Clustering;
- Selection of Cluster-heads and Clusters;
- Real-Time Operation;
- Synchronization;
- Data Aggregation;
- Repair Mechanisms;
- Quality of Service (QoS).

LEACH

LEACH (Low Energy Adaptive Clustering Hierarchy) is a self-organizing and adaptive clustering protocol proposed by (Heinzelman, Chandrakasan, & Balakrishnan, 2002) that uses randomization to distribute the energy load evenly among the sensor nodes. In the LEACH scheme, the nodes organize themselves into a local cluster and one node behaves as a local cluster head. LEACH includes a randomized rotation of the high energy cluster head position such that it rotates among the sensors. This feature leads to a balanced distribution of the energy consumption to all nodes and makes it possible to have a longer lifetime for the entire network.

TEEN

TEEN (Threshold sensitive Energy Efficient sensor Network) is a basic routing protocol of hierarchical clustered multi-hop routing protocol (Lee, Noh, & Kim,

2013). In the TEEN protocol, at every cluster setup phase, the cluster-head broadcasts to its cluster members, the following threshold values: a) Hard threshold: it is an absolute value for the sensed attribute. If the node senses this value, it turns on its transmitter and reports the data to the CH. b) Soft threshold: it is a small variation in the value of the sensed attribute, which causes the node to turn on its transmitter. The first time a parameter from the attribute set reaches its hard threshold value, the node transmits the sensed data. The sensed value is stored in a variable called sensed value. The node will transmit the data in the current cluster period only when both the following conditions are true: 1) the sensed attribute is greater than the hard threshold; 2) the sensed attribute differs from sensed value by an amount equal or greater than the soft threshold. Thereby, the hard threshold tries to reduce the number of transmission by sending only when the sensed attribute is in the range of interest. The soft threshold reduces the number of transmission by eliminating all the transmissions, which have little or no change in the sensed attribute (Lee, Noh, & Kim, 2013).

APTEEN

APTEEN (Adaptive Threshold sensitive Energy Efficient sensor Network protocol) (Manjeshwar & Agrawal, 2002) is an extension to TEEN and aims at both capturing periodic data collections and reacting to time critical events. The architecture is same as in TEEN. When the base station forms the clusters, the cluster heads broadcast the attributes, the threshold values, and the transmission schedule to all nodes. Cluster heads also perform data aggregation in order to save energy. APTEEN supports three different query types: historical, to analyze past data values; one-time, to take a snapshot view of the network; and persistent to monitor an event for a period of time.

HEED

HEED (Hybrid Energy-Efficient Distributed Clustering) is a multi-hop clustering algorithm for wireless sensor networks, with a focus on efficient clustering by proper selection of cluster-heads based on the physical distance between nodes. The main objectives of HEED are to (Younis & Fahmy, 2004):

- Distribute energy consumption to prolong network lifetime;
- Minimize energy during the cluster-head selection phase;
- Minimize the control overhead of the network.

The most important aspect of HEED is the method of cluster-head selection. Cluster-heads are determined based on two important parameters (Younis & Fahmy, 2004):

1. The residual energy of each node is used to probabilistically choose the initial set of cluster-heads. This parameter is commonly used in many other clustering schemes.
2. Intra-Cluster Communication Cost is used by nodes to determine the cluster to join. This is especially useful if a given node falls within the range of more than one cluster-head. In HEED it is important to identify what the range of a node is in terms of its power levels as a given node will have multiple discrete transmission power levels.

The power level used by a node for intra-cluster announcements and during clustering is referred to as cluster power level (Younis & Fahmy, 2004). Low cluster power levels promote an increase in spatial reuse (Younis & Fahmy, 2004) while high cluster power levels are required for inter-cluster communication as they span two or more cluster areas.

Therefore, when choosing a cluster, a node will communicate with the cluster-head that yields the lowest intra-cluster communication cost. The intra-cluster communication cost is measured using the Average Minimum Reachability Power (AMRP) measurement (Younis & Fahmy, 2004). The AMRP is the average of all minimum power levels required for each node within a cluster range R to communicate effectively with the cluster-head i. The AMRP of a node i then become a measure of the expected intra-cluster communication energy if this node is elevated to cluster-head. Utilizing AMRP as a second parameter in cluster-head selection is more efficient then a node selecting the nearest cluster-head (Younis & Fahmy, 2004).

PEGASIS

PEGASIS (Power-Efficient GAthering in Sensor Information Systems) is a chain-based protocol that is near optimal for data-gathering application in sensor networks (Lindsey, & Raghavendra, 2002). In PEGASIS protocol, the main idea is to form a chain among the sensor nodes so that each node communicates only with a close neighbor, the gathered data moves from node to node, get fused, and eventually a designated node transmits to the BS. Nodes take turns transmitting to the BS so that the average energy spent by each node per round is reduced. The sensor nodes will be organized to form a chain; nevertheless, building a chain to minimize the total length is similar to the traveling salesman problem, which is known to be intractable.

However, with the radio communication energy parameters can be built a simple chain with a greedy approach.

HEEP

HEEP (Hybrid Energy Efficiency Protocol) protocol combines two algorithms, LEACH and PEGASIS. HEEP suggest a new network self-organization approach, that join clusters-based and the chain-based approaches. This new approach is called chains clustering approach. Organizing the network nodes in chains clusters avoids the bad energy dissipation in LEACH protocol and reduces the routing delay generated by PEGASIS protocol (Boubiche, & Bilami, 2011). Based on the chains clustering approach, in each cluster adjacent chains node are formed and the most powerful node is selected to be the cluster head. All nodes will transmit their collected data to their CH using neighboring chains of nodes. Then CHs transmit the received data directly to the base station, or indirectly through the neighboring CHs. Transmitting collected data through the neighboring chains nodes can reduce transmission distances and optimize energy consumption. Data aggregation is applied by each node in a chain, to reduce the amount of exchanged data between nodes and their CH, which preserves energy reserves (Boubiche, & Bilami, 2011).

CLUSTER-BASED ROUTING PROTOCOLS FOR MSN

Most of routing protocols in WSNs considers all nodes are homogeneous with respect to energy which is not realistic approach. In particular round uneven nodes are attached to multiple Cluster-head; in this case cluster-head with large number of member ode will drain its energy as compare to cluster-head with smaller number of associated member nodes. Furthermore mobility support is another issue with routing protocol, to mitigate these issues, some protocols have proposed:

M-LEACH

(Nguyen, Defago, Beuran, & Shinoda, 2008) proposed M-LEACH (Mobile LEACH); M-LEACH allows mobility of non-cluster-head nodes and cluster-head during the setup and steady state phase. MLEACH also considers remaining energy of the node in selection of cluster-head. Some assumptions are also assumed in M-LEACH like other clustering routing protocols. Initially all nodes are homogeneous in sense of antenna gain, all nodes have their location information through GPS and base station is considered fixed in M-LEACH. Distributed setup phase of LEACH is modified by M-LEACH in order to select suitable cluster-head. In M-LEACH cluster-heads are elected on the basis of attenuation model (Heinzelman, Chandrakasan, & Balakrishnan, 2002).

Optimum cluster-heads are selected to lessen the power of attenuation. Other criteria of cluster-head selection are mobility speed. Node with minimum mobility and lowest attenuation power is selected as cluster-head in M-LEACH. Then selected cluster-heads broadcast their status to all nodes in transmission range. Non-cluster-head nodes compute their willingness from multiple cluster-heads and select the cluster-head with maximum residual energy.

In steady state phase, if nodes move away from cluster-head or cluster-head moves away from its member nodes then other cluster-head becomes suitable for member nodes. It results into inefficient clustering formation. To deal this problem MLEACH provides handover mechanism for nodes to switch on to new cluster-head. When nodes decide to make handoff, send DIS-JOIN message to current cluster-head and also send JOIN-REQ to new cluster-head. After handoff occurring cluster-heads reschedule the transmission pattern.

LEACH-ME

LEACH-ME (LEACH Mobile Enhanced) was proposed to enhance LEACH-M (Kumar, Vinu, & Jacob, 2008) by selecting the less mobile nodes relatively to its neighbors to be CHs. Each node contains cluster head transitions it has made during the steady state phase while transmitting data. Nodes transmit a transition count to its CH during the TDMA slot. The CH calculates the average transition count of its members for the few last cycles. As a result, an active slot will rise when the number of transition count is beyond the threshold value. During active slot, nodes broadcast their IDs and each node estimates the distance to all nodes and calculate mobility factor according to (1)

$$M_i\left(t\right) = \frac{1}{n-1} * \sum_{j=0}^{n-1} d_{ij}\left(t\right) \tag{1}$$

where $M_i\left(t\right)$ is the mobile factor based on remoteness of node *i* to all other nodes N is the number of neighbors of node *i*, and $d_{ij}\left(t\right)$ is the distance of node *i* from its neighbors *j*. After calculating the mobile factor, the nodes with least mobility factor value are selected to be CHs, taking into consideration the energy level of that node is not below a certain threshold. The steady state phase is the same for both LEACH-M and LEACH-ME.

LEACH-Mobile

(Kim & Chung, 2006) proposed a Self-Organization Routing Protocol Supporting Mobile Nodes for Wireless Sensor Network. In the proposed scheme, like LEACH is broken up into rounds, where each round begins with a set-up phase when the clusters are organized, followed by a steady-state phase when data transfers to the base station occur. In order to minimize overhead, the steady-state phase is long compared to the set-up phase. One of the basic ideas in LEACH-Mobile is to confirm the inclusion of sensor nodes in a specific cluster at the steady-state phase as the cluster head and non-cluster head node receives particular message at a given time slot according to TDMA time schedule that each sensor cluster has, and then to reorganize the cluster with minimum energy consumption.

LEACH-Mobile assumes that all the non-cluster head nodes of sensor network has to have data to send to cluster head necessarily at its time slot allocated in TDMA schedule.

While the cluster-head in LEACH protocol waits to receive sensed data according to TDMA schedule during steady-state phase, the cluster head in LEACH-Mobile transmits the request message for data transmission to non-cluster head node for gathering sensed data according to TDMA schedule at each time slot. As the data transfer takes place, the cluster head confirms with a time slot list of nodes whether the sensed data is received accordingly at an allocated TDMA time slot at every time when a frame ends, then marks the node on the list of non-receiving. If the sensed data is not received again from the node marked previously when the next frame ends, it removes the node and it may also assign this time slot to the newly joined node in TDMA schedule. It assumes for the cluster head that the nodes not responding to data-request message are moved and are located out of its cluster region. Then, TDMA schedule created by rescheduling is transmitted to all cluster members of nodes.

While cluster-head declares the membership of node within its own cluster region by data-request message, each mobile node confirms the cluster to which it will belong. After the clusters are organized and cluster heads are selected, the non-cluster head nodes transmit data to cluster head upon receiving data-request message. If data request message are not received until the frame ends with time slot allocated by TDMA schedule, the procedure of protocol operation goes to next frame. If mobile node does not receive data-request message even when next frame ends, it broadcasts cluster join-request message. Then the cluster-head upon receiving cluster join request message transmits cluster head advertisement message like a set-up phase to that node.

After this phase is completed, the mobile node decides the new cluster to which it will belong for this round as the mobile node moves. This decision is based on the received signal strength of the advertisement message.

E^2 MWSNRP

E^2MWSNRP (Energy Efficient Mobile Wireless Sensor Network Routing Protocol) proposed by (Sara, Kalaiarasi, Pari, & Sridharan, 2010) and expected to guarantee a longer network lifetime and better packet delivery ratio with less energy consumption. E^2MWSNRP is a multipath hybrid routing protocol that can be designed mainly for highly dynamic energy deficient mobile wireless sensor network where energy dissipation reduction and reliable transmission of data is a must. Despite the real shape of the sensor field, the entire area is assumed to be circumscribed into a big square and then divided into different square zones after the sensor nodes are deployed in the field.

- Routing inside the zone is called IntrA Precinct Routing (IAPR).
- The routing done outside a zone is called IntEr Precinct Routing (IEPR).
- **IntrA Precinct Routing (IAPR):** Every node in a precinct is assumed to be within the communication range of every other node in the precinct. So every sensor node can communicate with the fusion node using single hop communication. When an event is detected, the sensor node first communicates with the fusion node. The fusion node checks if the destination is within its precinct. If so, proactively the event is send to the destination. To forward the data to other precincts, the IntEr Precinct Routing is employed.
- **IntEr Precinct Routing:** It is a multipath reactive routing technique employed for communication among the fusion nodes. When a single path on demand routing protocol is used in such networks, a route rediscovery is needed in response to every route break (Sara, Kalaiarasi, Pari, & Sridharan, 2010). The E^2MWSNRP enables the selection of best paths from the computation of maximal nodal surplus energy. The RREQ message contains the following fields:

```
< Source address, source precinct id, sequence no., broadcast
id, hop count, destination address, maximum surplus energy >.
```

The broadcast id is incremented whenever the source issues a new RREQ. The sequence number denotes the freshness information about a route. As the RREQ travels from a source to various destinations, it automatically sets up the reverse path from all nodes to the source. These reverse path entries are maintained for at least enough time for the RREQ to traverse the network and produce a reply to the sender. If an intermediate fusion node has a current route to the destination and if the RREQ has not been processed previously, the node then unicasts a Route Reply Packet (RREP) back to its neighbor from which it received the RREQ. The RREP message contains the subsequent fields:

```
<Source address, destination address, destination precinct id,
sequence number, hop count, readiness factor, maximum surplus
energy, lifetime>.
```

If the readiness factor denotes 'Discard', a route error (RERR) message is propagated in the reverse path instead of RREP. Four important steps are involved:

- Energy Aware Selection Mechanism;
- Finding maximal nodal surplus energy along the best paths;
- Sorting the multipath in descending order using the nodal surplus energy;
- Forwarding the data packets through the path with maximal nodal surplus energy.

GBEER

Several researchers have focused to provide very energy efficient routing protocols for Wireless Sensor Network with mobile sinks. (Kweon, Ghim, Hong, & Yoon, 2009) have proposed the Grid Based Energy Efficient Routing (GBEER) for communication from multiple sources to multiple mobile sinks in wireless sensor network. With the global location information a permanent grid structure is built. Data requests are routed to the source along the grid and data is sent back to the sinks. The grid quorum solution is adopted to effectively advertise and request the data for mobile sinks. The communication overhead caused by sink's mobility is limited to the grid cell. There is no additional energy consumption due to multiple events because only one grid structure is built independently of the event.

Security in WSN

WSN is an emerging technology and have great potential to be employed in critical situations like battlefields and commercial applications such as building, one of the major challenges wireless sensor networks face today is security. While the deployment of sensor nodes in an unattended environment makes the networks vulnerable to a variety of potential attacks, the inherent power and memory limitations of sensor nodes makes conventional security solutions unfeasible. In the next sections we will discuss the security requirement in WSN, attacks on WSN also the mechanism of security in wireless sensors network.

Security Requirement in WSN

Wireless sensor networks share many characteristics with traditional networks, the security requirements of a wireless sensor network can be classified as follows:

- **Data Confidentiality:** Applications like surveillance of information, industrial secrets and key distribution need to rely on confidentiality. The standard approach for keeping confidentiality is through the use of encryption.
- **Data Authentication:** Authentication ensures the reliability of the message by identifying its origin. Attacks in sensor networks do not just involve the alteration of packets; adversaries can also inject additional false packets (Padmavathi & Shanmugapriya, 2009). Data authentication verifies the identity of the senders and receivers. Data authentication is achieved through symmetric or asymmetric mechanisms where sending and receiving nodes share secret keys. Due to the wireless nature of the media and the unattended nature of sensor networks, it is extremely challenging to ensure authentication.
- **Data Integrity:** Data integrity is a core requirement for secure sensor data in WSN. It is to ensure that information is not changed in transit, either due to malicious intent or by accident.
- **Data Freshness:** Even if confidentiality and data integrity are assured, there is a need to ensure the freshness of each message. Data freshness suggests that the data is recent, and it ensures that no old messages have been replayed. To ensure that no old messages replayed a time stamp can be added to the packet.
- **Availability:** This ensures that the desired network services are available even in the presence of denial-of-service attacks.
- **Authorization:** which ensures that only authorized sensors can be involved in providing information to network services
- **Non-Repudiation:** which denotes that a node cannot deny sending a message it, has previously sent.
- **Self-Organization:** a wireless sensor network is a typically an ad hoc network, which requires every sensor node be independent and flexible enough to be self-organizing and self-healing according to different situations. Due to random deployment of nodes no fixed infrastructure is available for WSN network management. Distributed sensor networks must self-organize to sup-

port multi-hop routing. They must also self-organize to conduct key management and building trust relation among sensors.

- **Secure Localization:** Often, the utility of a sensor network will rely on its Ability to accurately and automatically locate each sensor in the network. A sensor network designed to locate faults will need accurate location information in order to pinpoint the location of a fault. Unfortunately, an attacker can easily manipulate no secured location information by Reporting false signal strengths, replaying signals.

After we discuss the security requirement in WSN, in the next subsection we will classify and describe some attacks that can affect the network.

Attacks on WSN

Based on (Karlof, & Wagner, 2003) research, we can classify routing attacks into six categories:

- **Sink Hole Attack:** In the sinkhole attack, Attacker's goal is to lure all the traffic from a particular area to a compromise node. This attack may create also selective forwarding and black hole attacks.
- **Black Hole Attack:** Attacker node can create a black hole, by attracting and drop-ping all traffic in a specific zone.
- **Selective Forwarding:** In this kind of attacks, attacker may refuse to forward packets or drop them and act as a black hole.
- **Sybil Attack:** In a Sybil attack, a malicious node can represent multiple identities to the network. This kind of attacks is threatening to fault tolerant schemes such as distributed storage, multipath routing and topology maintenance.
- **Hello Flood Attack:** In a Hello Flood Attack, the attacker broadcasts hello packets to convince the nodes that the attacker is a neighbor.
- **Misdirection:** This kind of attacks can be done, by forwarding the message along with the wrong path or by sending false routing updates.

Mechanism of Security in WSN

To protect WSNs against different kinds of vulnerabilities, preventive mechanisms like cryptography and authentication can be applied to prevent some types of attacks. This kind of preventive mechanisms formed the first defense line for WSNs. However, some attacks like wormholes, sinkhole, could not be detected using this kind of preventive mechanisms. In addition, these mechanisms are only effective to prevent from outside attacks and failed to guarantee the prevention of intruders from inside the network (Silva *et al*, 2005). Because of that, it is necessary to use some mechanisms of intrusion detection.

- **Cryptography:** The encryption-decryption techniques devised for the traditional wired networks are not feasible to be applied directly for the wireless networks and in particular for wireless sensor networks. WSNs consist of tiny sensors which really suffer from the lack of processing, memory and battery power

(Pathan, Dai, & Hong, 2006). Applying any encryption scheme requires transmission of extra bits, hence extra processing, memory and battery power which are very important resources for the sensors' longevity. Applying the security mechanisms such as encryption could also increase delay, jitter and packet loss in wireless sensor networks (Pathan, Dai, & Hong, 2006).

- **Intrusion Detection Systems (IDS):** Are considered to act as the second defense line against network attacks that preventive mechanisms fail to address (Silva *et al*, 2005). An Intrusion detection system is defined in (Debar, Dacier, & Wespi, 1999) as "A system that dynamically monitors the events taking place on a system and decides whether these events are symptoms of an attack or constitute a legitimate use of the system". However, there are many challenges posed against the application of the IDS for WSNs. These challenges are due to the lack of resources like, energy, processing and storage.

In general, IDS schemes are categorized into misuse IDS and anomaly IDS. The former matches the new observations with the signatures stored in the database of the IDS. The later detects the abnormal activities from the predefined normal profile in order to identify possible attacks.

Add to the cryptography and intrusion detection system. There are a wide variety of security schemes can be invented to counter malicious attacks and these can be categorized as high level and low-level. Figure 2 (Padmavathi and Shanmugapriya, 2009) shows the order of security mechanisms.

OPEN ISSUES IN WIRELESS SENSORS NETWORK

Extremely energy-efficient solutions are required for each aspect of WSN design to deliver the potential advantages of the WSN phenomenon. There still exist several other grand challenges. In next subsections, we discuss these challenges and highlight open research issues for addressing them.

Figure 2. Security mechanisms
(Padmavathi and Shanmugapriya, 2009).

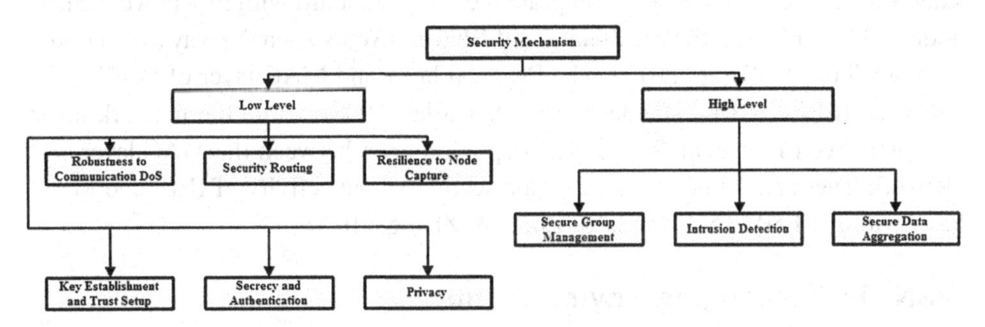

Integration of Sensor Networks and the Internet

In order to extend the applicability of WSN and provide useful information anytime and anywhere, their integration with the Internet is very important. Interconnecting WSN with the Internet is of great significance, while all-IP interconnection method is currently the most convenient and effective way. Considering that the conventional full instance of TCP/IP protocol stack is inappropriate for the sensor node (Zhou, & Zhang, 2013), design schemes of the protocol stack for all-IP WSN have been proposed by many institutions and scholars.

An Overview of How WSN Interconnects with IPv6 Network

According to the author in (Zhou, & Zhang, 2013), there are three major approaches to connect sensor networks to TCP/IP networks are as below:

- **Proxy Architecture:** In this way, all interaction between clients and sensor nodes is through the proxy by protocol conversion and protocol overlap. The drawback of the proxy approach is that it may create a single point of failure.
- **Delay Tolerant Networks:** The DTN method (Fall, 2003) is much like the proxy architecture, but it consists of a bundle layer that resides above the transport layer, which helps to avoid a single point of failure on the sink node. This approach is intended for challenged environments where network partitioning is frequent.
- **All-IP Architecture:** In all-IP WSN, all sensor nodes are required to implement with TCP/IPv6 protocol to be a network terminal, thus realizing a seamless integration of WSN and the TCP/IP Internet through direct communication.

The 6LoWPAN (IPv6 Over, Low power WPAN) (IETF Working Group, 2013) standard has been developed to integrate the IPv6, standard with low-power sensor nodes. This marks that the combination of IPv6 and WSN is on the way to standardization (Zhou, & Zhang, 2013). The Physical layer and MAC layer of 6LoWPAN adapt the IEEE802.15.4 standard (Hui, & Culler, 2008), while the network layer supports IPv6 (Yibo et al, 2011). An adaptation layer between the MAC layer and Network layer is designed to achieve the seamless connectivity of IPv6 and MAC layer with IEEE802.15.4 standard (Zhou, & Zhang, 2013).

WSNs in Challenging Environments

Wireless Underground Sensor Networks (WUSNs)

Wireless Underground Sensor Networks (WUSNs) are one important extension of the terrestrial wireless sensor networks, where the sensor nodes are buried underground and communicate wirelessly through soil. As a promising field, WUSNs enable a wide variety of novel applications, such as intelligent irrigation, sports field maintenance, border patrol, infrastructure monitoring, intruder detection, among of others (Akyildiz, Melodia, & Chowdhury, 2006; Akyildiz, Sun, & Vuran, 2009).

Although its deployment is mainly based on underground sensor nodes, a WUSN still requires aboveground devices for data retrieval, management, and relay functionalities. Accordingly, three different communication links exist in WUSNs based on the locations of the transmitter and the receiver (Silva, & Vuran, 2010):

- **Underground-to-Underground (UG2UG) Link:** Both the sender and the receiver are buried underground and communicate through soil. This type of communication is employed for multi-hop information delivery.
- **Underground-to-Aboveground (UG2AG) Link:** The sender is buried and the receiver is above the ground. Monitoring data is transferred to aboveground relays or sinks through these links.
- **Aboveground-to-Underground (AG2UG) Link:** Aboveground sender node sends messages to underground nodes. This link is used for management information delivery to the underground sensors.

Wireless Underwater Sensor Networks (UWSNs)

Wireless underwater sensor networks (UWSNs) are envisioned to enable applications for a wide variety of purposes such as, pollution monitoring, oceanographic data collection, assisted navigation, disaster prevention and tactical surveillance (Akyildiz, Pompili, & Melodia, 2005).

In the next subsections we will illustrate the difference between WSN terrestrial and UWSN; also we will discuss the UWSN architecture and finally the comparison between the kinds of UWSN architecture.

Differences between WSN Terrestrial and UWSN

Although WSN and UWSN are different, mainly due to the unique characteristics of water, certain aspects of WSN research can be applied to UWSN. The main differences between terrestrial and underwater sensor networks are as follows:

- **Communication Method:** UWSN uses acoustic signal while WSN uses radio waves.
- **Cost:** While terrestrial sensor nodes are expected to become increasingly inexpensive, underwater sensors are expensive devices. It is due to the UWSN's transceivers complexity and the increased protection required by the hardware.
- **Power:** UWSN needs more power because it uses acoustic signal and covers a longer distance. Compared to acoustic signal, RF needs less power, since the processing at receivers is not that complex.
- **Memory:** The connection of an acoustic signal can be disabled by special underwater situations, like shadow zones. Due to this fact, underwater sensors need to acquire more data to prevent the loss of data. However, this is not an issue for terrestrial sensors.
- **Density:** In terrestrial sensor application, like tracking system, sensors can be deployed densely. While an underwater sensor is more expensive than terrestrial sensor, it will cost more money to deploy densely. Even if money is not an issue, it is still not easy to deploy them.
- **Mobility:** Most sensor nodes in ground-based sensor networks are typically static, though it is possible to implement interactions between these static sensor nodes and a limit amount of mobile nodes (e.g., mobile data collecting entities like "mules" which may or may not be sensor nodes). In contrast, majority of underwater sensor nodes, except some fixed nodes equipped on surface-level buoys, are with low or medium mobility due to water current and other underwater activities. From empirical observations, underwater objects may move at the speed of 2-3 knots (or 3-6 kilometers per hour) in a typical underwater condition. Therefore, if a network protocol proposed for ground-based sensor networks does not consider mobility for majority of sensor nodes, it would likely fail when directly cloned for aquatic applications (Cui, Kong, Gerla, & Zhou, 2006).

Nowadays, many different protocols for terrestrial WSN have been developed. However, they cannot fit UWSN. Not only do the architectures of UWSN impact the development of a new protocol, but also the characteristics of underwater. It is another different place with terrestrial sensor network. Therefore, we may develop different kinds of protocol according to the architectures of UWSN. The following will discuss the idea that protocols should be designed according to the type of architecture.

UWSN Architectures

According to (Cui, Kong, Gerla, & Zhou, 2006), UWSN can be roughly classified into two broad categories:

- **Long-Term Non-Time-Critical Aquatic Monitoring:** This kind of UWSN can be work for a long time and the data collected by the sensors are not real-time data. For long-term monitoring, energy saving is a central issue to consider in the protocol design. Clearly, in the UWSNs for long-term aquatic monitoring, localization is a must-to-do task to locate mobile sensors, since usually only location-aware data is useful in aquatic monitoring. In addition, the sensor location information can be utilized to assist data forwarding since geo-routing proves to be more efficient than pure flooding. Furthermore, location can help to determine if the sensors float to the boundary of the interested area. If this happens, the sensors should have some mechanisms to relocate (self-propelled) or pop up to the water surface for manually redeployment. Lastly, reliable, resilient, and secure data transfer is required to ensure a robust observing system in the presented UWSN architecture.

In this type of network, sensor nodes are densely deployed to cover big and spacial continuous monitoring areas. Data are collected by local sensors, related by intermediate sensors, and finally reach the surface nodes (equipped with both acoustic and RF (Radio Frequency) modems), which can transmit data to the on-shore command center by radio.

- **Short-Time Time-Critical Aquatic Exploration**: Compare to Long-term non-time-critical UWSN, this kind of UWSN focus on real-time data. Therefore, how to make data transfer efficiently need to be more concern when designing network protocol. Also, this kind UWSN just work for a short term that means energy saving is not as important as long-term one. However, reliable, resilient, and secure data transfer is an always-desired advanced feature for both types of UWSNs.

In (Cui, Kong, Gerla, & Zhou, 2006) assume a ship wreckage & accident investigation team wants to identify the target venue.

Comparison of the Two Classifications

The difference between the two classifications is static and mobile. In (Cui, Kong, Gerla, & Zhou, 2006), long-term non-time-critical and short-term time-critical UWSN are based on mobile ability. That's why they concern the location aware in either way. Moreover, long-term and short-term did not distinguish 2D or 3D. Obviously, there are some differences in protocol design.

Next years, Underwater wireless sensor networks (UWNS) will become more and more important on the research of underwater world.

Wireless Multimedia Sensor Networks (WMSN)

The integration of low-power wireless networking technologies with inexpensive hardware such as complementary metal–oxide semiconductor (CMOS) cameras and microphones is now enabling the development of distributed networked systems that we refer to as wireless multimedia sensor networks (WMSNs) (Akyildiz, Melodia, & Chowdhury, 2006) (Akyildiz, Melodia, & Chowdury, 2007), that is, networks of wireless, interconnected smart devices that enable retrieving video and audio streams, still images, and scalar sensor data. In next sections we will describe/illustrate/discusses/write the important Factors influencing the design of multimedia sensor Networks, Reference architecture of WMSN and the security of WMSN.

IMPORTANT FACTORS INFLUENCING THE DESIGN OF WMSN

- **QoS Requirement:** One of the first and the most important challenge to the WMSNs design is to meet the application specific QoS requirements. The WMSNs are designed to address a range of application scenarios ranging from simple scalar application to multi-tier support involving heterogeneous sensors that includes multimedia sensor support besides the use of scalar sensors. Streaming multimedia content is generated over longer time periods and requires sustained information delivery. Hence, a strong foundation is needed in terms of hardware and supporting high-level algorithms to deliver QoS and consider application-specific requirements.
- **Resource Constraints:** Sensor devices are constrained in terms of battery, processing capability, memory, and achievable data rate.

- **High Bandwidth Demand:** In WMSNs the bandwidth requirement for multimedia data communication is order of magnitude higher than the bandwidth required for the existing WSNs. For example the scalar WSN architecture involving motes like TelosB or Micaz etc support Zigbee/802.15.4 radio standard that supports the data rate of up to 250Kbps.
- **Power Consumption:** Power consumption is a fundamental concern in WMSNs, even more than in traditional wireless sensor networks. An application-specific approach for energy conservation is introduced in (Gürses, & Akan, 2005) which can be used in state change detection of a hot spot.
- **Integration with Internet (IP) Architecture:** The WMSNs design should support various other wireless communication standards like Bluetooth, Wi-Fi etc and Internet Protocol (IP) suite. This may enables the user to pop the network's information from anywhere any time.

Add to these factors (QoS requirement, resources constraint … etc.) there are still others factors influencing the design of WMSN:

- Variable Channel Capacity.
- Cross-layer Coupling of Functionality.
- Multimedia Source Coding Techniques.

In the next section we will illustrate the Reference architecture of WMSN

Reference Architecture of WMSN

In (Akyildiz, Melodia, & Chowdury, 2007) introduce a reference architecture for WMSNs, where three sensor networks with different characteristics are shown:

1. Single-tier flat, homogeneous sensors, distributed processing, centralized storage.
2. Single-tier clustered, heterogeneous sensors, centralized processing, centralized storage.
3. Multitier, heterogeneous sensors, distributed processing, distributed storage.

Security of Wireless Multimedia Sensor Networks

As wireless multimedia sensor networks become more widely used, security issue in multimedia sensor networks also becomes a problem. While the use of stronger codes, watermarking techniques, encryption algorithms, amongst others, have resulted in secured wireless communication, there are altogether different considerations in WMSNs (Akyildiz, Melodia, & Chowdhury, 2006). To enhance the security, watermark techniques have been used (Wang, Peng, Wang, Sharif, & Chen, 2008). However, usually the enhanced security achieved by watermark techniques are at the cost of energy consumption because the computation calculation involved in watermark techniques cost energy (Luo, 2013), which is a precious resource for sensors. To address this problem, an adaptive energy-aware watermarking scheme was proposed in (Wang, Peng, Wang, Sharif, & Chen, 2008). Details can be found in (Wang, Peng, Wang, Sharif, & Chen, 2008). Another work about security in multimedia sensor networks can be found in (Wang, Peng, Wang, Sharif, & Chen, 2008).

FUTURE RESEARCH DIRECTIONS

The discussions above clearly show that, although a lot of WSN research is going on, still a lot of open issues exist. In this section, we present some of future research directions, summarized as following:

- **Heterogeneous WSNs:** Heterogeneous WSNs have emerged to solve the problems related to homogeneous WSNs such as the limited energy capacities. A heterogeneous WSN is generally formed of more than one type of nodes (Yu, Wang, Zhang, & Zheng, 2007) mostly homogeneous with limited calculating capacities and memory, and few heterogeneous nodes with higher capacities which gives an equilibrium to the tasks in the network and represents an effective way to increase the lifetime of the network (Kumar, Tsiatsis, & Srivastava, 2003) (Rhee, Seetharam, & Liu, 2004). Integrating heterogeneity in the network has many advantages. Indeed, the heterogeneity can improve the sensor networks scalability, reduce the energy needs without sacrificing performance, give a balance between the networks cost and functionality, encourage new broadband applications, and improve the security mechanisms with the use of more complex and energy consumer protocols. The heterogeneous WSNs also present disadvantages such as Heterogeneous nodes deployment problem which consists to define the number and the emplacement of the heterogeneous nodes. Also, introducing heterogeneous nodes in the network may present new security concerns. Indeed, the heterogeneous WSNs have to consider varying security capabilities. In heterogeneous WSNs, the homogeneous nodes are inexpensive and energy limited. Their main tasks consist in collecting and reporting data. The heterogeneous nodes have more complex and computation intensive tasks, power and memory. Three heterogeneity types are considered in wireless sensor networks (Yu, Wang, Zhang, Zheng, 2007; Katiyar, Chand, & Soni, 2010):

- **Computational Heterogeneity:** In computational heterogeneity, sensor nodes have more powerful processor and more memory which leads to additional complex calculations and data storage at longer terms.
- **Link Heterogeneity:** In link heterogeneity, sensor nodes consist of most powerful radio range allowing thus more reliable transmissions.
- **Energy Heterogeneity:** Nodes contain replaceable or directly fed power batteries, which allows network's longer lifetime.

- **Nano Sensor Networks (NSNs):** Due to the remarkable progress in nanotechnology, a new class of wireless sensor networks named nano-scale sensor networks (NSNs) is realized and more and more used. NSNs are generally composed of nano-scale devices which perform basic tasks like computing, sensing and communicating. These devices are not commercialized, but many projects are working on producing them in bulk. Some kinds of nano sensors are reported and survived like miniature hydrogen sensors (Villatoro, & Monzon-Herńandez, 2005), nano sensors for chemical and biological sensing (Yonzon, Stuart, Zhang, & McFarland, 2005), chemical and biological nano sensors for molecular level (Akyildiz, & Jornet, 2010) (Falconi, Damico, & Wang, 2007). In addition, communication possibilities at nano-scale are explored to connect the nano sensors to form a NSN. The paper proposed in (Akyildiz, Brunetti, Blazquez, 2008) confirms the possibility of communication at nano-scale either using electromagnetic or some form of molecular-based transceivers. New research activities are conducted in the recent years to understand the unique properties of nano materials that could be used for the communication between nano devices (Akyildiz, & Jornet, 2010), (Chou, 2012), (Atakan, & Akan, 2010) (Chou, 2013).

Due to its possible use at the atomic levels, a NSN can be exploited new kind of nanotechnology applications, where distributed communication between the nano sensors are envisaged to accomplish the application goal, which cannot be realized with conventional sensor networks. An example of the proposed applications for NSNs is the biomedical application (health monitoring and drug delivery), the environmental applications (plant monitoring and defeating insect plague), industrial (ultra-sensitive touch interfaces), and the military applications (biological and chemical defense) domains (Akyildiz, & Jornet, 2010).

- **Biosensor Network:** Researchers have recently focused on simulating animals behavior especially the rats behavior. Indeed, they have demonstrated that through the stimulation of regions of the brain, the animals can be guided though hostile environments (S. K. Talwar et al, 2002). They also can be trained to perform some tasks like finding targets such as explosives by stimulating brain regions to produce or reinforce stimulus cues for various commanded movements. According to the brain simulation, the animals allow the controller to guide them leading thus to a wireless communication.

Recently, search and rescue operation involve animals which are adept at negotiating difficult 3D terrain in both light and dark and which combine their natural functions such as olfactory, visual, auditory, and tactile senses to find chemicals or people. This makes them more effective than mechanical robots. Recently, and due to the advances in neurophysiology, it is possible to train a large number of rats which are remotely guidable at moderate cost. Also, the advances in low-power very large-scale integration (VLSI) and micro-electromechanical systems (MEMS) make it possible to design wireless communication and networking devices that could fit into a backpack carried by a rat.

In the applications dedicated to the rescue missions, group of rats can be autonomously guided and coordinated to form cooperative multi-hop wireless sensor network and complete a critical mission (Li, Panwar, Mao, Burugupalli, & Lee, 2005). The biosensor network is principally dedicated to natural disaster recovery applications (finding trapped people and hazards), homeland security applications (search for explosives, bio-agents, etc., in containers or cargo ships), military operations (e.g., reconnaissance and minesweeping), and law enforcement applications (e.g., collecting evidence from inaccessible regions) where it is important to develop wireless communication and networking technologies that enable the setup and operation of sensor network which is composed of a coordinated set of trained animals and some mechanical robots, all guided by a command center. In the future animals can be replaced with robots.

Biosensor networks present some problems that need to be addressed. Indeed, in such type of networks, the rats must work in three different types of roles: seekers, followers, and relays. The, it is important to more investigate cooperative control techniques to autonomously guide and reward a large set of animals performing different tasks, and generalize these techniques to include team of seekers. The search can aim to minimize the total search time by minimizing the distance traveled by each seeker. In this type of networks, the radio transmission system has to meet the needs of the sensors, the network management system for routing information, and the control system for status and remote guidance information which has its own throughput and signal quality requirements. The radio propagation characteristics must be explored in the physical layer.

- **Wireless Body Sensor Network (WBSN):** A wireless body sensor network (WBSN) is a type of WSNs based on radio frequency (rf). A WBSN main characteristic is that it allows the interconnection of tiny nodes with sensor or actuator capabilities in, on, or around a human body in a short range of about 2 m. Wireless Body Sensor Network is specially developed to monitor, manage and communicate different vital signs of the human body like temperature and blood pressure. For this, different sensors are installed on clothes or directly on the body or under the human skin. The actuators installed on the human body are used to inject controlling drugs or life-saving drugs. The communication between sensors, actuators and cellular phone is established through a central unit. The function of the cellular phone consists to transmit the information to and from the human body to the external world (physician, emergency). In WBSNs, IP addresses are assigned to each body. IEEE 802.15 Task Group 6 standard is used to develop energy-efficient devices and to develop applications for WBAN.

Many studies of the propagation of electromagnetic waves in on the body have been conducted by researches which have proposed some models for the physical layer. Even if the body movements may influence directly the received signal strength, the proposed models do not consider them. Recently, researches based on the galvanic coupling and transformation of information via the bones and which can offer promising results are conducted and more investigated. At the level of the data link layer and the network layer, some protocols have been proposed. This level still has a lot of open research issues. At the level of the data link layer, the researchers must take into account in developing specific MAC-protocols the body characteristics such as the movements, the human physiology such as the beating of the heart. They must also take into account the mobility of the nodes. The combination of thermal routing with more energy-efficient mechanisms represents an interesting research issue in this field. Also, Qos frameworks are essential and must be used. The mobility support embedded in the protocol, the security, the inter-operability and other issues represent promising research issues that must be investigated. Many of these mechanisms could be united in a cross-layer protocol to realize an optimal system.

CONCLUSION

Wireless sensor networks (WSN) have become one of the most attractive research areas in many scientific fields for the last years. WSNs are composed of several sensor nodes, where the main objective of a sensor node is to collect information from its surrounding environment and transmit it to one or more points of centralized control called base stations.

The WSN protocol stack is very similar to traditional protocol stack and integrates power and routing awareness, integrates data with networking protocols, communicates power efficiently through the wireless medium, and promotes cooperative efforts of sensor nodes.

WSNs are deployed in inaccessible areas, hostile environments, and have a variety of applications, such as: military applications, environmental monitoring, medical application, home application, industrial and commercial application.

In WSNs, three of the most important concerns are the low power consumption requirement, the mobility of sensor nodes to improve the sensing coverage and the vulnerability to many types of security attacks. In order to mitigate the above mentioned concerns, various hierarchical or cluster-based routing methods (LEACH, TEEN, APTEEN, HEED, PEGASIS, HEEP), protocols for mobile sensor network (M-LEACH, LEACH-Mobile, LEACH-ME, E2 MWSNRP, GBEER), and mechanisms of security (Cryptography, Authentication and Intrusion Detection Systems) are proposed in the literature.

In addition to energy efficiency, mobility and security there still exist some other open research issues, such as: integration of sensor networks and the Internet, WSNs in challenging environments (WUSNs & UWSNs) and wireless multimedia sensor networks.

The intent of this chapter is to present a practical survey on wireless sensor networks.

REFERENCES

Adams, J. (2004). *Designing with 802.15.4 and ZigBee*. Paper presented at the Industrial Wireless Applications Summit, San Diego, CA.

Akyildiz, I. F., Brunetti, F., & Blazquez, C. (2008). Nano networks: A new communication paradigm. *Computer Networks*, *52*(12), 2260–2279. doi:10.1016/j.comnet.2008.04.001

Akyildiz, I. F., & Jornet, J. M. (2010). Electromagnetic wireless nanosensor networks. *Nano Communication Networks*, *1*(1), 3–19. doi:10.1016/j.nancom.2010.04.001

Akyildiz, I. F., Melodia, T., & Chowdhury, K. R. (2006). A survey on wireless multimedia sensor networks. *Computer Networks*, *51*(4), 921–960. doi:10.1016/j.comnet.2006.10.002

Akyildiz, I. F., Melodia, T., & Chowdury, K. R. (2007). Wireless multimedia sensor networks: A survey. *IEEE Wireless Communications*, *14*(6), 32–39. doi:10.1109/MWC.2007.4407225

Akyildiz, I. F., Pompili, D., & Melodia, T. (2005). Underwater acoustic sensor networks: Research challenges. *Ad Hoc Networks*, *3*(3), 257–279. doi:10.1016/j.adhoc.2005.01.004

Akyildiz, I. F., & Stuntebeck, E. P. (2006). Wireless underground sensor networks: Research challenges. *Ad Hoc Networks Journal*, *4*(6), 669–686. doi:10.1016/j.adhoc.2006.04.003

Akyildiz, I. F., Su, W., Sankarasubramaniam, Y., & Cayirci, E. (2002, March). (2002-1). Wireless sensor networks: A survey. *Computer Networks*, *38*(4), 393–422. doi:10.1016/S1389-1286(01)00302-4

Akyildiz, I. F., Su, W., Sankarasubramaniam, Y., & Cayirci, E. (2002, August). (2002-2). A Survey on Sensor Networks. *IEEE Communications Magazine*, *40*(8), 102–114. doi:10.1109/MCOM.2002.1024422

Akyildiz, I. F., Sun, Z., & Vuran, M. C. (2009). Signal Propagation Techniques for Wireless nderground Communication Networks. *Physical Communication Journal*, *2*(3), 167–183. doi:10.1016/j.phycom.2009.03.004

Alrajeh, N. A., Khan, S., & Shams, B. (2013). Intrusion Detection Systems in Wireless Sensor Networks: A Review. *International Journal of Distributed Sensor Networks*, 1–7.

Atakan, B., & Akan, O. (2010). Carbon nano tube-based nano scale ad hoc networks. *IEEE Communications Magazine*, *48*(6), 129–135. doi:10.1109/MCOM.2010.5473874

Boubiche, D., & Bilami, A. (2011). HEEP (Hybrid Energy Efficiency Protocol) Based on Chain Clustering. *Int. J. Sensor Networks*, *10*(1/2), 25–35. doi:10.1504/IJSNET.2011.040901

Chou, C. T. (2012). Molecular circuits for decoding frequency coded signals in nano-communication networks. *Nano Communication Networks*, *3*(1), 46–56. doi:10.1016/j.nancom.2011.11.001

Chou, C. T.Chun Tung Chou. (2013). Extended Master Equation Models for Molecular Communication Networks. *IEEE Transactions on Nanobioscience*, *12*(2), 79–92. doi:10.1109/TNB.2013.2237785 PMID:23392385

Cui, J.-H., Kong, J., Gerla, M., & Zhou, S. (2006). Challenges: Building Scalable Mobile Underwater Wireless Sensor Networks for Aquatic Applications. *IEEE Network. Special Issue on Wireless Sensor Networking*, *20*(3), 12–18.

Debar, H., Dacier, M., & Wespi, A. (1999). Towards a taxonomy of intrusion-detection systems. *Computer Networks*, *31*(8), 805–822. doi:10.1016/S1389-1286(98)00017-6

Falconi, C., Damico, A., & Wang, Z. (2007). Wireless Joule nano heaters. *Sensors and Actuators*, *127*(1), 54–62. doi:10.1016/j.snb.2007.07.002

Fall, K. (2003). *A delay-tolerant network architecture for challenged internets*. Paper presented at the SIGCOMM'2003, Karlsruhe, Germany. doi:10.1145/863955.863960

Gao, T., Greenspan, D., Welsh, M., Juang, R., & Alm, A. (2005). *Vital SignsMonitoring and Patient Tracking over a Wireless Network*. Paper presented at IEEE 27th Annual International Conference of the Engineering in Medicine and Biology Society (EMBS), Shanghai, China.

Gürses, E., & Akan, Ö. B. (2005). Multimedia communication in wireless sensor networks. *Annales des Télécommunications*, *60*(7-8), 872–900.

Heinzelman, W. B., Chandrakasan, A. P., & Balakrishnan, H. (2002). An applicationspecific protocol architecture for wireless microsensor networks. *IEEE Transactions on Wireless Communications*, *1*(4), 660–670. doi:10.1109/TWC.2002.804190

Hui, J. W., & Culler, D. E. (2008). *IP is dead, long live IP for wireless sensor networks*. Paper presented at the sixth ACM conference on embedded network Sensor System (SenSys '08), Raleigh, USA. doi:10.1145/1460412.1460415

IETF Working Group. (2013). *IPv6 over low power WPAN working group*. Available from http://tools.ietf.org/wg/6lowpan/

Karlof, C., & Wagner, D. (2003). Secure routing in wireless sensor networks: Attacks and coun-termeasures. *Ad Hoc Networks*, *1*(2-3), 293–315. doi:10.1016/S1570-8705(03)00008-8

Katiyar, I., Chand, N., & Soni, S. (2010). Clustering Algorithms for Heterogeneous Wireless Sensor Network: A Survey. *International Journal of Applied Engineering Research*, *1*(2), 273–287.

Kifayat, K., Merabti, M., Shi, Q., & Llewellyn, D. (2010). Security in Wireless Sensor Networks. In Handbook of Information and Communication Security. Springer. doi:10.1007/978-3-642-04117-4_26

Kim, D. S., & Chung, Y. J. (2006). *Self-organization routing protocol supporting mobile nodes for wireless sensor network*. Paper presented at the First Int. Multi-Symp. on Computer and Computational Sciences, Hangzhou, China. doi:10.1109/IMSCCS.2006.265

Kumar, G. S., Vinu, P. M. V., & Jacob, K. P. (2003). *Mobility Metric based LEACH-Mobile Protocol*. Paper presented at the 16th International Conference on Advanced Computing and Communications (ADCOM'08), Chennai, India.

Kumar, R., Tsiatsis, V., & Srivastava, M. B. (2003). *Computation Hierarchy for In-Network Processing*. Paper presented at the 2nd Intl. Workshop on Wireless Networks and Applications, San Diego, CA.

Kweon, K., Ghim, H., Hong, J., & Yoon, H. (2009). *Grid-Based Energy-Efficient Routing from Multiple Sources to Multiple Mobile Sinks in Wireless Sensor Networks*. Paper presented at the 4th International Symposium on Wireless Pervasive Computing (ISWPC'09), Melbourne, Australia. doi:10.1109/ISWPC.2009.4800585

Lee, S., Noh, Y., & Kim, K. (2013). Key Schemes for Security Enhanced TEEN Routing Protocol in Wireless Sensor Networks. *International Journal of Distributed Sensor Networks*, *2013*, 1–8. doi:10.1155/2013/374796

Li, Y., Panwar, S. S., Mao, S., Burugupalli, S., & Lee, J. (2005). A Mobile Ad Hoc Bio-Sensor Network. *Proceedings of the IEEE, ICC*, 2005.

Lindsey, S., & Raghavendra, C. (2002). *PEGASIS: Power-Efficient Gathering in Sensor Information Systems*. Paper presented at the IEEE Aerospace Conference, Montana, USA. doi:10.1109/AERO.2002.1035242

Loo, C. E., Yong, M., Leckie, C., & Palaniswami, M. (2006). Intrusion Detection for Routing Attacks in Sensor Networks. *International Journal of Distributed Sensor Networks*, *2*(4), 313–332. doi:10.1080/15501320600692044

Luo, Z. (2008). Survey of Networking Techniques for Wireless Multimedia Sensor Networks. *International Journal of Recent Technology and Engineering*, *2*(2), 182–183.

Manjeshwar, A., & Agrawal, D. P. (2002). *APTEEN: A Hybrid Protocol for Efficient Routing and Comprehensive Information Retrieval in Wireless Sensor Networks*. Paper presented at the 2nd International Workshop on Parallel and Distributed Computing Issues in Wireless Networks and Mobile Computing, Ft. Lauderdale, FL. doi:10.1109/IPDPS.2002.1016600

Maraiya, K., Kant, K., & Gupta, N. (2011). Application based Study on Wireless Sensor Network. *International Journal of Computers and Applications*, *21*(8), 9–15. doi:10.5120/2534-3459

Ng, J. W. P., Lo, B. P. L., Wells, O., Sloman, M., Peters, N., Darzi, A., . . . Yang, G.-Z. (2004). *Ubiquitous Monitoring Environment for Wearable and Implantable Sensors*. Paper presented at the International Conference on Ubiquitous Computing (UbiComp), Tokyo, Japan.

Nguyen, L. T., Defago, X., Beuran, R., & Shinoda, Y. (2008). *An Energy Efficient Routing Scheme for Mobile Wireless Sensor Networks*. Paper presented at the IEEE International Symposium on Wireless Communication Systems (ISWCS'08), Reykjavik, Iceland.

Padmavathi, G., & Shanmugapriya, D. (2009). A Survey of Attacks, Security Mechanisms and Challenges in Wireless Sensor Networks. *International Journal of Computer Science and Information Security*, *4*(1-2), 1–9.

Pathan, A.-S. K., Dai, T. T., & Hong, C. S. (2006). A Key Management Scheme with Encoding and Improved Security for Wireless Sensor Networks. In S. Madria et al. (Ed.), *International Conference on Distributed Computing and Internet Technology (ICDCIT'06) (LNCS)* (Vol. 4317, pp. 102-115). Berlin, Germany: Springer. doi:10.1007/11951957_10

Rhee, S., Seetharam, D., & Liu, S. (2004). Techniques for Minimizing Power Consumption in Low Data-Rate Wireless Sensor Networks. In *Proc. of IEEE Wireless Communications and Networking Conference*. Atlanta, GA: IEEE.

Sara, G. S., Kalaiarasi, R., Pari, S. N., & Sridharan, D. (2010). Energy Efficient Mobile Wireless Sensor Network Routing Protocol. In N. Meghanathan et al. (Ed.), *Recent Trends in Networks and Communications: Proceedings of the International Conferences, NeCoM 2010, WiMoN 2010, WeST 2010*, (LNCS) (Vol. 90, pp 642-650). Berlin, Germany: Springer. doi:10.1007/978-3-642-14493-6_65

Silva, A. P. R. D., Martins, M. H. T., Rocha, B. P. S., Loureiro, A. A. F., Ruiz, L. B., & Wong, H. C. (2005). *Decentralized intrusion detection in wireless sensor networks*. Paper presented at the 1st ACM international workshop on Quality of service & security in wireless and mobile networks, Montreal, Canada. doi:10.1145/1089761.1089765

Silva, A. R., & Vuran, M. C. (2010). *Communication with Aboveground Devices in Wireless Underground Sensor Networks: An Empirical Study*. Paper presented at the IEEE International Conference on Communications (ICC), Cape Town, South Africa. doi:10.1109/ICC.2010.5502315

Talwar, S. K., Xu, S., Hawley, E. S., Weiss, S. A., Moxon, K. A., & Chapin, J. K. (2002). Behavioural Neuroscience: Rat Navigation Guided by Remote Control. *Nature*, *417*(6884), 37–38. doi:10.1038/417037a PMID:11986657

Villatoro, J., & Monzon-Hernandez, D. (2005). Fast detection of hydrogen with nano fiber tapers coated with ultra thin palladium layers. *Optics Express*, *13*(13), 5087–5092. doi:10.1364/OPEX.13.005087 PMID:19498497

Wang, H., Peng, D., Wang, W., Sharif, H., & Chen, H. (2008). *Energy-Aware Adaptive Watermarking for Real-Time Image Delivery in Wireless Sensor Networks*. Paper presented at theIEEE International Conference on Communications (ICC '08), Beijing, China. doi:10.1109/ICC.2008.286

Wood, A., Virone, G., Doan, T., Cao, Q., Selavo, L., Wu, Y., . . . Stankovic, J. (2006). *ALARM-NET: Wireless Sensor Networks for Assisted-Living and Residential Monitoring*. Technical Report CS-2006-11. University of Virginia.

Yibo, C., Hou, K.-M., Zhou, H., Shi, H.-L., Liu, X., Diao, X., . . . De Vaulx, C. (2011). *6LoWPAN stacks: a survey*. Paper presented at the 7th International Conference on Wireless Communications, Networking and Mobile Computing (WiCOM), Wuhan, China.

Yonzon, C. R., Stuart, D. A., Zhang, X., McFarland, A. D., Haynes, C., & Vanduyne, R. (2005). Towards advanced chemical and biological nano sensors-An overview. *Talanta*, *67*(3), 438–448. doi:10.1016/j.talanta.2005.06.039 PMID:18970187

Younis, O., & Fahmy, S. (2004). HEED: A Hybrid Energy-Efficient Distributed Clustering Approach for Ad Hoc Sensor Networks. *IEEE Transactions on Mobile Computing*, *3*(4), 366–379. doi:10.1109/TMC.2004.41

Yu, L., Wang, N., Zhang, W., & Zheng, C. (2007). *Deploying a Heterogeneous Wireless Sensor Network*. Paper presented at the International Conference on Wireless Communications, Networking and Mobile Computing (WiCom'07), Shanghai, China.

Zhang, P., Sadler, C., Lyon, S., & Martonosi, M. (2004). *Hardware design experiences in ZebraNet*. Paper presented at the ACM SenSys'04, Baltimore, MD. doi:10.1145/1031495.1031522

Zhou, Q., & Zhang, R. (2013). *A Survey on All-IP Wireless Sensor Network*. Paper presented at the 2nd International Conference on Logistics, Informatics and Service Science, Beijing, China. doi:10.1007/978-3-642-32054-5_105

KEY TERMS AND DEFINITIONS

E²MWSNRP: Energy Efficient Mobile Wireless Sensor Network Routing Protocol is a multipath hybrid routing protocol that can be designed mainly for highly dynamic energy deficient mobile wireless sensor network where energy dissipation reduction and reliable transmission of data is a must.

GBEER: Grid Based Energy Efficient Routing is a very energy efficient routing protocol for communication from multiple sources to multiple mobile sinks in wireless sensor network.

Intrusion Detection Systems: Is defined as a system that monitors computer system or network activities to detect signs of violations of security policies. In general, IDS schemes are categorized into misuse IDS and anomaly IDS.

LEACH-ME: Low Energy Adaptive Clustering Hierarchy-Mobile Enhanced is an enhanced version of M-LEACH. In LEACH-ME, the selection of cluster-head is based on the less mobile nodes in relation to its neighbors. The steady state phase is the same for both LEACH-M and LEACH-ME.

LEACH-Mobile: Low Energy Adaptive Clustering Hierarchy-Mobile is a self-organization routing protocol supporting mobile nodes for WSN. In the proposed scheme, like LEACH is broken up into rounds, where each round begins with a setup phase when the clusters are organized, followed by a steady state phase when data transfers to the base station occur. In order to minimize overhead, the steady state phase is long compared to the setup phase. One of the basic ideas in LEACH-Mobile is to confirm the inclusion of sensor nodes in a specific cluster at the steady state phase.

M-LEACH: Mobile-Low Energy Adaptive Clustering Hierarchy is a clustering routing protocol which allows mobility of cluster-head and non-cluster-head nodes during the setup and steady state phase. The setup phase of LEACH is modified by M-LEACH in order to select suitable cluster-head based on attenuation model and mobility speed. In steady state phase, if nodes move away from cluster-head or cluster-head moves away from its member nodes, MLEACH provides an handover mechanism for nodes to switch on to new cluster-head in order to avoid an inefficient clustering formation.

UWSN: Wireless underwater sensor network is an extension of the terrestrial wireless sensor networks, where the sensor nodes are underwater and communicate by acoustic signals through water.

Wireless Sensor Network: An infrastructure comprised of several sensor nodes, where the main objective of a sensor node is to collect information from its surrounding environment and transmit it to one or more points of centralized control called base stations.

WMSN: Wireless Multimedia Sensor Network is composed of several wireless interconnected smart nodes which are equipped with cameras, microphones, and other sensors producing multimedia data content.

WUSN: Wireless Underground Sensor Network is an important extension of the terrestrial wireless sensor networks, where the sensor nodes are buried underground and communicate wirelessly through soil.

Chapter 7
Design, Architecture, and Security Issues in Wireless Sensor Networks

Piyush Kumar Shukla
UIT RGPV, India

Kirti Raj Bhatele
UIT RGPV, India

Lokesh Sharma
Chang Gung University (CGU), Taiwan

Poonam Sharma
MITS RGPV, India

Prashant Shukla
SIRT RGPV, India

ABSTRACT

Wireless Sensor Networks (WSNs) provide a new paradigm for sensing and disseminating information from various environments, with the potential to serve many and diverse applications. In this chapter, we report the latest trends in WSN research, focusing on middleware technology and related areas, and including application design principles. We give an overview of WSNs and design aspects of applications, including existing research prototypes and industry applications. We describe the technology supporting these sensor applications from the view of system architecture and network communication. We then highlight outstanding issues and conclude with future perspectives on middleware technology.

INTRODUCTION

In recent years, advances in miniaturization, yet simple low power circuit design and improved low cost, small size batteries have made a new technological vision possible: wireless sensor network (WSN) (You, Lieckfeldt, Salzmann, Timmer-

DOI: 10.4018/978-1-4666-8687-8.ch007

mann, 2009). Wireless Sensor Networks (WSNs) are designed by sensor nodes that communicate each other and also processing data and sensing environment (Li & Halpern, June 2001). A conventional Wireless Sensor is illustrated with the help of Figure 1. A sensor node is basically a device that converts a sensed attribute (such as temperature, vibrations) into a form understandable by the users. A functional block diagram of a typical sensor node is given in Figure 2. WSNs, which can be considered as a special case of ad-hoc networks with reduced or no mobility, are expected to find increasing deployment in coming years, as they enable reliable monitoring and analysis of unknown and untested environments These networks are "data centric", i.e., unlike traditional ad-hoc networks where data is requested from a specific node, data is requested based on certain attributes such as, "which area has temperature over 35C or 95F". Therefore A large numbers of sensors need to be deployed to accurately reflect the physical attribute in a given area. Due to lack of a better word, typical sensor consists of transducer to sense a given physical quantity with a predefined precision, an embedded processor for local processing, small memory unit for storage of data and a wireless transceiver to transmit or receive data and all these devices run on the power supplied by an attached battery.

It is interesting to note that precise specifications of various components, may depend on the type of applications in hand, but the basic characteristics are essentially present to fulfil desired application functionalities. There are few integrated sensors commercially available and can be used directly as plug and play unit to monitor and control some specific physical parameters as decided by the user.

Figure 1. Wireless sensor diagram

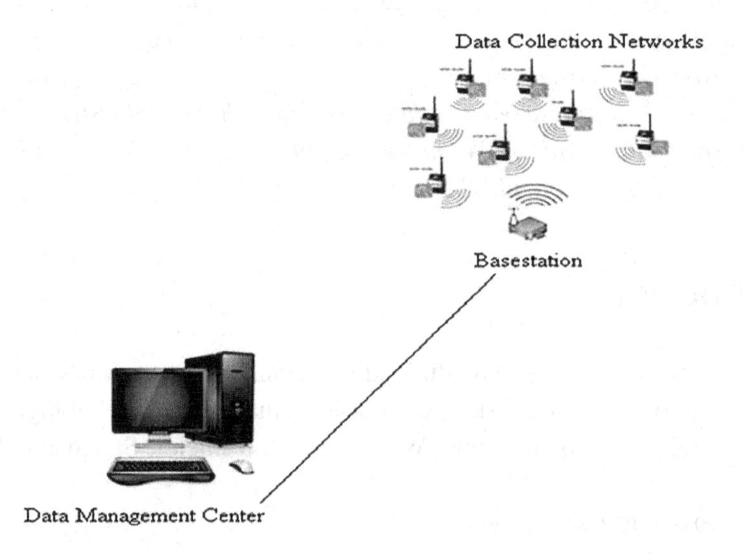

Figure 2. Functional block diagram of a typical sensor node

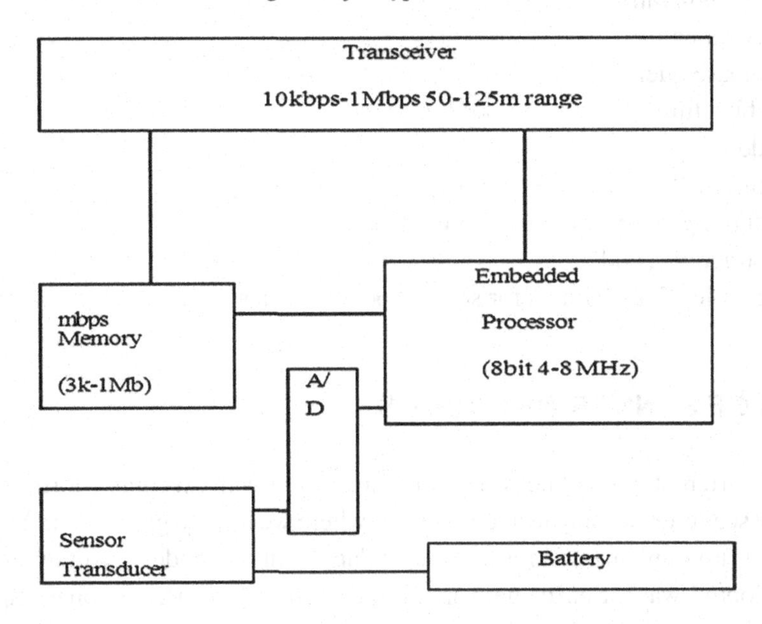

But, there are many basic sensors transducers that could convert many problems quantities such as temperature, pressure, velocity, acceleration, stress and strain, fatigue, tilt, light intensity, sound, humidity, gas - sensors, biological, pollution, nuclear radiation, civil structural sensors, blood pressure, sugar level, white cell count and many others .These basic generic transducers need to be interfaced and connected to other devices and such custom made unit can be used for a given specific application.

OVERVIEW OF KEY ISSUES

Lately there has been a developing enthusiasm toward Wireless sensor networks (WSN). Recent headways inside the field of sensing, processing and correspondences have pulled in research deliberations and enormous ventures from changed quarters in the field of WSN. Conjointly sensing systems can uncover previously surreptitiously phenomena. The different zones wherever major research exercises happening inside the field of WSN are:

- Deployment,
- Localization,
- Synchronization,

- Data aggregation,
- Dissemination,
- Database querying,
- Architecture,
- Middleware,
- Security,
- Designing less power consuming devices,
- Abstractions, and
- Higher level algorithms for sensor specific issues.

TYPES OF SENSOR NETWORKS

There are different types of network which are currently in executable forms. Sensor nodes are scattered densely in a region from where we have to gather data. Multihop techniques are implemented for transfer of data from one node to another node via different opted ways. Sensor networks are designed considering some important principles that needed to be focused. Power efficiency is one the important parameter, primarily energy has to be saved so that network will remain in active mode for long duration. Sensor network will collect data into one centralized body where all the data is stored and can be accessed easily. Location awareness mechanism has to be implemented so that the address of sensor node can be determined. There are different approaches which can be implemented for energy efficient routes are following.

- **Maximum Available Power Route:** In this the route which has maximum energy to serve will be preferred. Before this power energy has to be calculated for each node.
- **Minimum Energy Route:** The route that consumes minimum energy to serve will be preferred between source and sink.
- **Minimum Hop Route:** The route that has lesser number of hopes between source and sink will given higher priority.
- **Maximum Minimum Power Available Route:** The route along which the minimum available power is larger than the minimum available power of other routes will be served.

TYPES OF SENSOR NETWORKS BASED ON DIFFERENT EFFECTIVE STRATEGIES

- **Small Minimum Energy Communication Network (SMECN):** It a protocol which compute energy efficient subnet work where the communication network is given (Rodoplu & Meng, 1999; Li & Halpern, 2001). In this sub graph designed will have lesser number of edges only if the broadcast region is circular. Flooding is a technique which is used for routing in sensor networks. In this the data packets are broadcast to all nodes by Multihop technique they will reach up to destination. It has certain disadvantage such as implosion, overlap, resource blindness. Gossiping is like selecting randomly node for transmission of data it reduces the serious issues of flooding (Hedetniemi & Liestman, 1988).
- **Sensor Protocols for Information via Negotiations (SPIN):** It is an adaptive protocol based on negotiations and resource adaptations working on the deficiencies of Flooding. It has three types of messages Advertisement(ADV), data(DATA) and request(REQ) .It also broadcast the messages but by advertising and then data will transmitted. SPIN is data-centric routing (Heinzelman, Kulik & Balakrishnan, 1996).
- **Sequential Assignment Routing (SAR):** It creates multiple trees in which root of tree is one hop neighbor from the sink. Trees grow outward and avoiding nodes with very low QOS and energy reserves (Sohrabi, Gao, Ailawadhi & Pottie, 2000). There are two algorithms based on SAR, LEACH and Direct diffusion. In former Low energy adaptive clustering hierarchy (LEACH) is clustering based protocol that minimizes energy dissipation in sensor networks(Heinzelman, Chandrakasan & Balakrishnan, 2000). LEACH selects cluster-heads with high energy dissipation in communication with the base station is spread to all sensor nodes in network. It having two phases, setup and steady phase. Steady phase have longer durability for network communication. Direct Diffusion setup gradients for data to flow from source to sink during interest dissemination (Intanagonwiwat, Govindan & Estrin, 2000).

WIRELESS SENSOR NETWORKS APPLICATIONS

Now-a-days, Wireless sensor network is used in various fields with different environments for determining strategic location in automobile, airplanes vehicle etc. to constantly monitor device, these wired sensors are large to cover as much area is desirable. The organization of such a network should be pre-planned to find a strate-

Figure 3. Deployments of nodes

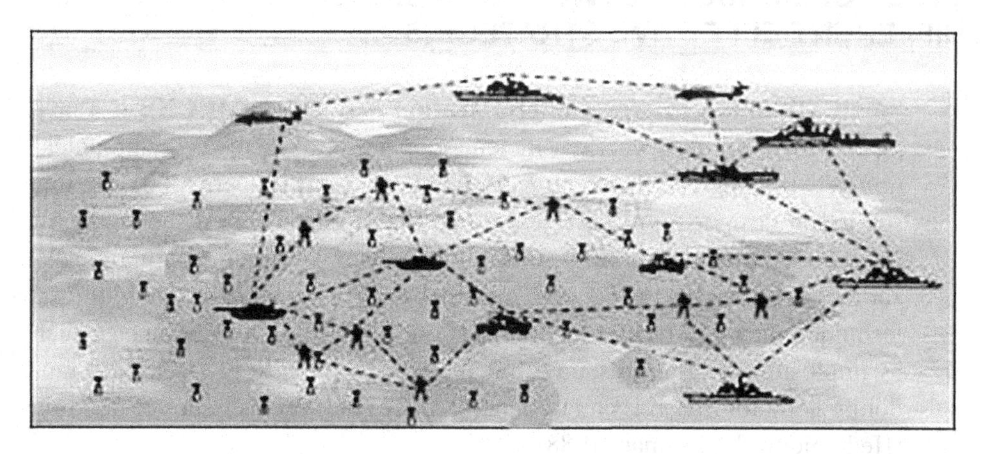

gic position to place these nodes and then should be installed appropriately. Figure 3 simply depicts a model for the deployments of Nodes in a battle field. Wireless sensor network have been conceived with military application in battlefield surveillance and tracking of enemy activities. There are innumerable application do exist for sensor network such as weather monitoring, security, and tactical surveillance, distributed computing, fault detection and diagnosis in machinery, large bridges and tall structures, detecting ambient conditions such as temperature, movement, sound, light, radiation, vibration, smoke, gases or the presence of certain biological and chemical objects.

- **Military Applications:** Military has implemented wireless sensor network in command, control, communications, computing, intelligence, surveillance, reconnaissance and targeting (C4ISRT) systems. For military operations precision is the important aspect to destruct opponent, so sensors has to deploy in densely, since sensor network is low cost and dense deployment is required from wherever the information is needed. Sensor networks can be served in targeting systems to track and monitor the opponent paths. Intelligent ammunition is one of the applications of sensor network which move itself according to the sink location. In chemical attack, nuclear and biological weapons sensors can be easily deployed without any loss of human and maintaining the secrecy of intruder.

Wireless sensor networks can be an integral part of military command, control, communications, computing, intelligence, surveillance, reconnaissance and targeting (C4ISRT) systems. The rapid deployment, self-organization and fault tolerance characteristics of sensor networks make them a very promising sensing technique for military C4ISRT. Since sensor networks are based on the dense deployment of disposable and low-cost sensor nodes, destruction of some nodes by hostile actions does not affect a military operation as much as the destruction of a traditional sensor, which makes sensor networks concept a better approach for battlefields. Some of the military applications of sensor networks are monitoring friendly forces, equipment and ammunition; battlefield surveillance; reconnaissance of opposing forces and terrain; targeting; battle damage assessment; and nuclear, biological and chemical (NBC) attack detection and reconnaissance. Monitoring friendly forces, equipment and ammunition: Leaders and commanders can constantly monitor the status of friendly troops, the condition and the availability of the equipment and the ammunition in a battlefield by the use of sensor networks. Every troop, vehicle, equipment and critical ammunition can be attached with small sensors that report the status. These reports are gathered in sink nodes and sent to the troop leaders. The data can also be forwarded to the upper levels of the command hierarchy while being aggregated with the data from other units at each level. Battlefield surveillance: Critical terrains, approach routes, paths and straits can be rapidly covered with sensor networks and closely watched for the activities of the opposing forces. As the operations evolve and new operational plans are prepared, new sensor networks can be deployed anytime for battlefield surveillance. Reconnaissance of opposing forces and terrain: Sensor networks can be deployed in critical terrains, and some valuable, detailed, and timely intelligence about the opposing forces and terrain can be gathered within minutes before the opposing forces can intercept them. Targeting: Sensor networks can be incorporated into guidance systems of the intelligent ammunition. Battle damage assessment: Just before or after attacks, sensor networks can be deployed in the target area to gather the battle damage assessment data. Nuclear, biological and chemical attack detection and reconnaissance: In chemical and biological warfare, being close to ground zero is important for timely and accurate detection of the agents. Sensor networks deployed in the friendly region and used as a chemical or biological warning system can provide the friendly forces with critical reaction time, which drops casualties drastically. We can also use sensor networks for detailed reconnaissance after an NBC attack is detected. For instance, we can make a nuclear reconnaissance without exposing a team to nuclear radiation.

- **Environmental Applications:** Sensor networks have wide application in environmental aspects such as temperature, humidity, pressure, water fall, pollution etc. Forest fire can be easily detected by the help of densely deployed sensors in forest. Flood detection can be informed by the help of sensors and take some measures.
- **Health Applications:** In medical field sensors are widely used such as in surgical equipments for detecting, diagnosing the patients. Some of the health applications for sensor networks are providing interfaces for the disabled;

integrated patient monitoring; diagnostics; drug administration in hospitals; monitoring the movements and internal processes of insects or other small animals; telemonitoring of human physiological data; and tracking and monitoring doctors and patients inside a hospitals (Bulusu, Estrin, Girod and Heidemann, 2001) (Kahn, Katz and Pister, 1999).

- **Home Applications:** Home automation: As technology advances, smart sensor nodes and actuators can be buried in appliances, such as vacuum cleaners, micro-wave ovens, refrigerators, and VCRs (Petriu, Georganas, Petriu, Makrakis, Groza, 2000). These sensor nodes inside the domestic devices can interact with each other and with the external network via the Internet or Satellite. They allow end users to manage home devices locally and remotely more easily.

The design of smart environment can have two different perspectives, i.e., human-centered and technology-centered (Abowd, Sterbenz, 2000). For human-centered, a smart environment has to adapt to the needs of the end users in terms of input/ output capabilities. For technology-centered, new hardware technologies, networking solutions, and middleware services have to be developed. A scenario of how sensor nodes can be used to create a smart environment is described in (Herring, Kaplan, 2000). The sensor nodes can be embedded into furniture and appliances, and they can communicate with each other and the room server. The room server can also communicate with other room servers to learn about the services they offered, e.g., printing, scanning, and faxing. These room servers and sensor nodes can be integrated with existing embedded devices to become self-organizing, self-regulated and adaptive systems based on control theory models as described in (Herring, Kaplan, 2000). Another example of smart environment is the ''Residential Laboratory'' at Georgia Institute of Technology (Essa, 2000). The computing and sensing in this environment has to be reliable, persistent, and transparent.

ARCHITECTURE OF PROTOCOL STACK

A wireless sensor network is a specially appointed course of action for multifunctional sensor nodes in a sensor field. Sensor nodes are thickly dispersed over an outsized, even remote, space regardless if a number of nodes malfunction, the network will continue to function. There are two fundamental formats for wireless sensor networks, the essential could be a star design wherever the nodes convey, in an exceedingly single hop, on to the sink at whatever point feasible and shared correspondence is most minimal. In second, it is directed once again to the sink by means of learning passing between nodes. This multi-hop correspondence is anticipated to devour less power than single-hop correspondence as a consequence of nodes is relatively close

Figure 4. Protocol stack

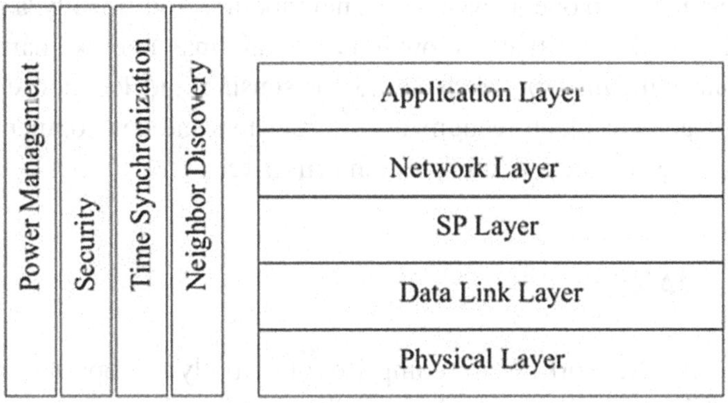

Power Management
Security
Time Synchronization
Neighbor Discovery

Application Layer
Network Layer
SP Layer
Data Link Layer
Physical Layer

to each other. While focusing on building standard protocol architecture for communication in sensor networks must be resource-aware and compo-sable framework.

There have been very few efforts in the past on protocol layering in sensor networks. A typical protocol stack is depicted in figure 4. The research proposal on the introduction on Sensor net Protocol (SP) as a common protocol layer for sensor networks was the first consolidated effort in this direction.

Sensor net Protocol is a best-effort single hop broadcast protocol which is situated between the network and the link layer in the protocol stack. The placement of SP between the network and link layers is quite obvious. As any sensor network application is closely tied to the network layer routing protocols, the common layer of SP cannot be at the network layer. And similarly there are numerous link layer protocols and each Sensor-net application developer might prefer to be able to choose a protocol which will be best suited for the application.

Power and routing algorithms integrate information with the networking protocols efficiently to communicate in wireless medium and help the efforts of sensor nodes. There are energy aware routing algorithms for efficiently implement routing protocols and data integration for network protocols. Protocol layer consist of physical layer, data link layer, Sensor net Protocol network layer, transport layer, application layer, power management plane, mobility management plane and task management plane.

The first layer of protocol stack, physical layer used to addresses the robust, transmission and receiving techniques. The network layer can perform routing and transfer data to transport layer, whereas transport layer maintain flow of data. Depending on the sensing tasks, different types of application Software can be set up and used on the application layer. The power management plane manages how a sensor node uses its power and manages its power consumption among the three operations (sensing, computation, and wireless communications). For instance, to avoid getting duplicated messages, a sensor node could put off it collector once accepting a message from one in every of its neighbours. Additionally, a sensor node telecasts to its neighbours that it's low in power and may not take part in steering messages. The remaining power is saved for sensing and location errands. The mobility management plane recognizes and registers the development/versatility of sensor nodes as a system administration primitive.

WSNs VS. MANETs

A Wireless Sensor Network (WSN) comprises of spatially appropriated independent sensors to watch physical or ecological conditions, in the same way as temperature, sound, vibration, weight, movement or toxins and to deliver glove pass their insight through the system to a principle area.

A Mobile ad-hoc network (MANET) is generally delineated as a system that has a few free or self-ruling nodes, normally made out of cell phones or option versatile things, that may sort out themselves in changed ways that and work while not strict top-down system organization. There square measure numerous different setups that would be known as MANETs and in this way the potential for this sort of system keeps on being contemplated. A comparison table is given below.

INTERNAL SENSOR SYSTEMS

System platform and OS support, to provide the functionality described above, a sensor node is composed of one or more sensors, a signal conditioning unit, an analog-to-digital conversion module (ADC), a processing unit with memory, a radio transceiver, and a power supply. Hardware platform Categories: Sensor node hardware platforms can be divided into three categories (Zhao & Guibas, 2004).

- **Adapted General Purpose Computers:** Adapted general-purpose computers. These platforms are low-power personal computers (PCs), embedded PCs, and personal digital assistants (PDAs). These platforms mainly run on Windows CE, Linux, or other operating systems developed for mobile devices. Because of the high processing ability and the high bandwidth communication, these platforms offer the opportunity to use higher level programming languages, which makes it easier to develop and implement software components.
- **Embedded Sensor Modules:** These platforms are assembled from commercial off the-shelf (COTS) Chips. Using COTS offers several benefits. These

Table 1. A comparison highlighting the different characteristic, problems, similarity and differences in between the WSN and MANET

	WSN (Wireless Sensor Network)	MANET (Mobile Ad Hoc Network)
Characteristics	• Batteries for Power consumption • Ability to cope with node failures • Mobility • Dynamic network topology • Communication failures • Heterogeneity of nodes • Scalability • Ability to withstand harsh environmental conditions • Unattended operation	• No centralized control. • Self-organizing and self-restoring. • Multiple hops Communication. • Rapid Link Wreckage.
Applications	• Security in military operations. • Environmental monitoring. • Seismic and structural monitoring. • Industrial automation. • Bio-medical applications • Health and Wellness Monitoring • Inventory Location Awareness	• Battlefield communication • Sensor networks • Personal area networking using PDAs, laptops and hand phones, etc. • Search-and-rescue • Cellular network and wireless Hot Spot extension
Problems		• No central infrastructure. • Limited range of communication. • Mobility of participants Charging.
Similarity	• Both are Distributed Wireless networks. • Multihop Routing. • Both Ad hoc and sensor nodes are usually battery-powered. • Both networks use a wireless channel placed in an unlicensed spectrum.	
Differences	• Density of deployment in sensor network is higher than ad hoc network. • In MANET unicast and multicast is common whereas in sensor network Multihop routing applicable. • The network size of MANET depends upon no. of active users whereas in WSNs it depend upon observed area. • WSN having no such problem of traffic congestion it exists in MANET. • WSNs the sensor nodes are state of art computation devices, simple and cheap elements. • WSNs are conceivable with different network densities, but in MANET it is not quite large. • Energy is critical in WSN for network lifetime but in MANET charging resources are there. • QOS services vary, in WSN energy optimization is major and in MANET low jitter. • WSN is data centric but no such protocol in MANET.	

components are widely used, making them cheap because of big production quantities, and are well supported by the manufacturers and communities.

- **System on Chip(SoC):** These platforms integrate micro electromechanical systems (MEMS) sensors, microcontroller, and wireless transceiver technologies on one chip, an application-specific integrated circuit (ASIC).

In a sensor network, application-particular necessities drive the whole hardware design, from methodology abilities to radio data measure and sensor modules, to obliging the fittings to be standard. Then again, these requirements led to a huge kind of equipment parts, making wireless sensor network hardware not only modular, but also heterogeneous (Frohlich).

While designing WSN Operating system these are major issues which has to consider are Architecture, Programming model, Scheduling, Memory Management and protection, Communication protocol and support, and resource sharing (Farooq). These WSN software frameworks are not full-blown operating systems, since they lack a powerful scheduler, memory management, and file system support. However, these frameworks are widely referred to asWSN operating systems. Such as TinyOS, Contiki, Mantis and SoS (Bischoff).

STANDARDS

It belongs to the standard family of IEEE 802.15.The two network layer which majorly associated with Wireless sensor network are physical and MAC layer. Physical layer assist frequency, bitrate spectrum handling, modulation and channel for transmission whereas MAC layer do packet formats, operational modes and timing aspects of packets.

Storage

Storing, collecting and querying data across miniaturized battery powered Wireless Sensor Networks (WSN) is a key research focus today. Distributed Data-Centric Storage (DCS), an alternate to External Storage (ES) and Local Storage (LS), is a promising efficient storage and search mechanism (Ahmed). Reliable data management raises the interest of nonstop data accessibility, even when nodes fizzle, e.g., as a consequence of power loss, to deliver the robustness, region unit connected inside the space of sensing component systems. Data is recorded and transmitted to substitutes, that territory unit associated with mains power frameworks (Siegmund).

Testbeds

As WSNs have the potential to operate in diverse environments, simulation tools and mathematically modeling do not capture physical channel characteristics accurately. This results in in-accurate performance evaluations. Therefore, some researchers have focused on designing WSN testbeds, so that system and protocol performance can be evaluated in a more realistic setting (Farooq). Simulations and even emulations are not considered sufficient for the deployment of new technologies as they often lack realism. WSN test beds to offer numerous users access to their testbeds in a standardized flexible way that matches these requirements (Ellbruck).

NETWORK SERVICES

Localization

Localization is that the strategy by that Associate in determining item decides its special directions amid a given field. The most crucial test for versatile WSN is that the need of restriction. In order to understand a special connection, or for right route all through a sensing locale, sensor position ought to be well known. As an after-effect of sensor nodes is likewise sent alterably, or could revision position all through run-time, there is likewise no technique for knowing the position of each node at any given time. limitation conspires that give high-correctness situating insight in WSN's can't be utilized by portable sensors, as an after-effect of they for the most part need incorporated handling, take excessively long to run, or make suppositions concerning the setting or design that don't matter to dynamic networks (Amundson).

Synchronization

Time synchronization is a critical piece of infrastructure for any distributed system. Distributed, wireless sensor networks make extensive use of synchronized time, but often have unique requirements in the scope, lifetime, and precision of the synchronization achieved, as well as the time and energy required to achieve it (Elson).

Coverage

Coverage problem is an important and fundamental issue in sensor networks, which reflects how well a sensor network is monitored or tracked by sensor. Two types of sensor coverage are investigated: area coverage and target coverage (Deying). Sensor placement and density control are two different issues; they both boil down to the issue of determining a set of locations either to place the sensors or to activate sensors in the vicinity, with the objective of fulfilling the following two requirements coverage and connectivity (Hou).

Compression and Aggregation

Energy efficiency of data collection is one of the dominating issues of wireless sensor networks (WSNs). Data aggregation techniques have been heavily investigated, there are still imperfections to be improved on such as simple aggregation function are used (ex Min/Max/Sum) (Xiang). Aggregation includes compressing-based and forecasting-based method; compressing aggregation focus on compress the data packets accompanied with transmitting based on spatial correlation; while forecasting aggregation tends to use mathematical model to fit the time series and predict the new value due to highly temporal correlation (CUI).

SECURITY

Security could be a comprehensively talking utilized term including the attributes of confirmation, trustworthiness, protection, no repudiation, and hostile to playback (Undercoffer, Joshi, & Pinkston, 2002). For the secure transmission of arranged sorts of data over networks, several cryptographic, steganographic and elective methods square measure utilized that are well known.

- **Cryptographic:** The craft of defensive information by renovating it into partner degree indecipherable configuration alluded to as figure content. Singularly those that have a mystery key will disentangle the message into plain content. Encrypted messages can sometimes be broken by cryptanalysis, also called code breaking.
- **Steganographic:** The hiding of a secret message within an ordinary message and the extraction of it at its destination. Steganography takes cryptography a step farther by hiding an encrypted message so that no one suspects it exists.
- **Data Confidentiality:** Is whether the information continues a framework is ensured against unplanned or unapproved access. Since frameworks region unit by and large usual oversee delicate insight, data Confidentiality is normally a live of the force of the framework to defend its data. Thus, this could be a fundamental component of Security.
- **Data Integrity:** The assurance that information can only be accessed or modified by those authorized to do so. Measures taken to ensure integrity include controlling the physical environment of networked terminals and servers, restricting access to data, and maintaining rigorous authentication practices.

- **Data Freshness:** Some piece of hunt algorithms that gives greater weight to fresher content over more seasoned substance for a couple of pursuit questions. The freshness issue counteracts later, to a great degree hierarchal pages from demonstrating beginning once new substance is a considerable measure of adequate to the inquiry.

Security Goals for Sensor Network

Wireless Sensor Network has its own distinct properties which makes it unique. Their unusual properties like their volume, pattern of distribution and resource constraints give rise to few securities necessities (Malik). Security goals are categorized as primary and secondary goals (Paul, Walters, Liang, Shi, W, Chaudhary, 2006). The primary goals are standard security goals like Confidentiality, Integrity, Authentication and Availability (CIAA) whereas secondary goals are Data freshness, Self- Organization, Time Synchronization and Secure Localization (Malik) (Padmavathi, and Shanmugapriya, 2009). The primary goals are:

Data Confidentiality is the most important issue in sensor network security. Confidentiality means the ability to hide messages from a passive attacker so that any messages transferred among sensor nodes remain confidential. Sensor nodes should not leak any data to its neighbours. Data can only be accessed by only those sensor nodes that are authorized for it.

Provision of data confidentiality does not allow revealing data but it is not helpful when data is inserted in the original message by an attacker. In the wireless sensor network, Data Integrity is needed to assure the reliability of the data. It refers to the ability to ensure that the message has not been altered or tempered with and new data has not been inserted in the packet during transmission (Padmavathi, and Shanmugapriya, 2009). The integrity of the sensor network will be in trouble, when false data will be injected by a malicious node present within the sensor network and Unstable conditions because of wireless channel cause damage and loss of data (Padmavathi, and Shanmugapriya, 2009).

Data authentication assures reliability of transmitting message by indentifying communicating entities. In wireless sensor network, Authentication is used to verify identities of sending and receiving nodes. Sending and receiving nodes use shared secret Key for transmitting messages among sensor nodes. Symmetric and Asymmetric schemes are used for achieving data authentication among communicating parties. It is a challenging task within the sensor network to ensure authentication because of unattended nature of sensor network and wireless nature of the transmission media.

Data availability means desired service will be available whenever required. Availability finds whether a sensor node is able to use required resources and whether the sensor network is available to transmit the message within the network. However, failure of the sensor node or base station availability will damage the whole sensor network. Thus, the data availability is the important parameter for maintaining an operational sensor network (Padmavathi, and Shanmugapriya, 2009) (Paul, Walters, Liang, Shi, W, Chaudhary, 2006).

In the sensor network, Freshness of message is required together with the confidentiality and data integrity of the message (Akykildiz, Su, Sankarasubramaniam & Cayirci, 2002). Data Freshness ensures that the data is fresh and it also provides assurance that the no old messages have been replayed within the sensor network. To provide assurance of data freshness a nonce or time counter can be included into the packet during transmission.

Wireless Sensor Network is one type of ad-hoc network. In this type of network, each sensor node should be self-Organizing and self–healing to handle diverse situations. In this, no fixed infrastructure is available for network management. This feature of the sensor network provides a great challenge for the security of the sensor network (Padmavathi, & Shanmugapriya, 2009). Due to dynamic nature of the sensor network, sometimes deployment of the sensor nodes using pre- installed shared key mechanisms among the sensor nodes and base station becomes impossible task. It is desirable that sensor nodes within the sensor network should be self-organizing for achieving multi-hop routing and also for establishing shared keys among nodes for developing trust (Jaydip, 2009).

Applications of the sensor network needs time synchronization. In WSN, Security Mechanisms should be time-synchronized. A more collaborative WSN may need group synchronization among a group of sensor nodes for tracking various applications.

The utility of the wireless sensor network will depend on its ability to locate each sensor node in the network. A sensor network designed to identify faults requires exact location information for indentifying the location of the fault. If the location information of any sensor node is not secured then an adversary can easily change the unsecured location information by reporting false signal strengths within the network and by replaying signals (Padmavathi, & Shanmugapriya, 2009) (Jaydip, 2009).

Taxonomy of Attacks

Wireless sensor networks are vulnerable to security attacks because of the broadcast nature of the transmission medium of the sensor network. Sensor networks are more vulnerable because nodes are deployed in the hostile environment where these nodes are not physically protected. It becomes impossible to monitor each sensor node deployed in the large scale wireless sensor network and protect them from logical or physical attacks. Attackers may perform different types of security attacks in sensor network to make the network unstable. Based on the capability of an Attacker, the attacks are classified as:

- **Outsides Attacks:** Those attacks that are performed by the nodes that are not the part of the network whereas Insider attacks are performed by the legitimate nodes of the sensor network which behaves in unintended and in unauthorized ways. Robustness is required to overcome outside attacks and resilience to Insider attacks (Chaudhari & Kadam,).
- **Passive Attack:** Defined as monitoring and eavesdropping on the communication channel within the wireless sensor network by the unauthorized user whereas the active attack involves modification of the original data stream and creation of false stream (Chaudhari & Kadam).
- **Mote-Class Attacks:** In the mote-class attacks an attacker attacks sensor nodes within the WSN using some nodes of similar capabilities whereas in the laptop-class attacks, an attacker can use more powerful devices like laptop to attack a sensor network that have large transmission range, processing power.

Attacks Based On the Protocol Stack

The Physical Layer has the following responsibilities such as frequency selection, data encryption, signal detection, carrier frequency generation, modulation etc. (Akykildiz, Su, Sankarasubramaniam, & Cayirci, 2002). Attacks on the physical layer are:

- *Jamming is one type of Denial of Service (DoS) attack.* In this attack, an adversary attempts to unstable the operations of the network through the broadcast of the high energy signal. It interferes with the radio frequencies used by the nodes within the network to disturb network operations. In the WSN, Jamming attack can be classified as Constants (It corrupts the packet during transmission), Deceptive (This type of attack injects a constant stream of bytes within the network so that it behaves like legitimate traffic in the network), random (It randomly performs sleep and jamming operation to save energy in the communication) and reactive (In this type of attack, when it sense that the network has more traffic, it sends the jam signal to disturb the network). Spread-spectrum techniques which is used for radio communication is used to defend against jamming.

- *Wireless Sensor networks work in the outdoor environments.* Sensor nodes in these types of networks are greatly susceptible to physical attacks because of unattended nature of the network. An adversary can extract cryptographic keys or other confidential data from the sensor nodes memories for destroying the whole sensor network. To provide defence against physical attacks, one can tamper proof the physical packages of the sensor nodes.
    - **Self-Destruction:** When an adversary tries to access the nodes physically, the memory contents of the sensor nodes should be vaporized so that leakage of the confidential information is prevented (Chaudhari & Kadam).
    - **Fault Tolerant Protocols:** Wireless Sensor Network Protocols must be resilient to these types of attacks.

Data link Layer is used for multiplexing of data streams, Medium access control and Error control. Attacks which are performed at the link layer include collisions, unfairness in the allocation and interrogation.

- **Resource Exhaustion:** A malicious node disturbs the MAC protocol through continuously sending request or transmitting data over the transmission medium. This type of attack leads starvation in the network so that other nodes cannot access transmission medium. To provide defence against this attack, Rate limiting to the MAC admission control technique scheme can be used in which network can avoid excessive transmission request to prevent energy drain of the medium. Time division multiplexing technique can also be utilized for it.
- **Collision:** A collision happens in the channel when two nodes try to send their data on the same frequency simultaneously (Jaydip, 2009). . When packets collide, data portion of the packets may be altered that will cause the mismatch of the checksum at the receiving side. So the data packet will be discarded. Collision can be prevented by using the Error Correcting Codes (Chaudhari & Kadam).
- **Unfairness:** Unfairness is caused by repeated application of the Resource Exhaustion and Collision attacks or abusively utilizing cooperative MAC layer priority mechanisms. Unfairness is a partial Denial of Service (DoS) attack but it highly degrades performance of the channel. To provide protection against unfairness, small frame should be utilized in the channel so the any node seizes the channel for short duration only (Chaudhari & Kadam).
- **Interrogation:** This type of attack utilizes two way (RTS/CTS) handshake protocols that numerous MAC layer protocols use to alleviate hidden node

terminal problem. In this, an adversary can repeatedly transmit RTS frames for getting CTS responses from a targeted neighboring node.

In the WSN, Network Layer is vulnerable to numerous types of attack such as spoofed routing information, selective packet forwarding, sinkhole, Sybil, wormhole, hello flood etc.

- **Spoofed Routing Information:** The most direct attack against a routing protocol tries to get the routing information while is being exchanged between the nodes within the network (Chaudhari & Kadam) (Jaydip, 2009). . An adversary may spoof, alter or replay the routing information for disrupting network traffic (Karlof & Wagner, 2003). These disruption comprise the creation of routing loops, attracting or repelling network traffic from select nodes, extending and shortening source routes, generating fake error messages, partitioning the network, and increasing end-to-end latency (Chaudhari & Kadam)(Jaydip, 2009).
- **Selective Forwarding:** Wireless sensor networks are multi-hop networks. It works on the principle that the participating nodes will transmit the messages with loyalty. Attacking nodes may refuse to forward certain messages in the network or drop them. If the attacking nodes drops all the messages forwarded through them then it know as Black Hole Attack. However, if they selectively forward some messages in the network and drop others then it is called Selective Forwarding Attack (Karlof & Wagner, 2003).
- **Sinkhole:** In this attack, an adversary forged the routing information for making a compromised node an attractive choice for its neighbours. As a result, the neighbour nodes will select the compromised nodes as next hop for transmitting their data. Sinkhole attack is used to make selective forwarding attack very simple as it lures all the traffic towards the compromised node and creates a metaphorical sinkhole with an attacker at the centre (Chaudhari & Kadam).
- **Sybil Attack:** In many WSN application, sensor nodes work together for accomplishing a goal. To complete a task, sensor nodes can use distribution of subtask and redundancy of information. In this case, a sensor node can keep multiple identities of the other legitimate nodes in the WSN is called Sybil Attack. Sybil attack is used to degrade integrity of data, security and resource utilization, whereas distributed algorithms are used to achieve this. This attack can also be performed for defeating routing algorithms, data aggregation, voting, fair resource allocation and foiling misbehaviour detection.
- **Wormhole Attack:** A wormhole is a low latency link between two network portions. This link is used by an adversary for replaying network messages

over it. Wormhole attack is most vulnerable attack in the network in which adversaries records the message at one portion of the network and tunnel those messages to the other network portions. This attack is a significant threat to the WSN because it doesn't need to compromise a sensor node whereas it can be performed at the starting stage when sensor node starts to discover neighbouring information.

- **Hello Flood:** In this Attack, An adversary sends or replays Hello packet that are needed in various protocols to announce the sensor nodes to their neighbours. Sensor nodes which are receiving Hello Packets may consider that they are in the radio range of the source node. A laptop-class adversary can transmit Hello Packet to all the other sensors to convince them that compromised node belong to their neighbours. Because of this large number of sensor nodes will transmit their packets to this neighbour that will cause oblivion.
- **Node Capture Attack:** It is analyzed that capturing a single node is enough for an adversary to get complete control over the whole network. In this attack, an adversary can extract cryptographic information and get unlimited access to the confidential information stored on the memory of the physically captured node. A reasonable solution to node capture attack issues will definitely constitute a groundbreaking work in the wireless sensor network.
- **Acknowledgement Spoofing:** Routing algorithms which are used in the WSN sometimes needs acknowledgement of the transmitted packets. An attacking node can overhear packet transmission which is destined for neighbouring nodes to spoof acknowledgements for conveying false information to these neighbouring nodes. To provide protection against this attack, authentication via encryption of all transmitted packets and also packet headers must be utilized.

Attacks launched on the transport layer are as follows:

- **Flooding:** An adversary may repeatedly send new connection request until the resources needed by each connection are exhausted or reached a maximum limit. In both the cases, further legitimate request will be avoided (Jaydip, 2009). One solution to provide defence against this attack is each connecting node shows its commitment to the requested connection by solving a puzzle or maximum limit of number of connections for a particular node can be increased.
- **De-Synchronization Attacks:** De-synchronization means disruption of existing connections within the network. In this attack, an adversary can repeatedly spoof messages to an end of the network causing the other end to request

retransmission of lost frames. Hence, these messages are retransmitted and if an attacker maintains a correct timing, it can prevent the source and destination node from being exchanging any meaningful information. This attack will cause significant wastage of energy of legitimates nodes within the network in an endless synchronization recovery protocol (Chaudhari & Kadam).

CONSTRAINS IN SECURITY OF WIRELESS SENSOR NETWORK

A WSN consists of large number of sensor nodes, these devices require some form of resources such as power, storage, bandwidth, communication channels, network etc. The size of nodes is very small so we require optimize devices that can perform the above functionalities. Some of the major constraints are following.

- **Energy Constrains:** Energy has been an important concern among wireless sensor network community. Applications usually require WSN to have a long lifetime. While traditionally sensor nodes are powered by batteries with limited energy and changing battery is a bad idea since its costly, infeasible even impossible (hazardous place like volcano), people turn their attention to other solutions. Low energy designs so that the battery efficiently minimizes energy consumption or maximize lifetime while meeting required performance constraints.
- **Node Constrains:** Each node can only send a finite number of bits. For this minimized sending each bit very slowly and the delay versus energy tradeoff for each bit. Short range networks must consider both transmit and processing energy of nodes. Even sleep modes can also save energy but in complicated networking. There are issues of limited memory and energy constraints in nodes.
- **Network Constrains:** Network has Low-power microscopic sensors with wireless communication capability is the biggest challenge to accomplish. Miniaturization of computer hardware and intelligence network to move congestion efficiently. The physical devices to be embedded with environments so that reliability of devices retain for longer lifetime in network.

HOLISTIC APPROACH FOR SECURITY ENHANCEMENT

This methodology is utilized to help the execution of remote identifier system with respect to security, life span, and property underneath regularly changing ecological

Figure 5. Security layers in WSN

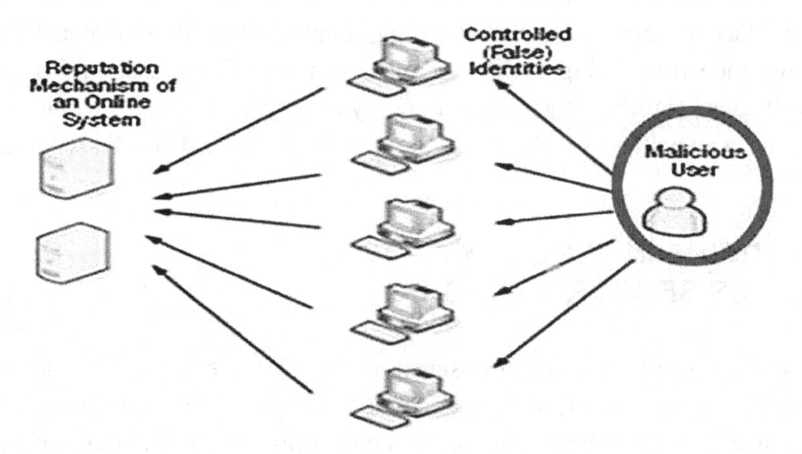

conditions. The holistic approach of security contemplations with respect to including all the layers for ensuring general security in an exceedingly arrange. One security solutions for one layer won't be an efficient employing approach could be the best option. The Figure 5 depicts the security layers in WSN.

SENSOR NETWORKS SIMULATORS

To study sensor networks or to test application and protocols, simulators are required to execute algorithms. As motes having very little communication and energy resources so there is need to implement virtual environment for researchers. A WSN simulator consists of various modules namely events, medium, environment, node, transceiver, protocols, and applications. There are various sensor network simulators such as NS2, TOSSIM, GLoMoSim(Global Mobile Information system simulator), UWSim(Under water sensor network), Avrora, SENS(A Sensor Enviroment and Network Simulator), COOJA(COntikiOs Java), Castalia, Shawn, EmStar, JSim, SENSE, VisualSense, (J)Prowler .

- **NS2:** It is event (unique ID) simulator wherever the development of your time relies on upon the worldly request of occasions that region unit kept up by a scheduler. Scheduler keeps up requested framework with the occasions to be dead and flames them one by one, summoning the handler of the occasion.
- **TOSSIM, TinyOS:** A component-based runtime environment designed to provide support for deeply embedded systems which require concurrency intensive operations while constrained by minimal hardware resources.

- **Global Mobile Information System Simulator (GloMoSim):** Network protocol simulation software that simulates wireless and wired network systems. GloMoSim is designed using the parallel discrete event simulation capability provided by Parsec, a parallel programming language. GloMoSim currently supports protocols for a purely wireless network.

CONCLUSION

In this chapter we provided you a brief introduction of wireless sensor network. The reader should be able envision number of application and the technologies that are used in sensor network. Today wireless sensor network having great potential to explore the universe. In most of the technologies sensors has started playing a bigger role in collecting data from remote places which is quite impossible to manually collect at low cost for long time with sustainable results. In future years remote sensing component actuation interfaces can provide extra compelling for recognition structure for damage. A noteworthy point of interest of wireless sensor network over old link basically based recognition framework is that the collocation of strategy force with the sensing transducer. WSN's are intended for particular applications. Applications epitomize, however aren't confined to, ecological perception, modern machine perception, reconnaissance frameworks, and military target interest each application contrasts in choices and necessities. To help these differing qualities of uses, the occasion of new correspondence conventions, calculations, outlines, and administrations are needed.

REFERENCES

Abowd, G.D., & Sterbenz, J.P.G. (2000). Final report on the interagency workshop on research issues for smart environments. *IEEE Personal Communications*, 36–40.

Ahmed, K. (n.d.). *Techniques and challenges of data centric storage schemes in wireless sensor network*. Academic Press.

Akykildiz, I. F., Su, W., Sankarasubramaniam, Y., & Cayirci, E. (2002). A Survey on Sensor Networks. *IEEE Communications Magazine*.

Amundson, I. (n.d.). *A survey on localization for mobile wireless sensor network*. Academic Press.

Bischoff, R. (n.d.). *Wireless sensor network platforms*. Retrieved from http://www.ieee802.org/15/

Bulusu, N., Estrin, D., Girod, L., & Heidemann, J. (2001). *Scalable coordination for wireless sensor networks: self-configuring localization systems.* Paper presented at the International Symposium on Communication Theory and Applications (ISCTA 2001), Ambleside, UK.

Cui, J. (n.d.). *Data Aggregation in Wireless sensor networks: Compressing and forecasting*. Academic Press.

Dasgupta, S., Bhattacharya, I., & Bose, G. (2009). Energy-aware cluster based node scheduling algorithm in wireless sensor network for preserving maximum network life time. In *Proceeding of International Conference on Methods and Models in Computer Science (ICM2CS).*

Deying, L. (n.d.). *Sensor Coverage in Wireless sensor network.* Academic Press.

Ellbruck, H. (n.d.). *Using and operating wireless sensor network testbeds with WISEBED*. Academic Press.

Elson, J. (n.d.). *Time synchronization for wireless sensor network.* Academic Press.

Essa, I. A. (2000). Ubiquitous sensing for smart and aware environments. *IEEE Personal Communications.*

Farooq, M. O. (n.d.). *Operating system for wireless sensor networks: A survey.* Academic Press.

Farooq, M. O. (n.d.). *Wireless Sensor networks Testbeds and state of the art multimedia sensor nodes*. Academic Press.

Frohlich, A. A. (n.d.). *Operating system support for wireless sensor network.* Academic Press.

Hedetniemi, S., & Liestman, A. (1988). A survey of gossiping and broadcasting in communication networks. *Networks, 18*(4), 319–349. doi:10.1002/net.3230180406

Heinzelman, W. R., Chandrakasan, A., & Balakrishnan, H. (2000). Energy-efficient communication protocol for wireless microsensor networks. *IEEE Proceedings of the Hawaii International Conference on System Sciences*, 1–10.

Heinzelman, W. R., Kulik, J., & Balakrishnan, H. (1999). Adaptive protocols for information dissemination in wireless sensor networks. In *Proceedings of the ACM MobiCom'99*. doi:10.1145/313451.313529

Herring, C., & Kaplan, S. (2000). Component-based software systems for smart environments. *IEEE Personal Communications,* 60–61.

Hou, J. C. (n.d.). *Coverage in Wireless sensor networks*. Academic Press.

Intanagonwiwat, C., Govindan, R., & Estrin, D. (2000). Directed diffusion: a scalable and robust communication paradigm for sensor networks. In *Proceedings of the ACM Mobi-Com*. doi:10.1145/345910.345920

Jaydip, S. (2009, August). Security on Wireless Sensor Networks Security. *International Journal of Communication Networks and Information Security, 1*.

Kahn, J. M., Katz, R. H., & Pister, K. S. J. (1999). Next century challenges: mobile networking for smart dust. *Proceedings of the ACM MobiCom'99*. doi:10.1145/313451.313558

Karlof, C., & Wagner, D. (2003). Secure routing in wireless sensor networks: Attacks and countermeasures. In *Proceedings of the 1st IEEE International Workshop on Sensor Network Protocols and Applications*. doi:10.1109/SNPA.2003.1203362

Li, L., & Halpern, J. Y. (2001). *Minimum-energy mobile wireless networks revisited.* Paper presented at the IEEE International Conference on Communications ICC'01, Helsinki, Finland.

Malik, M.Y. (n.d.). An Outline of Security in Wireless Sensor Networks: Threats, Countermeasures and Implementations. *Wireless Sensor Networks and Energy Efficiency: Protocols, Routing and Management.* DOI: 10.4018/978-1-4666-0101-7.ch024

Noury, N., Herve, T., Rialle, V., Virone, G., Mercier, E., Morey, G., & Porcheron, T. et al. (2000). Monitoring behavior in home using a smart fall sensor. *IEEE-EMBS Special Topic Conference on Microtechnologies in Medicine and Biology*. doi:10.1109/MMB.2000.893857

Padmavathi, G. & Shanmugapriya, D. (2009). A Survey of Attacks, Security Mechanisms and Challenges in Wireless Sensor Networks. *International Journal of Computer Science and Information Security, 4.*

Petriu, E. M., Georganas, N. D., Petriu, D. C., Makrakis, D., & Groza, V. Z. (2000). Sensor-based information appliances. *IEEE Instrumentation & Measurement Magazine*, 31–35.

Rappaport, T. (1996). *Wireless Communications: Principles and Practice*. Englewood Cliffs, NJ: Prentice-Hall.

Rodoplu, V., & Meng, T. H. (1999). Minimum energy mobile wireless networks. *IEEE Journal on Selected Areas in Communications*, *17*(8), 1333–1344. doi:10.1109/49.779917

Siegmund, N. (n.d.). *Towards Robust data storage in wireless sensor networks*. Academic Press.

Sohrabi, K., Gao, J., Ailawadhi, V., & Pottie, G.J. (2000). Protocols for self-organization of a wireless sensor network. *IEEE Personal Communications*, 16–27.

Sundani, H. (n.d.). *Wireless Sensor network simulators A survey and comparison*. Academic Press.

Undercoffer, J., Avancha, S., Joshi, A., & Pinkston, J. (2002). Security for Sensor Networks. *In Proceedings of CADIP Research Symposium*.

Walters, L., & Shi, C. (2006). Wireless Sensor Network Security: A Survey. Security in Distributed, Grid and Pervasive Computing.

Warneke, B., Liebowitz, B., Pister, K.S.J. (2001). Smart dust: communicating with a cubic-millimeter computer. *IEEE Computer*.

Xiang, L. (n.d.). *Compressed data Aggregation for energy efficient Wireless sensor network*. Academic Press.

You, J., Lieckfeldt, D., Salzmann, J., & Timmermann, D. (2009). *Connectivity Aware Topology Management for Sensor Networks*. GAF & Co.

Zeng, X. (n.d.). *GloMoSim A library for parallel simulator of large scale wireless networks*. Academic Press.

Zhao, F., & Guibas, L. (2004). *Wireless Sensor Networks: An Information Processing Approach*. San Francisco, CA: Morgan Kaufmann.

KEY TERMS AND DEFINITIONS

Energy Constrains: Energy has been an important concern among wireless sensor network community. Applications usually require WSN to have a long lifetime.

MANET: A Mobile ad-hoc network (MANET) is generally delineated as a system that has a few free or self-ruling nodes, normally made out of cell phones or option versatile things, that may sort out themselves in changed ways that and work while not strict top-down system organization

Steganography: The hiding of a secret message within an ordinary message and the extraction of it at its destination

System on Chip (SoC): These platforms integrate micro electromechanical systems (MEMS) sensors, microcontroller, and wireless transceiver technologies on one chip, an application-specific integrated circuit (ASIC).

WSN: Wireless Sensor Networks (WSNs) are designed by sensor nodes that communicate each other and also processing data and sensing environment.

Chapter 8
Authentication in Next Generation Network

G. C. Manna
BSNL, India

Vishnu Suryavanshi
GHRCE Nagpur, India

ABSTRACT

Next generation Network has been evolved based on the concept of integration of voice, data and mobile networks, single point control for provisioning of all services and single account for all services of an user or corporate over an all IP platform. User authentication of next generation network depends on Session Initiation Protocol but existing networks do not directly support this protocol. The new networks authentication policy together with existing networks authentication policy completes the authentication process. Authentication steps for each type of network with trace routes involving associated protocols has been discussed in the chapter.

INTRODUCTION

Telephone network has been evolved from basic voice based services. Till 1980's, Data communication was added to voice based telephone as access network for narrow band and directed to a server through modem. During 1990's, customer data path was in telephone access network but filtered out at central office to follow IP based data network switched through LAN switch/Router, popularly known as broadband network. Mobile voice telephony grew in parallel which also paved way

DOI: 10.4018/978-1-4666-8687-8.ch008

to mobile data communication through mobile switch. During 2000s, the mobile access technology was able to support broadband mobile services through separate core network for data communication. Lot of value added services were loaded on these networks. Thus the network based concept of telephony was evolved to service based unified communication system based on all IP nodes and protocols and popularly known as next generation network.

Next section deals with the background of different telephone networks and evolution of backbone transmission systems. Logging in to a network is the first step to get services which is termed as authentication process. Authentication process for basic telephone, broadband, and mobile networks using Public Switched Telephone Network (PSTN) and Public Land Mobile Network (PLMN) has been discussed in subsequent sections. Authentication of User Equipment (UE) based on USIM/ISIM, without USIM/ISIM and for 2G based services for Next Generation Network (NGN) has been discussed later followed by conclusion.

BACKGROUND

Early telephones were hardwired to communicate with only a single other telephone using overhead lines. Later telephone exchange switchboards located at central offices were used to provide the switching or interconnection of two or more individual subscriber lines for calls made between them, rather than requiring direct lines between subscriber stations. Common battery non multiple/multiple switch boards with line jacks were extended to each customer premises on overhead lines. Such switch boards were managed by operators to connect the customers with each other and there was no secrecy of communication& there were time delay as well.

Automatic exchanges, or dial service, came into existence. Automation replaced human operators with electromechanical systems and telephones were equipped with a dial by which a caller transmitted the destination telephone number to the automatic switching system. Such electromagnetic auto exchanges like Strowger & Pentaconta Cross bar technologies lived the network for a long time. Initial electronic switching systems gradually evolved from electromechanical hybrids with stored program control (SPC) to the fully digital new technology switches viz. OCB from Alcatel, EWSD from Siemens etc. This Central office concept introduced Main switching unit (MSU) & Remote switching unit (RSU) architecture by which telephone line faults were drastically reduced. These digital switches use time & space switching concepts and they work by connecting two or more digital circuits, according to the dialed telephone number. These automatic exchanges were provided with tone generators to generate different tones like Dial tone / Ring back tone / busy tone etc. Also the exchanges have elements called Registers / Translators / Controllers / Line

& trunk interface units / switching network etc. When the caller (A number) lifts the phone, the loop condition is recognized by the exchange & dial tone is fed to him. On dialing, the dial tone is cut off & a register is assigned to collect the dialed digits of the called (B number). Then the dialed digits are analyzed by the translator and charger to decide the routing & charging and the call is put through to the B number by reserving time slots on either side of the switching network. Ringing current is fed to B number & it rings. When the B party answers, it is recognized by the exchange and it switches the time slots on both sides. When either party releases the call, the exchange terminates the session & effects charging to the calling party. This formed the basis of public switched telephone network (PSTN).For long distance, interexchange calls are set up between switches and controlled using the Common Channel Signaling System 7 (CCS7) protocols (Bosse, 1998). This CCS7 concept helped to carry the signaling messages of different calls on a separate path than that of voice path. CCS7 signaling paved the way for remote signaling, signal transfer points (STP), overlay networks such as intelligent network (IN) etc. CCS7 ISDN user part (ISUP) messages are widely used to exchange the signaling information like calling number, calling line category, dialed digits, answer, release conditions of A / B numbers. Service functions were enhanced through Intelligent Network by setting up a separate Service control point (SCP) and the PSTN exchanges act as a Service Switching Point (SSP). The SSP & SCP communicate with each other on well-defined set of IN Application Protocol (INAP). The example of INAP messages are Initial Detection Point (IDP), Apply charging (AC), Apply charge report (ACR) etc. The examples of IN services are Virtual Card Calling (VCC), prepaid cards, Account Card Calling (ACC), Free phone (FPH), Virtual private network (VPN), Reverse charging, etc. These PSTN digital exchanges provide integrated services digital network (ISDN) by which voice, data& image communication turned to reality. An ISDN (2B+D) subscriber can establish two independent calls and one data call simultaneously on the existing path of single telephone line. For business customer with large volume of traffic, a possibility of 30 simultaneous calls (30B+D) based Private Branch Exchange (PBX) also prepared. Apart from basic services, ISDN offers supplementary services such as Calling line Identification Presentation (CLIP), Calling Line Identification Restriction (CLIR), Closed User Group (CUG), COLP, COLR, etc. An element called NT (Network Termination) is mandatory at the customer premises for ISDN service. On introduction of digital technology switches, lots of local exchanges, Tandem, & TAX exchanges were introduced in the PSTN cloud to cater to the growing STD, ISD traffic needs.

Broadband data services are used by Telecom Companies to leverage their existing investment in copper in the local loop for which Asymmetric Digital Subscriber Loop (ADSL) modulation technique is used. Asymmetric Digital Subscriber Loop (ADSL) has found favour as a broadband delivery mechanism, in view of its high downstream bandwidth. Downstream refers to data flowing from the service provider to the user. Most of the popular applications, like web browsing, video streaming, FTP downloads, etc., require much higher downstream bandwidth than upstream bandwidth (Forouzan, 2011). ADSL manages to extract high data rates in this direction. With all the local telecom companies edging closer to the customer with their distributed access mechanisms like DLCs and RSUs, the distances lie well within the range of most customer premises, and thus broadband delivery can be quite effective, with a richer user experience. To provide broadband service to the customer, there are few elements need to be introduced in the network, viz. DSLAM, TIER I LAN switch, TIER II LAN switch, BNG, AAA, Billing system, CPE etc. CPE is the Customer Premises Equipment modem which uses existing telecom line to provide composite telephone and data services. For many such telephone lines, Digital Subscriber Line Access Multiplexer (DSLAM) is deployed at the exchange premises where the speech path & data path are multiplexed/demultiplexed. DSLAMs which work using Ethernet technologies in local areas are concentrated using LANs of different capacities and Broadband Network Gateway Routers (BNG) where it is connected to the service provider's network internet backbone. The broadband users are authenticated for at the AAA server by their user name & password. Their authorization is also fetched from AAA server and at the end of the session, it generates a ticket with volume of upload/download data for accounting purpose. The Broadband services are provided in an integrated manner by the broadband service provider. Some of these are, High Speed Internet Services, Video on Demand, Multicast Video Streaming, Interactive 'e'-Learning, Interactive Gaming. Very High data rate Digital Subscriber Line (VDSL) with speed of 24Mb downlink are also provided with introduction of Line Media Gateway (LMG), DSLAMs in Next Generation Network (NGN).

GSM network consists of different elements like HLR, MSC, GGSN, SGSN, BSC, RNC, BTS, and Node –B etc. There are two different cores viz. circuit switch (CS) to carry voice traffic & packet switch (PS) to carry data traffic. There are different protocols like Mobile application protocol (MAP), BSS Application part (BSSAP), Customised Application for Mobile Enhanced Logic (Camel) application part (CAP), Transaction capabilities application part (TCAP), Signaling connection control part (SCCP), ISDN user part (ISUP) etc. The MAP messages are used to query the remote repository like HLR. CAP messages are involved in mobile IN based prepaid call scenario. It was possible to introduce Mobile number portability (MNP) only with the help of these SCCP messages. Initially at the access part, BSC/BTS were introduced to cater to voice traffic originated from Mobile. Later by plugging a Pocket Controller unit (PCU)/BTS to the same BSC along with GGSN/SGSN, low speed data traffic was introduced with specifications in the interfaces and this is called GPRS or 2G services. General Packet Radio Services (GPRS) is a new technology for GSM networks that will allow GPRS compatible handsets to transfer data at much higher rates than the current 9.6 or 14.4 kb/s available to GSM handsets. GPRS has been designed to enhance existing GSM networks to provide end-to-end transport of packet based IP data. This technology has effectively make mobile Internet access a reality and is seen by many as a stepping-stone between 2nd Generation (2G) and 3rd Generation (3G) networks. New GPRS network elements use packet-switched technology to transfer data. Packet-switched networks allocate bandwidth in a dynamic way, when required, to transfer data to/from terminal equipment that makes much more efficient use of the available resources. GPRS can also offer asymmetrical transfer rates where the downlink to the mobile has greater bandwidth than the uplink. Universal Mobile telecommunication system (UMTS) is an evolution from GSM and other second generation (2G) mobile systems. UMTS is an International Mobile Telecommunications - 2000 (IMT-2000) 3G system. UMTS architectures provide a smooth transition from second generation telecommunications systems by slowly phasing in new software and new network elements. The main technological difference between 2G and 3G systems is the new Code Division Multiple Access (CDMA) and wideband CDMA techniques in the Radio Access Network (RAN) that increases bandwidth and efficiency. With introduction of split architecture in the core network, and by using Wideband CDMA (WCDMA) radio technology, wireless broadband speed of 14.4 mbps has become true. The emergence of fourth generation (4G) of mobile services uses Orthogonal Frequency Division Technology (OFDM) which allows users to connect to different networks depending on their location. Mobile handset can connect a wireless local area network(WLAN) when inside a building, switch to a 3G mast when outside, and connect to a standard 2G network in areas with no 3G coverage.4G is an all IP Network with higher bit rates, enhanced Multimedia services, and smooth streaming videos, worldwide access/roaming capability uses intelligent software as driver technique. The debate of wired vs wireless will end in the NGN era through Fixed Mobile Convergence (FMC).

Next generation network (NGN) is a packet based network which is able to provide multimedia telecom services and able to make use of multiple broadband, QoS enabled transport technologies in which service related functions are independent from underlying transport related technologies. NGN is a collection of technologies which provides convergence for voice, data and video services. Voice shall also be transported through packet switching. In NGN basically the switching and call intelligence functions are separated. It consists of different layers, viz.

1. **Access Layer:** Combines all the access technologies like POTS, ISDN, GSM/UMTS, ADSL, etc.

2. **Transport Layer:** The backbone network and the techniques for transport. It is IP backbone.
3. **Control Layer:** This layer controls call handling. It comprises of the equipment that manages signaling (SGW) and call handling (Media Gateway Controller).
4. **Service Layer:** This layer comprises of the equipment that centralizes the service logic and data. It is also called as Application Server or Media Server.
5. **Management Layer:** It spans over all the other layers and it comprises of all the management equipment.

The major network elements of NGN are:

- Media gateway controller (MGC),
- Trunk media gateway (TMG),
- Line Media gateway (LMG),
- Signaling gateway (SGW),
- Application servers etc.

There are different set of rules to be used for communication between these entities which are called protocols. Major protocols of NGN are:

1. **Session Initialed Protocol (SIP):** The Session Initiation Protocol (SIP) is an application layer Signaling Control Protocol used to establish, maintain and terminate multimedia sessions between the two components viz. User Agent (End system/terminal) & Network Servers (Proxy, Redirect, Registrar)
2. **MEGACO/H.248:** It is based on Master/Slave relationship. It is fundamentally based on two key concepts of termination and context. This protocol is used between MGC & TMG/LMG.
3. **SIGTRAN:** These protocols are used to handle ss7 stack over IP which comprises of protocols viz. M3UA, M2UA, M2PA, SUA, and IUA. These protocols are transported through SCTP (Stream Control Transport Protocol) instead of TCP. This happens between MGCs or with MGW or with external signaling elements.
4. **Real-Time Transport Protocol (RTP):** It is a protocol for end to end delivery for the real time bearer data. Encapsulates audio or video inside RTP Packet, use UDP as transport protocol and delver to the other end. The examples for enhanced services in NGN are prepaid service, Video phone, & fixed mobile convergence etc.
5. **DIAMETER:** For signaling exchange between Packet Switch and Voice Switch networks for AAA purposes.

Transmission system evolution took place in four phases:

1. **Open Wire Systems:** Till fifties the long distance voice communication was almost entirely transported over Open Wire Carrier system. The voice signals for these systems were modulated to a higher frequency and carried through open wire systems. These open wire systems are capable of carrying traffic of three to twelve subscribers at a time.
2. **Coaxial Cables:** With the introduction of symmetrical pair cable carrier system followed by the Coaxial Cable system, the simultaneous voice channel carrying capacity to 960 voice channels had become a reality. Over the years this system was improved and developed to carry 2,700 simultaneous voice channels. These Coaxial Cable systems were provided with number of repeaters in between at a spacing of 4.5 km or 9 km depending upon the system used viz. 12 MHz systems or 4 MHz systems.
3. **Digital Transmission System:** Till eighties, Coaxial cable and UHF transmission media were used to provide connectivity with the introduction of Digital Transmission Systems viz. Digital UHF, Digital Microwave, and Digital Coaxial Systems. Underground coaxial cable was initially used for the connectivity of large and medium cities. Media diversity is provided through Radio Relay Systems. These Radio relay systems were very reliable and beneficial where laying and maintenance of underground cable is extremely difficult. In late eighties, Synchronous Digital Hierarchy (SDH) standard was introduced. SDH in conjunction with optical fiber cables, high capacity transmission telecom network was achieved. The basis of Synchronous Digital Hierarchy (SDH) is synchronous multiplexing where data from multiple tributary sources is byte interleaved. SDH brought the advantages to network providers viz. High transmission rates, Simplified add & drop function, High availability, capacity matching and Reliability. The product family includes Add-drop multiplexers (ADM), Terminal Multiplexers (TM), Regenerators (REG) and synchronous Digital Cross connects (DXC).In SDH, there are various network topologies and protection of path are possible e.g. Point to point, Point to multipoint, Ring Topology, Mesh topology etc.
4. **Optical Fiber Cable:** Optical Fiber Cable finds its way to provide connectivity in early 1980s spanning long distances between local phone systems as well as providing the backbone connectivity for many network systems. These systems are capable of carrying large number of voice channels compared to others and offer the circuit at low cost per kilometer of circuit. To cope up the growing demand of bandwidth, OFC is used to connect exchanges between different cities for STD traffic. Subsequently, intra-city connectivity between exchanges and finally most of the small rural exchanges were also connected

through the OFC systems. In recent years OFC is now started replacing copper wire as an appropriate means of communication signal transmission in local loop. The advantages of OFC are bandwidth, electromagnetic immunity, low loss of signal, high security.

Following two sections discusses with authentication mechanisms observed in PSTN voice, broadband data networks and mobile telephone networks as no separate authentication is done by next generation networks in addition for incoming traffic. Next section discusses different scenarios of authentication for user equipment of NGN with or without USIM/ISIM followed by conclusion.

PUBLIC SWITCHED TELEPHONE NETWORK (PSTN)

Authentication in Telephone Exchange

The basic telephone exchange components are shown in Figure 1 (Manna, 2010).A subscriber line consists of pair of copper conductors extended from a port of an electronic board in Local Subscriber Access unit. The port is called Network End (NE). The subscriber line is identified by a Directory Number (ND). When the telephone is on hook, the loop is normally kept at very high resistance and during an incoming call, a battery is connected which drives the ringing bell. When the telephone goes off-hook, the indication is extended by offerring a low resistance path, comparatively large current flows which is interpreted as acceptance of an incoming call or a new call as in appropriate case. For a new call, a query is send from access unit to Translator through an internal link for authentication of the ND. For an ALCATEL type of exchange, see Figure 1.

The appropriate command/response is as shown in Figure 2.

On interrogation of subscriber characteristics with directory number ND, NE is returned as check of authentication as shown in Figure 2. The subscriber calls from a Dual Tone Multi Frequency (DTMF) phone and subject to malicious call observation (IAM+CAM3). The subscriber is authorised for making all outgoing calls except ISD (SR3), enjoys own phone locking/unlocking facility (SRC), can bar outgoing calls (DF1) etc. Once authenticated, dial tone is fed to subscriber line from tone generator unit. The subscriber dialed digits are collected at Register unit. After receipt of first two or three digits, Register sends them to Translator unit for outgoing route determination and a route is pre-selected towards another exchange or Trunk Automatic Exchange (TAX) for long distance call. When all the digits are received in register, input line is put through by Marker unit to output line or trunk of PCM unit. The digits are send to terminating exchange by building a complete

Figure 1. Basic public switched telephone network block diagram

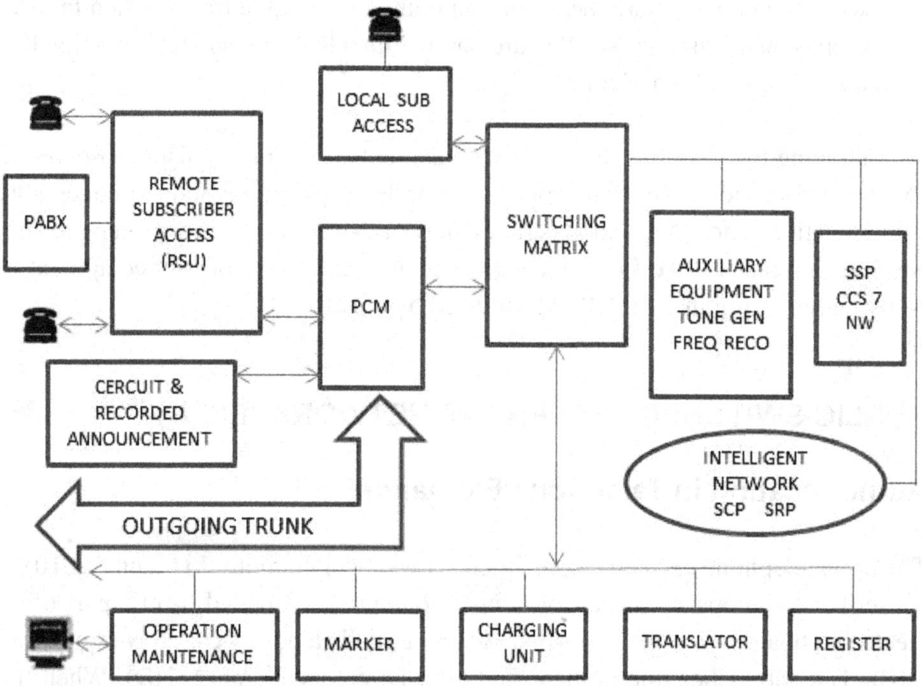

path and after authentication alarm is sent to the called party. When called party lifts handset, the route is through for communication. If the exchange is equipped with CCS7 signalling system, the dialed digits are sent to CCS7 unit which, in turn, send to distant CCS7 unit of terminating exchange through Initial Address Message (IAM). The called party is first authenticated and Address Complete Message (ACM) is returned to calling party CCS7 unit. The exchange sends ringing tone to the called party and if the handset is lifted, Answer and Charge (ANC) message is returned

Figure 2. Subscriber authority string for PSTN switch

```
ABOIN:
     CEN1/14-11-26/1700-16-11/SUBS CHARASTERISTIC INTERROGATION

ND=28220123
     PROCESSING TGLAIN ACC

          ND = 28220123   NE = 3-1-17
          TAX  = 00007263  + 00000000+ 00000000+ 00000000
          TYPE = SR3  + DTMF  + KLA
          CAT  = SRC  + RVT  + DF1+ IAM  + CAM2

     TGLAIN EXEC
```

to calling exchange. Marker unit of corresponding exchanges put through the links and the call is completed.

Authentication in Intelligent Network

Telephone exchange network authentication and authorisation are limited within its own subscriber base. With introduction of server based Intelligent Network, all the local exchanges have been upgraded as Service switching point (SSP), and Service control points (SCP) and new elements like Service Resource Point (SRP) are added. This helped to provide services like Free Phone Service (FPH),Virtual Private Network (VPN),Virtual Card Calling Service (VCC),Account Card Calling service (ACC),Premium Rate Service (PRM),Universal Access Number service (UAN),Tele Voting (VOT) etc. All the SSPs are virtually linked by creating signaling links/link sets/route sets and are able to communicate with SCP by exchanging Signaling connection control part (SCCP) messages like Initial detection point (IDP) / Basic call state model (BCSM). Most of the services are defined with logical access code like VCC - 1602 345, Free phone - 1600 345 1600 etc. When A party number is making a VCC call, there will be authentication for A number first for dial tone and calling card is also authenticated by IN subsequently. A party will be dialing 1602 345. The SSP will send an IDP message in which A number, B number as 1602 345 & service key will be sent to SCP. SCP based on the service key will decide and instruct SSP to connect to Service resource point (SRP). The SRP will be able to collect the digits dialed by the customer & able to play announcement. Customer will be asked to dial the secret code of the VCC card (16 digits hidden secret number (HRN).Such HRN will be collected by the SRP & presented to the SCP.SCP is having the data base for authentication verification of the HRN & Check the balance amount & validity period. Customer is prompted to dial the B number of his choice & the B number will be presented to the SCP. SCP will check the charging matrix & if balance is sufficient, it grants duration (say 5 minutes) & instruct SSP to connect the call to B party by message AC. SSP will connect the call to B number on ISUP & report the event on BCSM message to SCP. When either party releases the call, such event will be reported by SSP to SCP & its balance in VCC is reduced & a Call Detail Record (CDR) with proper A number & B number is generated.

AUTHENTICATION IN BROADBAND NETWORK

Telephone and PC are connected to ADSL modem using RJ11 and USB ports respectively (see Figure 3). The voice part is placed in a lower frequency band and

Figure 3. Block schematic of broadband network

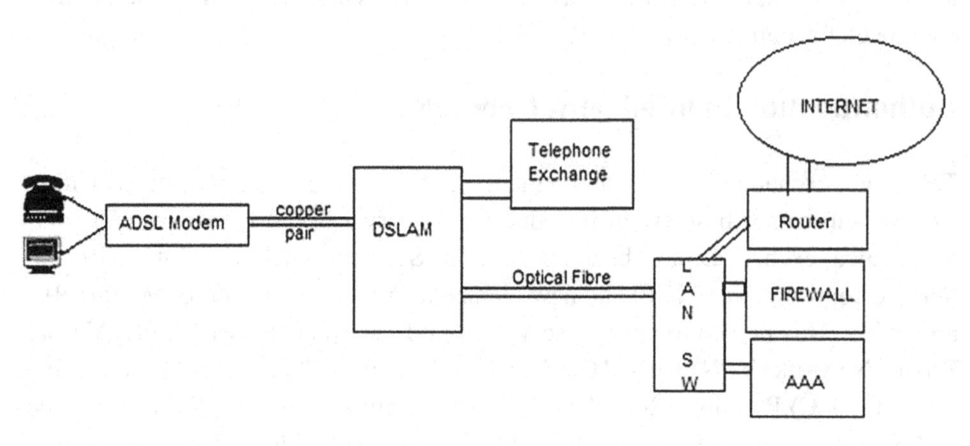

data part in higher frequency band deploying Discrete MultiTone (DMT), combined and transported through existing telephone copper wire to DSLAM usually located in telephone exchange. DSLAM filters the composite signal, puts lower frequency part to telephone exchange. The data is extracted from higher frequency component, multiplexed with other users data and send on copper line using Ethernet technology. LAN Switch is used as aggregator for DSLAM traffic. The service provider LAN SW at local end is carried through MPLS VPN network to service provider's nodal service centre. After Authentication and Authorisation, a user is allowed to access the internet.

Challenge Handshake Authentication Protocol (CHAP) is used to grant access to the network Forouzan, 2011). The AAA server stores the password of the user. During authentication, the server sends few bytes as challenge packet to the user. The user applies a predefined function that takes the challenge bytes and user password as input and creates a result. The user sends the result as response. The server computes the same and matches the result for authentication.

Protocols involved sequentially in authentication process of broadband network has been illustrated in Figure 4. The PC client software Quantaco sends a PPPoE Active Discovery Initiation(PADI) request packet as broadcast to initiate the session which is rebroadcasted by DSLAM and responded by Montreal _RAS Server through PPPoE Active Discovery Offer(PADO) message. The negotiation takes place at MAC layer. The client PC responds with a PPPoE Active Discovery (PADR) Request .Server responds by generating an unique Id for the PPP session and sends it in a PPPoE Active Discovery Session (PADS). The PC open PPP Link Control Protocol (PPP LCP) configuration request which is sometimes rejected or code rejected, however repeated request lead to PPP LCP configuration acknowledge by server. PPP LCP message again generated by PC for IP address identification

Figure 4. Protocols for authentication in broadband network

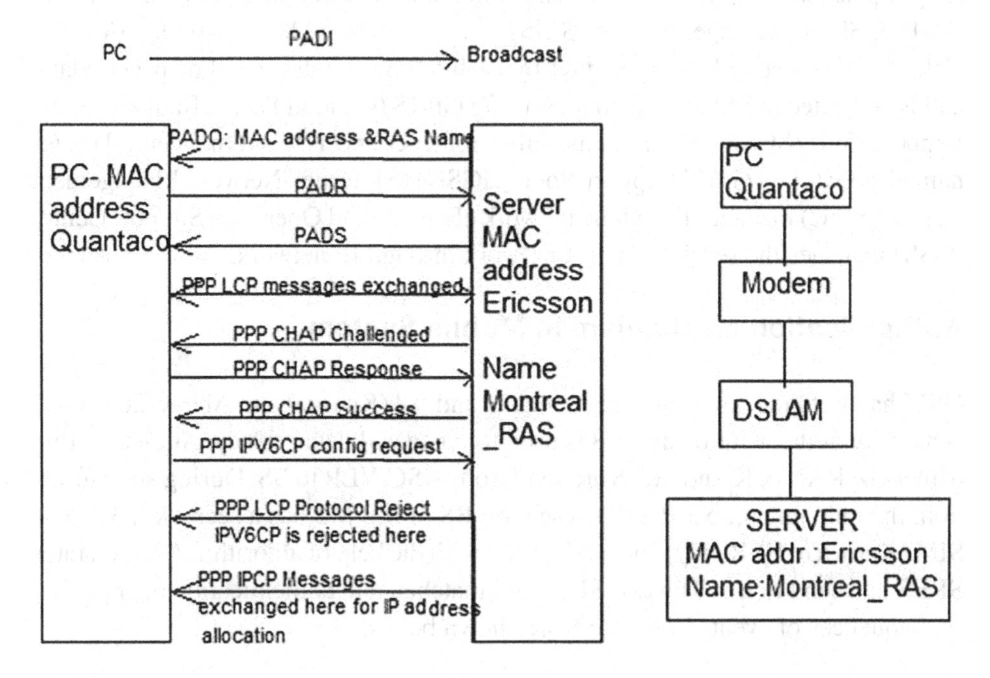

which is replied by PPP CHAP message by Server. PPP CHAP response message is send by PC to access server which in turn forwarded to authentication server. On receipt, PPP CHAP success is returned. PPP IPCP messages are exchanged after which an IP address is allocated for further communication with the external world.

AUTHENTICATION IN MOBILE SYSTEM

Generic block diagram of a mobile network is shown in Figure 5 (Garg and Wikes, 2005) . The user handset i.e. Mobile Station (MS) is housed with a detachable Subscriber Identification Module (SIM) containing customer mobile number and security information. GSM is Time Division Multiple Access (TDMA) based technology. When a customer's MS makes a registration request, his call is routed through Base Transceiver Station (BTS), Base Station Controller (BSC) and Mobile Switching Centre (MSC) where his identity, authority, security and subscription information are verified from Home Location Register (HLR), Authentication Centre (AuC) and Equipment Identity Register (EIR). The registration information of customer is made available in Visitor Location Register (VLR) for call handling. A Mobile Originated (MO) call, if from post-paid customer, is handled by MSC/HLR whereas for pre-paid, subscription information is obtained from mobile Intelligent Network (M-IN). Short Message Service (SMS) is handled by SMS controller (SMS-C). Calls to other networks are handled by mobile/PSTN gateways. For packet data, call is separated at BSC and send to Serving GPRS (General Packet Radio Service) Support Node (SGSN) which is basically a router connects to another central router named as Gateway GPRS Support Node (GGSN) to Internet. Network Management Centre (NMC) manages the whole network elements and Operation Support Centre (OSS) manages the service support network through IP network.

Authentication Mechanism in Mobile System

GSM handset holds two algorithms viz A3 and A8 (Kaushal and Abhay, 2008). A3 is used for authentication and A8 is used for voce and data privacy. Authentication triplet viz. RAND, K_i and SRES are send from MSC/VLR to BS. During authentication, the random number RAND is sent by BS to MS.MS has a secret key K_i in its SIM (Known to HLR only) and RAND and with the help of algorithm A3, generates SRES and send to BS. The two SRES are matched and authentication is granted.

Sequences of events in Figure 6 are shown below.

1. MS sends a location update request to MSCS/VLR. After receiving location update request from MS, MSCS/VLR checks correctness of subscriber data and judges location update category, to determine subsequent operations. (Trace Event NO 1)
2. According to some conditions, MSCS/VLR determines whether to send location update request to HLR. In general, this operation is originated in the following cases:
    a. Subscriber starts the system for the first time.
    b. Subscriber roams across the MSCS/VLR system.
    c. Subscriber data is incorrect or inconsistent with that in the HLR due to MSC/VLR reset or individual reasons.

In this case, the trace of first instance is taken.

3. Authentication parameter is checked from HLR and MS. Firstly. MSC/VLR send the authentication info request to HLR (Trace Event No 2 to 6) via
    a. MAP open request -open the dialogue for authentication.
    b. Send authentication info-request to send Authentication triplet rand, Kc and SRES.

Figure 5. Generic block diagram of mobile network

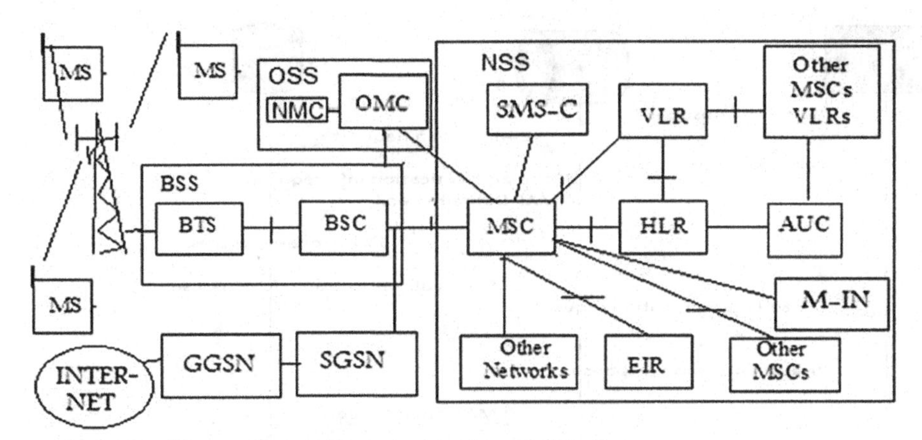

MS=Mobile Station BSS=Base Station Subsystem

BTS=Base Transceiver Station, BSC=Base Station Controller

OSS=Operation Support System, OMC=Operation & management Centre

MSC=Message Switching Centre, HLR=Home Location Register,

VLR=Visitor Location Register, AUC=Authentication Centre

M-IN=Mobile Intelligent Network, SGSN=Serving GPRS Support Node

GGSN=Gateway GPRS support Node,GPRS=General Packet Radio Service

c. HLR responds to the above message by sending authentication parameter to MSC/VLR.

4. MSC/VLR requests MS to send authentication info and MS responds by sending SRES to MSC/VLR (TraceN o 7 to 8).
5. If both Sres matched then the authentication of SIM is successful.
6. MSCS/VLR sends a location update request to HLR, which causes HLR to originate subscriber data insertion operation, to transmit subscriber data from HLR to MSC/VLR.MSCS/VLR acknowledges that all subscriber data from HLR is correct and returns acknowledgement primitive to HLR (Trace event No:9 to 19). This is done by following steps:
   a. MAP Open Request…Open the dialogue for Location update.
   b. Update location for the subscriber.
   c. HLR acknowledges the MAP open request.
   d. HLR then sends the subscriber data to MSC/VLR (i.e. subscription data) for storage.
   e. MSC/VLR stores the data and acknowledges the same to HLR
   f. HLR sends the acknowledgement for location update.

Figure 6. Authentication process transactions in GSM mobile communication system

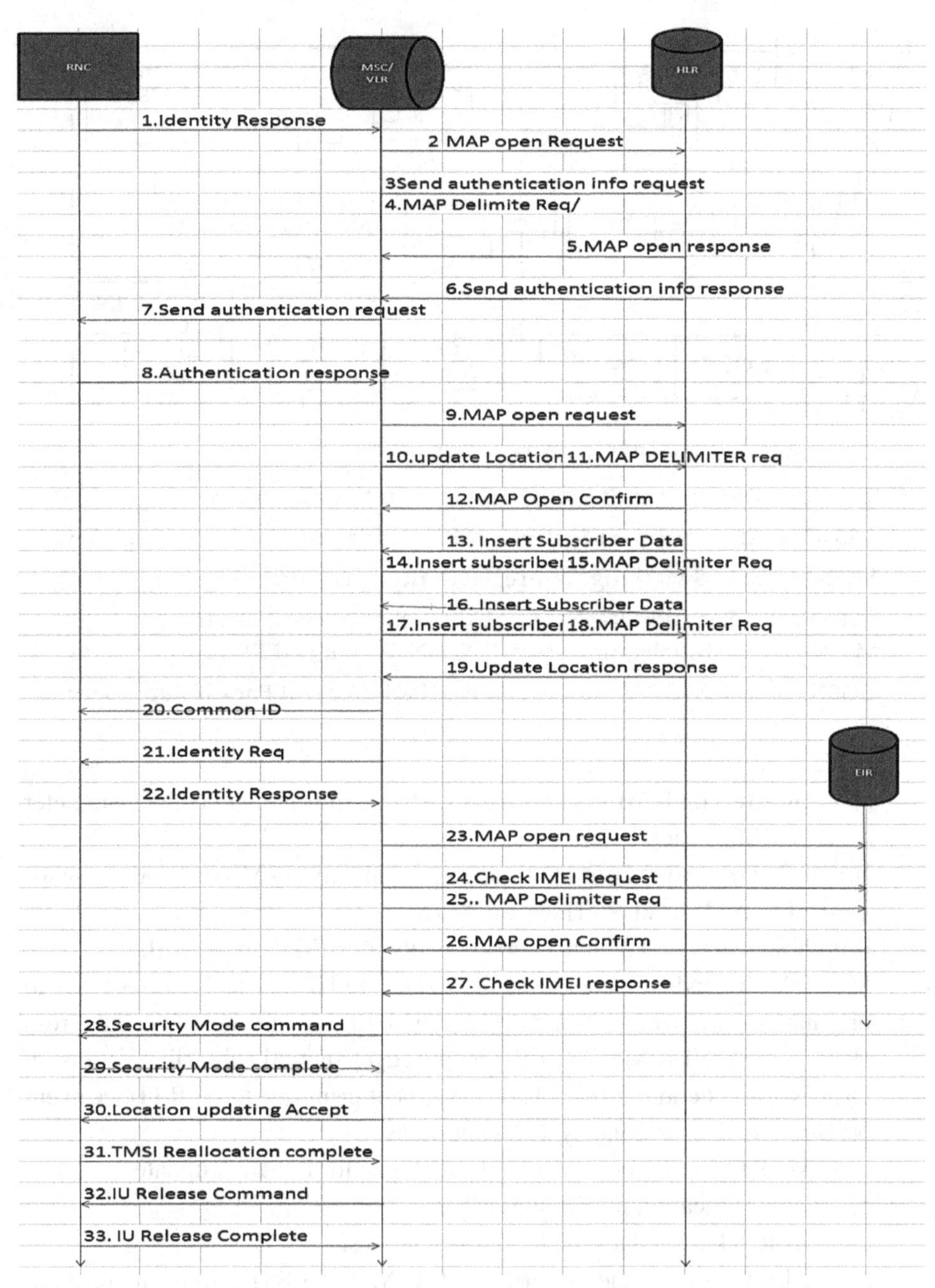

7. Then the Equipment Identity is to be checked by the MSC/VLR from EIR .For this MSC send MS an identity request via Common ID and Identity request message. (Trace No 20 to 21)

8. MS responds MSC/VLR by sending the IMEI via Identity response message. (trace event No:22)
9. Then IMEI is to be checked by the MSC/VLR from EIR (Equipment Identity Register) to confirm the IMEI status (whether it is white or Grey or Black listed). This is done by as follows. (Trace Event No: 23 to 27)
   a. MAP open request to EIR to Open dialogue
   b. MSC/VLR sends Check IMEI request to EIR
   c. EIR sends IMEI status to MSC/VLR (as white list)
10. The MSC/VLR sends the MS the security mode command to allocate TMSI to MS. In this command the Encrypted data is send to MS. MS responds by sending the Security command complete after choosing the Proper decryption algorithms. (Trace Event No 28 to 29)
11. MSC/VLR sends update location accepted to MS. (Trace event No: 30)
12. MS sends TMSI allocation complete to MSC/VLR to inform that the MS is successfully able to store TIMSI. (Trace Event No: 31)
13. The location update complete by releasing the air inter channel. (Trace Event No 32 to 33)

AUTHENTICATION IN NGN

NGN is designed to aggregate all types of telephone and data services e.g. PSTN, Broadband, PLMN, etc. (Kumar and Upadhyay, 2008). The block schematic is given in Figure 8.

The P-CSCF meant for broadband to access the IMS network through the Line Media Gateway (LMG) or SIP Access Gateway in general. The S-CSCF is for all fixed-line, mobile and NGN customers. The I-CSCF and S-CSCF implement routing and addressing. The S-CSCF obtains subscription data from the HSS and interacts with the application servers to provide services for subscribers. The P-CSCF is the first contact point for subscribers visiting the IMS network. A UE obtains the P-CSCF address based on configuration information or using the P-CSCF discovery mechanism. One CSCF unit is used for storing customer's contribution for supplementary services and termed as Advanced Telephony Server (ATS).Users from 2G and 3G mobile network (PLMN) and fixed line network (PSTN) communicates with NGN through media gateway (MGW) where MGCF plays vital role for all signaling co-ordinations. For users of PSTN and PLMN network, NGN relies upon the modality of authentication done by individual networks and do not revalidate. For subscribers of NGN network, authentication is carried out (Singh, 2007).

Figure 7. Events trace for mobile authentication

No.	Time	Type	Event	Direction	OfficeId	IMSI
1	11/13/2014 18:41:30	RANAP	Identity Response	RECEIVE	1101(RNCCEN1)	404811260420131.00
2	11/13/2014 18:41:30	VLRMAP	MAP OPEN REQUEST	SEND		404811260420131.00
3	11/13/2014 18:41:30	VLRMAP	Send Authentication Info Request	SEND		404811260420131.00
4	11/13/2014 18:41:30	VLRMAP	MAP DELIMITER REQUEST	SEND		404811260420131.00
5	11/13/2014 18:41:30	VLRMAP	MAP OPEN CONFIRM	RECEIVE		404811260420131.00
6	11/13/2014 18:41:30	VLRMAP	Send Authentication Info Response	RECEIVE		404811260420131.00
7	11/13/2014 18:41:30	RANAP	Authentication Request	SEND	1101(RNCCEN1)	404811260420131.00
8	11/13/2014 18:41:31	RANAP	Authentication Response	RECEIVE	1101(RNCCEN1)	404811260420131.00
9	11/13/2014 18:41:31	VLRMAP	MAP OPEN REQUEST	SEND		404811260420131.00
10	11/13/2014 18:41:31	VLRMAP	Update Location Request	SEND		404811260420131.00
11	11/13/2014 18:41:31	VLRMAP	MAP DELIMITER REQUEST	SEND		404811260420131.00
12	11/13/2014 18:41:31	VLRMAP	MAP OPEN CONFIRM	RECEIVE		404811260420131.00
13	11/13/2014 18:41:31	VLRMAP	Insert Subscriber Data Request	RECEIVE		404811260420131.00
14	11/13/2014 18:41:31	VLRMAP	Insert Subscriber Data Response	SEND		404811260420131.00
15	11/13/2014 18:41:31	VLRMAP	MAP DELIMITER REQUEST	SEND		404811260420131.00
16	11/13/2014 18:41:31	VLRMAP	Insert Subscriber Data Request	RECEIVE		404811260420131.00
17	11/13/2014 18:41:31	VLRMAP	Insert Subscriber Data Response	SEND		404811260420131.00
18	11/13/2014 18:41:31	VLRMAP	MAP DELIMITER REQUEST	SEND		404811260420131.00
19	11/13/2014 18:41:31	VLRMAP	Update Location Response	RECEIVE		404811260420131.00
20	11/13/2014 18:41:31	RANAP	Common ID	SEND	1101(RNCCEN1)	404811260420131.00
21	11/13/2014 18:41:31	RANAP	Identity Request	SEND	1101(RNCCEN1)	404811260420131.00
22	11/13/2014 18:41:31	RANAP	Identity Response	RECEIVE	1101(RNCCEN1)	404811260420131.00
23	11/13/2014 18:41:31	MSCMAP	MAP OPEN REQUEST	SEND		404811260420131.00
24	11/13/2014 18:41:31	MSCMAP	Check IMEI Request	SEND		404811260420131.00
25	11/13/2014 18:41:31	MSCMAP	MAP DELIMITER REQUEST	SEND		404811260420131.00
26	11/13/2014 18:41:31	MSCMAP	MAP OPEN CONFIRM	RECEIVE		404811260420131.00
27	11/13/2014 18:41:31	MSCMAP	Check IMEI Response	RECEIVE		404811260420131.00
28	11/13/2014 18:41:31	RANAP	Security Mode Command	SEND	1101(RNCCEN1)	404811260420131.00
29	11/13/2014 18:41:31	RANAP	Security Mode Complete	RECEIVE	1101(RNCCEN1)	404811260420131.00
30	11/13/2014 18:41:31	RANAP	Location Updating Accept	SEND	1101(RNCCEN1)	404811260420131.00
31	11/13/2014 18:41:32	RANAP	Tmsi Reallocation Complete	RECEIVE	1101(RNCCEN1)	404811260420131.00
32	11/13/2014 18:41:32	RANAP	Iu Release Command	SEND	1101(RNCCEN1)	404811260420131.00
33	11/13/2014 18:41:32	RANAP	Iu Release Complete	RECEIVE	1101(RNCCEN1)	404811260420131.00

Figure 8. Next generation network with interfaces for PSTN and PLMN

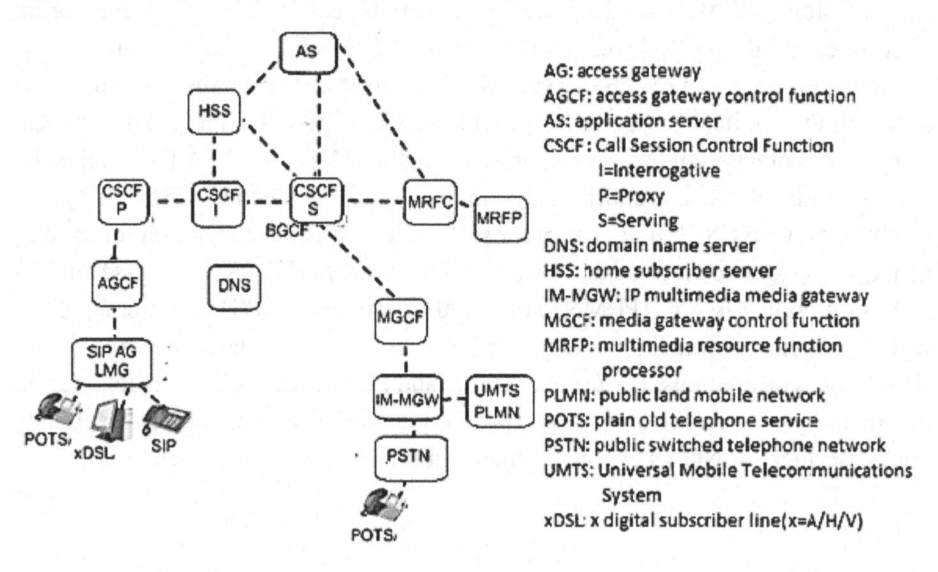

The S-CSCF supports the following authentication modes:

1. **Authentication of UE with USIM/ISIM:** IMS Authentication and Key Agreement (IMS AKA) authentication is used when UEs with IMS Subscriber Identity Module (ISIM) cards or Universal SIM (USIM) to access the IMS network.
2. **Authentication of PC Nodes without USIM/ISIM:** Digest authentication is used when SIP UEs access the IMS network. A SIP UE does not have SIM, USIM, or ISIM cards. It uses HTTP Digest authentication and accesses the IMS network using a user name and password (Franks et al., 1999; fielding and Reschke, 2014).
3. **Authentication with SIM from 2G Mobile Network:** It is used when packet switched (PS) subscribers of 2G network access the IMS network.

Authentication in NGN Network with SIM/USIM

Case 1: Call Flow Scenario from GSM to LMG POTS

Let us analyse a trace-route when a call is made from GSM to LMG POTS as shown in Figure 9.

As the call is made from outside network of IMS, the call request comes to IMS from MGCF as MGCF is responsible for signalling and media inter-working.

1. MGCF sends INVITE message to Core side (I-CSCF) to start the session.
2. I-CSCF queries HSS using LIR (Location-Info-Request) message to get the name of S-CSCF to which I-CSCF has to forward the request.
3. HSS responds to LIR with Location-Info-Answer (LIA) command, containing the name of the S-CSCF.
4. Request for session is forwarded to S-CSCF
5. S-CSCF communicates with ATS (Advanced Telephony Server). Supplementary services are facilitated by ATS.
6. The S-CSCF obtains the address of the called P-CSCF during registration of the callee.
7. The INVITE message is forwarded from S-CSCF to LMG where the called party is connected via P-CSCF.
8. After reception of INVITE, the Called Party (LMG) responds with "180 RINGING" message if the called party is not busy.
9. Calling Party MGCF (as GSM is routed via MGCF) acknowledges "180 RINGING" message with "PRACK" message.

Figure 9. Call trace for authentication from GSM to NGN

No.	Source Address	Source Eqpt Name	Source Port	Destination Address	Dest Eqpt Name	Destination Port	Message Type
1	55.191.54.22	MGCF	5060	55.191.54.7	Core side (I-CSCF)	31685	INVITE
2	55.191.54.7	Core side (I-CSCF)	31701	55.191.54.22	MGCF	5060	100 TRYING
3	55.191.54.7	Core side (I-CSCF)	3540	55.191.54.68	HSS Core side (FE)	3868	LIR
4	55.191.54.68	HSS Core side (FE)	3868	55.191.54.7	Core side (I-CSCF)	3540	LIA
5	55.191.54.8	Core side (S-CSCF)	31700	55.191.54.18	Core side (ATS)	5060	INVITE
6	55.191.54.18	Core side (ATS)	5060	55.191.54.8	Core side (S-CSCF)	31700	100 TRYING
7	55.191.54.18	Core side (ATS)	5060	55.191.54.8	Core side (S-CSCF)	31700	INVITE
8	55.191.54.8	Core side (S-CSCF)	31700	55.191.54.18	Core side (ATS)	5060	100 TRYING
9	55.191.55.134	Access side (P-CSCF)	31698	55.191.84.16	LMG Signalling IP	5060	INVITE
10	55.191.84.16	LMG Signalling IP	5060	55.191.55.134	Access side (P-CSCF)	31698	100 TRYING
11	55.191.84.16	LMG Signalling IP	5060	55.191.55.134	Access side (P-CSCF)	31698	180 RINGING
12	55.191.54.8	Core side (S-CSCF)	31700	55.191.54.18	Core side (ATS)	5060	180 RINGING
13	55.191.54.18	Core side (ATS)	5060	55.191.54.8	Core side (S-CSCF)	31700	180 RINGING
14	55.191.54.7	Core side (I-CSCF)	31701	55.191.54.22	MGCF	5060	180 RINGING
15	55.191.54.22	MGCF	5060	55.191.54.7	Core side (I-CSCF)	31702	PRACK
16	55.191.54.8	Core side (S-CSCF)	31700	55.191.54.18	Core side (ATS)	5060	PRACK
17	55.191.54.18	Core side (ATS)	5060	55.191.54.8	Core side (S-CSCF)	31700	PRACK
18	55.191.55.134	Access side (P-CSCF)	31698	55.191.84.16	LMG Signalling IP	5060	PRACK
19	55.191.84.16	LMG Signalling IP	5060	55.191.55.134	Access side (P-CSCF)	31698	200 OK
20	55.191.54.8	Core side (S-CSCF)	31700	55.191.54.18	Core side (ATS)	5060	200 OK
21	55.191.54.18	Core side (ATS)	5060	55.191.54.8	Core side (S-CSCF)	31700	200 OK
22	55.191.54.7	Core side (I-CSCF)	31702	55.191.54.22	MGCF	5060	200 OK
23	55.191.84.16	LMG Signalling IP	5060	55.191.55.134	Access side (P-CSCF)	31698	200 OK
24	55.191.54.8	Core side (S-CSCF)	31700	55.191.54.18	Core side (ATS)	5060	200 OK
25	55.191.54.18	Core side (ATS)	5060	55.191.54.8	Core side (S-CSCF)	31700	200 OK
26	55.191.54.7	Core side (I-CSCF)	31701	55.191.54.22	MGCF	5060	200 OK
27	55.191.54.22	MGCF	5060	55.191.54.7	Core side (I-CSCF)	31702	ACK
28	55.191.54.8	Core side (S-CSCF)	31700	55.191.54.18	Core side (ATS)	5060	ACK
29	55.191.54.18	Core side (ATS)	5060	55.191.54.8	Core side (S-CSCF)	31700	ACK
30	55.191.55.134	Access side (P-CSCF)	31698	55.191.84.16	LMG Signalling IP	5060	ACK
31	55.191.84.16	LMG Signalling IP	5060	55.191.55.134	Access side (P-CSCF)	31698	INVITE
32	55.191.55.134	Access side (P-CSCF)	31698	55.191.84.16	LMG Signalling IP	5060	100 TRYING
33	55.191.54.8	Core side (S-CSCF)	31700	55.191.54.18	Core side (ATS)	5060	INVITE
34	55.191.54.18	Core side (ATS)	5060	55.191.54.9	Core side (MRFC)	31625	INVITE
35	55.191.54.9	Core side (MRFC)	31625	55.191.54.18	Core side (ATS)	5060	100 TRYING
36	55.191.54.18	Core side (ATS)	5060	55.191.54.8	Core side (S-CSCF)	31700	100 TRYING
37	55.191.54.9	Core side (MRFC)	2945	55.191.54.19	Core side (MRFP)	2953	ADD_REQ
38	55.191.54.19	Core side (MRFP)	2953	55.191.54.9	Core side (MRFC)	2945	ADD_REPLY
39	55.191.54.9	Core side (MRFC)	31625	55.191.54.18	Core side (ATS)	5060	200 OK
40	55.191.54.18	Core side (ATS)	5060	55.191.54.8	Core side (S-CSCF)	31700	INVITE
41	55.191.54.8	Core side (S-CSCF)	31700	55.191.54.18	Core side (ATS)	5060	100 TRYING
42	55.191.54.7	Core side (I-CSCF)	31702	55.191.54.22	MGCF	5060	INVITE
43	55.191.54.22	MGCF	5060	55.191.54.7	Core side (I-CSCF)	31702	100 TRYING
44	55.191.54.22	MGCF	5060	55.191.54.7	Core side (I-CSCF)	31702	200 OK
45	55.191.54.8	Core side (S-CSCF)	31700	55.191.54.18	Core side (ATS)	5060	200 OK
46	55.191.54.18	Core side (ATS)	5060	55.191.54.9	Core side (MRFC)	31626	ACK
47	55.191.54.18	Core side (ATS)	5060	55.191.54.9	Core side (MRFC)	31626	INFO
48	55.191.54.9	Core side (MRFC)	31626	55.191.54.18	Core side (ATS)	5060	200 OK
49	55.191.54.18	Core side (ATS)	5060	55.191.54.8	Core side (S-CSCF)	31700	ACK
50	55.191.54.7	Core side (I-CSCF)	31702	55.191.54.22	MGCF	5060	ACK
51	55.191.54.9	Core side (MRFC)	2945	55.191.54.19	Core side (MRFP)	2953	MOD_REQ
52	55.191.54.18	Core side (ATS)	5060	55.191.54.8	Core side (S-CSCF)	31700	200 OK
53	55.191.55.134	Access side (P-CSCF)	31698	55.191.84.16	LMG Signalling IP	5060	200 OK
54	55.191.84.16	LMG Signalling IP	5060	55.191.55.134	Access side (P-CSCF)	31698	ACK
55	55.191.54.8	Core side (S-CSCF)	31700	55.191.54.18	Core side (ATS)	5060	ACK
56	55.191.54.19	Core side (MRFP)	2953	55.191.54.9	Core side (MRFC)	2945	MOD_REPLY
57	55.191.54.9	Core side (MRFC)	2945	55.191.54.19	Core side (MRFP)	2953	MOD_REQ
58	55.191.54.19	Core side (MRFP)	2953	55.191.54.9	Core side (MRFC)	2945	MOD_REPLY
59	55.191.54.22	MGCF	5060	55.191.54.7	Core side (I-CSCF)	31702	BYE
60	55.191.54.8	Core side (S-CSCF)	31700	55.191.54.18	Core side (ATS)	5060	BYE
61	55.191.54.18	Core side (ATS)	5060	55.191.54.9	Core side (MRFC)	31626	BYE
62	55.191.54.9	Core side (MRFC)	31626	55.191.54.18	Core side (ATS)	5060	200 OK
63	55.191.54.18	Core side (ATS)	5060	55.191.54.8	Core side (S-CSCF)	31700	BYE
64	55.191.55.134	Access side (P-CSCF)	31698	55.191.84.16	LMG Signalling IP	5060	BYE
65	55.191.54.18	Core side (ATS)	5060	55.191.54.8	Core side (S-CSCF)	31700	200 OK
66	55.191.54.7	Core side (I-CSCF)	31702	55.191.54.22	MGCF	5060	200 OK
67	55.191.54.9	Core side (MRFC)	2945	55.191.54.19	Core side (MRFP)	2953	SUB_REQ
68	55.191.84.16	LMG Signalling IP	5060	55.191.55.134	Access side (P-CSCF)	31698	200 OK
69	55.191.54.8	Core side (S-CSCF)	31700	55.191.54.18	Core side (ATS)	5060	200 OK
70	55.191.54.19	Core side (MRFP)	2953	55.191.54.9	Core side (MRFC)	2945	SUB_REPLY

10. Subsequently when Called party (LMG) answers the call, it sends "200 OK" message to MGCF (Calling Party).
11. The Calling client acknowledges that the Called party has answered by issuing a "ACK" message.
12. At this point media (voice, data, video) are exchanged. Different messages (like ADD_REQ, ADD_REPLY, MOD_REQ, MOD_REPLY, etc.) are exchanged among MGCF, MRFP, MRFC and core elements like S-CSCF, ATS of IMS to establish and control the media streams.
13. Once the conversation is over, one of the parties hangs up, which causes a "BYE" message to be sent. In the present case, calling party (GSM) hangs up and hence MGCF sends "BYE" message to LMG.
14. The called party (LMG) acknowledges "BYE" message by sending "200 OK" response.

The Multimedia Resource Function Processor (MRFP) controls the bearer on the Mb interface including processing the media streams (e.g. audio transcoding). The Multimedia Resource Function Controller (MRFC) interprets signaling information from an S-CSCF or a SIP-based Application Server and controls the media streams resources in the MRFP accordingly.

Case 2: IMS Authentication and Key Agreement (IMS AKA)

The Authentication process shown in Figure 10 with trace events in figure 11 are explained below:

1. User Equipment (UE) initiates a register request to P-CSCF (Proxy Call Session Control Function). P-CSCF check the IMPU (IP Multimedia Public Id) and IMPI (IP Multimedia Private ID) home network. The main function of P-CSCF is to obtain the Address of I-CSCF. This is done by sending a query request to DNS server and subsequent reception of response from the DNS. The register request (SIP register) is forwarded to I-CSCF. (Trace Event No: 1)
2. The I-CSCF makes query to HSS (Home Subscriber Server) to obtain the serving CSCF i.e. S-CSCF. This message is called UAR (User Authentication Request).If there are more than one HSS is present, the UAR is send to SLF (Subscriber Location Function) (Trace event no 2) to obtain the actual HSS .This address is received vis UAA (user authentication Answer). (Trace Event NO 3).Then the UAR message is send to the particular HSS and the response is obtained thereafter. (Trace Event No 4 and 5). The I-CSCF sends the Register message to S-CSCF.

Figure 10. Authentication process for NGN UE with ISIM in NGN network

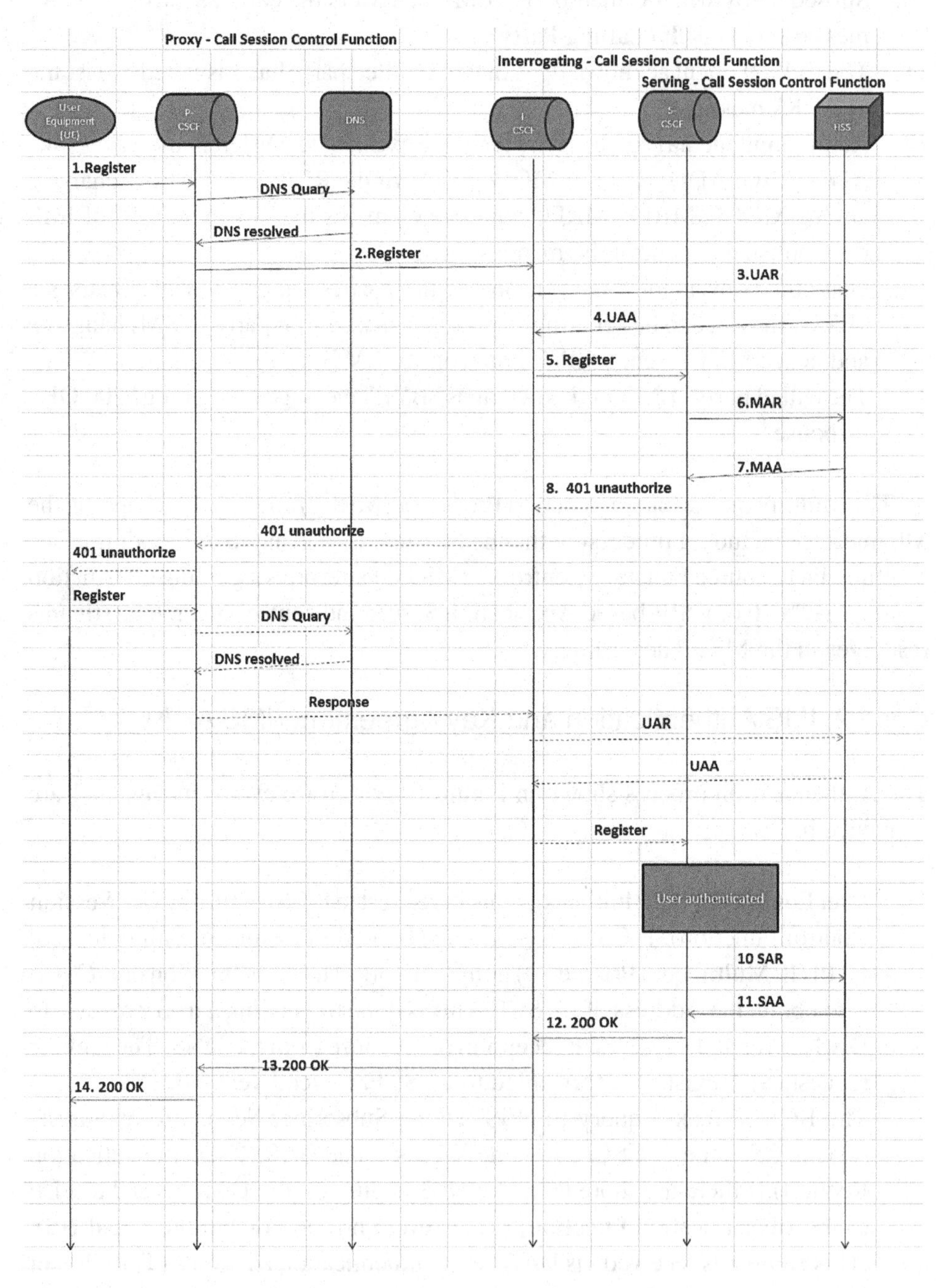

3. The S-CSCF then sends Multimedia Authentication Request MAR to HSS for Multimedia Authentication Request to obtain the Authentication vectors (AV). (trace Event No 6)

a. RAND Random number used to generate XRES, AUTN, CK, IK .
b. AUTN authentication Token
c. CK Cipher Key
d. IK integrity key
e. Xres Expected result from UE.
f. HSS give response with AV to S-CSCF.

4. S-CSCF send an Unauthorized (401 Unauthorized message) to I-CSCF along with AV and the I-CSCF sends this message to P-CSCF. The P-CSCF removes the CK and IK and sends the rand to UE. (Trace event No 7)
5. The UE then resend the register message with authentication parameter Like Xres to P-CSCF, I-CSCF and S-CSCF via HSS. (same process as first message but along with Xres).Trace No 8 to 13
6. The Xres is checked at S-CSCF to authenticate the UE.
7. In Trace No. 14 '423 Interval too brief' error message is encountered and Registration process failed. This is due to some network or other issues. (Trace event No 14)
8. The actual registration process reinitiated after this and step from 1 to 6 is repeated for trace event no 15 to 25
9. The User Authentication is done at S-CSCF and SCSCF resend SAR (Server Assignment Request) (Trace Event No 26) to HSS .In response HSS returns SAA (Server Assignment Answer) in which the service profile is send. This Service profile is downloaded to S-CSCF. (Trace Event No 30)
10. The Registration acknowledgement in the form of 200OK is send to UE from S-CSCF via I-CSCF and P-CSCF. (Trace event No 27).
11. S-CSCF sends a register message to AS (application Server) and a 200 Ok i.e. acknowledgement message received from AS. (Trace Event No28-29)

Exchange of message types between different elements are shown in Figure 11.

AUTHENTICATION PROCESS FOR SIP FOR SESSION WITHOUT USIM/ISIM

For understanding, Authentication process for SIP, Let us analyse the trace route when a call is made from 24441114 to 24441113.

In a SIP based network, Authentication can take place between a user agent and a server (proxy, registrar and user agent server) where a server requires a user agent to authenticate itself before processing the request. Similarly a user agent can request authentication of sever (known as Mutual Authentication).

In the present case, mutual authentication takes place between client and server.

Figure 11. Authentication trace for IMS UE in IMS network

No.	TimeStamp	Source Address	Source Port	Destination Address	Destination Port	Message Type
1	11:29:56.783	55.191.116.36	5060	55.191.103.134	31633	REGISTER
2	11:29:56.811	55.191.102.7	3540	55.191.102.77	3868	UAR
3	11:29:56.895	55.191.102.77	3868	55.191.102.7	3540	UAA
4	11:29:56.895	55.191.102.7	3543	55.191.102.68	3868	UAR
5	11:29:56.979	55.191.102.68	3868	55.191.102.7	3543	UAA
6	11:29:56.979	55.191.102.8	4541	55.191.102.68	3868	MAR
7	11:29:57.039	55.191.102.68	3868	55.191.102.8	4541	MAA
8	11:29:57.039	55.191.103.134	31649	55.191.116.36	5060	401 UNAUTHORIZED
9	11:29:57.071	55.191.116.36	5060	55.191.103.134	31633	REGISTER
10	11:29:57.075	55.191.102.7	3541	55.191.102.77	3868	UAR
11	11:29:57.159	55.191.102.77	3868	55.191.102.7	3541	UAA
12	11:29:57.159	55.191.102.7	3540	55.191.102.68	3868	UAR
13	11:29:57.231	55.191.102.68	3868	55.191.102.7	3540	UAA
14	11:29:57.231	55.191.103.134	31649	55.191.116.36	5060	423 INTERVAL TOO BRIEF
15	11:29:57.287	55.191.116.36	5060	55.191.103.134	31633	REGISTER
16	11:29:57.292	55.191.102.7	3540	55.191.102.77	3868	UAR
17	11:29:57.375	55.191.102.77	3868	55.191.102.7	3540	UAA
18	11:29:57.375	55.191.102.7	3541	55.191.102.68	3868	UAR
19	11:29:57.447	55.191.102.68	3868	55.191.102.7	3541	UAA
20	11:29:57.447	55.191.103.134	31649	55.191.116.36	5060	401 UNAUTHORIZED
21	11:29:57.483	55.191.102.7	3540	55.191.102.77	3868	UAR
22	11:29:57.479	55.191.116.36	5060	55.191.103.134	31633	REGISTER
23	11:29:57.555	55.191.102.77	3868	55.191.102.7	3540	UAA
24	11:29:57.555	55.191.102.7	3541	55.191.102.68	3868	UAR
25	11:29:57.639	55.191.102.68	3868	55.191.102.7	3541	UAA
26	11:29:57.640	55.191.102.8	4540	55.191.102.68	3868	SAR
27	11:29:57.640	55.191.103.134	31649	55.191.116.36	5060	200 OK
28	11:29:57.640	55.191.102.8	31652	55.191.102.18	5060	REGISTER
29	11:29:57.651	55.191.102.18	5060	55.191.102.8	31652	200 OK
30	11:29:57.711	55.191.102.68	3868	55.191.102.8	4540	SAA

SIP provides a challenge-based mechanism for Authentication known as Digest Authentication. By digest authentication, when the client initializes a connection with a proxy server, the proxy responds with 407 Proxy Authentication Required to challenge the UAC (User agent client).

The above scenario can be observed in the present trace route where the client (55.191.84.19 - IP Address of LMG) sent INVITE request message to server(55.191.55.134 - IP Address of Access side CSCF) to initiate the session and the server responds with 407 Proxy Authentication Required to challenge the client.

SIP uses headers for authentication. The Authenticate header field in 407 Proxy Authentication Required response includes the following parameters.

- **Digest:** Indicator of Authentication scheme.
- **Realm:** Associated Protection domain (In present case "dartec.ims.ca")
- **QoP:** Specifying quality of Protection, in present case its value is "auth" (authentication).
- **Nonce:** Unique string specified by server. SIP defines a nonce value to authenticate the client to the server. A recommended digest implementation should generate nonce value with at least a digest of Client IP Address and Time-stamp. The digest containing time-stamp as described above allows the same nonce value to be used repeatedly in the subsequent transactions as long as the time-stamp is expired.
- **Algorithm:** The algorithm used for checksum calculation. The default value is MD5.

After the receipt of 407 Proxy Authentication the Client acknowledges the receipt of 407 challenge by sending ACK message to the server.

Figure 12. Authentication in call scenario between POTS in NGN core network

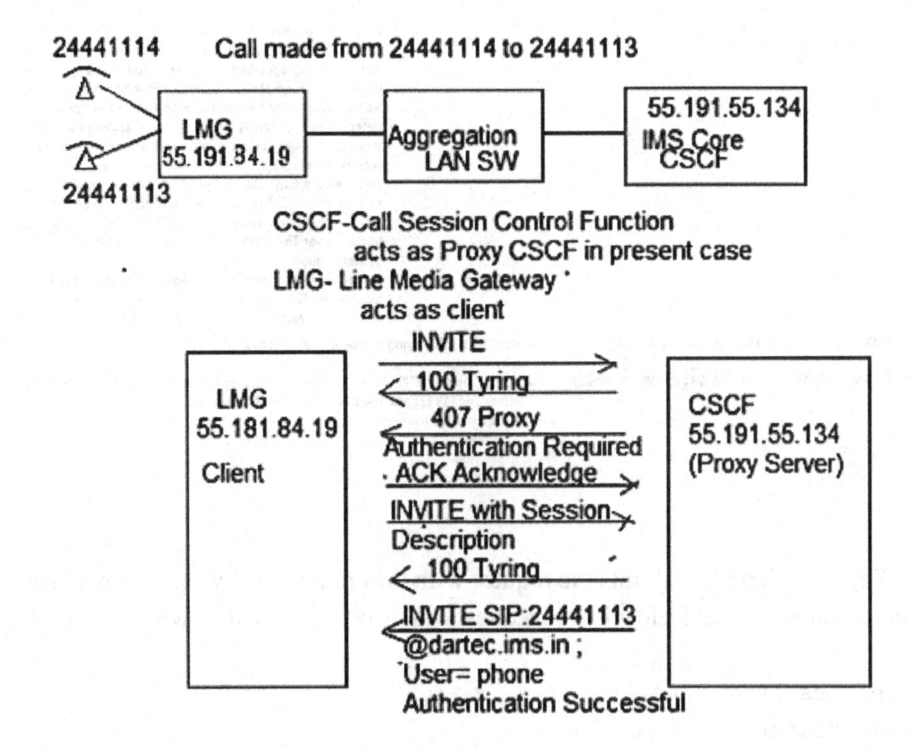

Figure 13. Simplified steps of Authentication process for POTS in NGN

No.	Time	Source	Destination	Protocol	Length	Info	Explanation
3	8.392289	55.191.84.19	55.191.55.134	SIP/SDP	1111	Request: INVITE tel:24441113, with session description	Client initiates the session
4	8.403073	55.191.55.134	55.191.84.19	SIP	302	Status: 100 Trying	Request for session is in process
5	8.404468	55.191.55.134	55.191.84.19	SIP	586	Status: 407 Proxy Authentication Required	When a client initializes a connection with proxy server,the proxy responds with 407 Proxy Authentication Required to challenge the UAC (User Agent Client). This 407 Proxy Authentication Required contains header which includes following Parameters: Digest: Indicator of Authentication scheme. realm: Associated Protection domain(In present case "dartec.ims.ca") qop: Specifying quality of Protection, In present case its value is "auth" (authentication). nonce: Unique string specified by server. SIP defines a nonce value to authenticate the client to the server.A recommended digest implementation should generate nonce value with atleast a digest of Client IP Address and Time-stamp.The digest containing time-stamp as described above allows the same nonce value to be used repeatedly in the subsequent transactions as long as the time-stamp is expired. algorithm: The algorithm used for checksum calculation. The default value is MD5.
6	8.415346	55.191.84.19	55.191.55.134	SIP	375	Request: ACK tel:24441113	Client acknowledges the receipt of 407 challenge.
7	8.420388	55.191.84.19	55.191.55.134	SIP/SDP	1338	Request: INVITE tel:24441113, with session description	After the client receives 407 challenge from server,it re-submits the request with the credentials by including a Proxy-Authorization header field with the request which includes the following parameters: Digest username=24441114 realm="dartec.ims.ca" nonce =String specified by server in authentication header. nc = Count of no of requests that client has sent with the nonce value. nonce-count(nc) is to enable the server to detect message replay attacks by maintaining own copy of this count. If the same nc value appears twice,then the request is a replay. cnonce= If a qop value is specified ,Client should specify cnonce value. SIP defines optional cnonce parameter whose value is generated and stored by the client and sent to the server.This parameter is used for server authentication. response = The string 32 hex digits calculated checksum algorithm=MD5 (explained in previous message) qop=auth (explained in previous message)
8	8.426832	55.191.55.134	55.191.84.19	SIP	302	Status: 100 Trying	Request is in process
9	8.620609	55.191.55.134	55.191.84.19	SIP/SDP	1455	Request: INVITE sip:24441113@dartec.ims.ca;user=phone, with session description	The client is authenticated successfully.The server forwards the INVITE request to next level.

Then the client re-submits the request with the credentials by including a Proxy-Authorization header field with the request which includes the following parameters:

```
Digest username=24441114
realm="dartec.ims.ca"
nonce =String specified by server in authentication header.
```

```
nc = Count of no of requests that client has sent with the
nonce value. nonce-count(nc) is to enable the server to detect
message replay attacks by maintaining own copy of this count.
If the same nc value appears twice, then the request is a re-
play.
cnonce= If a qop value is specified, Client should specify
cnonce value. SIP defines optional cnonce parameter  whose
value is generated  and stored by the client and sent to the
server. This parameter is used for server authentication.
response = The string 32 hex digits calculated checksum
algorithm=MD5
qop=auth
```

Once the Client is authenticated successfully, the server forwards the invite request to the next level.

The salient points of the authentication trace is given in Figure 13.

CONCLUSION

User authentication in Next Generation Network has been discussed above from application viewpoint. The protocols implemented at background and transport mechanism are required to follow strict regulation guidelines i.e. 6 sec for real-time (voice) functions and 20 sec for data service functions. Further robust authentication procedures are possible with higher speed of hardware implementation and development of efficient algorithms.

REFERENCES

Fielding, R., & Reschke, J. (Eds.). (2014). *Request for Comments: 7235 Hypertext Transfer Protocol (HTTP/1.1): Authentication*. Internet Engineering Task Force (IETF). Retrieved from https://tools.ietf.org/html/rfc7235

Forouzan, B. (2011). *Data Communications and Networking*. New Delhi, India: Tata McGraw Hill Education Private Limited.

Franks, J., Hallam-Baker, P., Hostetler, J., Leach, P., Lawrence, S., Luotonen, A., & Stewart, L. (1999). *Request for Comments: 2617 HTTP Authentication: Basic and Digest Access Authentication.* Available from http://tools.ietf.org/html/rfc2617

Garg, V. K., & Wikes, J. E. (2005). *Principles and Applications of GSM*. Singapore: Pte. Ltd.

Kaushal, R., & Abhay, V. (2008). *HLR Interface to AUC*. Stockholm: Ericsson AB.

Kumar, V., & Upadhyay, P. P. (2008). Next generation Network. *Telecommunications, 58*(2), 18–27.

Manna, G.C. (2010, August). *Quality of Service Monitoring*. Course material conducted at Information and Communication Authority (ICTA), Port Louis, Mauritius.

Singh, P. K. (2007). *NGN Security. In Compendium on Next Generation Networks* (pp. 27–28). Delhi, India: The Gondals Press.

van Bosse, J. G. (1998). *Signaling in Telecommunication Networks*. New York, NY: John Wiley & Sons, Inc.

Chapter 9
Cryptographic Algorithms for Next Generation Wireless Networks Security

Vishnu Suryavanshi
GHRCE Nagpur, India

G. C. Manna
BSNL, India

ABSTRACT

At present a majority of computer and telecommunication systems requires data security when data is transmitted the over next generation network. Data that is transient over an unsecured Next Generation wireless network is always susceptible to being intercepted by anyone within the range of the wireless signal. Hence providing secure communication to keep the user's information and devices safe when connected wirelessly has become one of the major concerns. Quantum cryptography algorithm provides a solution towards absolute communication security over the next generation network by encoding information as polarized photons, which can be sent through the air security issues and services using cryptographic algorithm explained in this chapter.

INTRODUCTION

Does increased security provide comfort to paranoid people? Or does security provide some very basic protections that we are naive to believe that we don't need? During this time when the Internet provides essential communication between tens of millions of people and is being increasingly used as a tool for commerce, security

DOI: 10.4018/978-1-4666-8687-8.ch009

becomes a tremendously important issue to deal with. There are many aspects to security and many applications, ranging from secure commerce and payments to private communications and protecting passwords. One essential aspect for secure communications in next generation wireless network is that of cryptography, which the focus of this chapter is.

Security in computer world determines the ability of the system to manage, protect and distribute sensitive information (Abdel-Karim R. Al Tamimi 2006). The most attractive and fast growing network is 802.11 in wireless networks. In 1997IEEE 802.11 introduced standards for wireless local network (WLAN) communication, some of these standards are:

- Using the 2.4 GHz radio spectrum and 11 Mbps max data rate is 802.11b.
- Using the 5 GHz radio spectrum and 54 Mbps max data rate is 802.11a.
- Using the 2.4 GHz radio spectrum and 54 Mbps max data rate is 802.11g.

Wireless Robust Security Network is 802.11i (Quality of service). It is used in quality of service for traffic prioritization to give delay sensitive application such as multimedia and voice communication priority(SANS, 2005).Next generation wireless technology 3G, 4G and more has been gaining rapid popularity in recent years.

They have ubiquitous wireless communications and services as Integration of multi-networks is using IP technology; similar technology to the wired Internet where users are freed from their local networks, not just IP end-to-end but over-the-air packet switching, high bandwidth / high-speed wireless and highly compatible with wired network infrastructures like ATM, IP.

These technologies are facing security problems in the software products used to access the vast Internet, operating systems, www browsers and e-mail programs(Chandra, et al., 2008).For secure data transformation cryptographic algorithms plays a key role.

A cryptographic technique provides three forms of security namely confidentiality, data integrity and authentication. Confidentiality refers to protection of information from unauthorized access (Daemean & Rijmen, 1999). Information has not been manipulated in any unauthorized way is ensured by data integrity. Authentication can be explained in two groups as entity authentication and message authentication. Detecting any modifications to the message provides message authentication. Entity authentication assures the receiver of a message, about both the identity of the sender and his active participation (Kumar & Purohit, 2010)

Need of a standard depends on the ease of use and level of security which it provides. Here, the distinction between wireless usage and security standards show that the security is not maintained well up to with the growth past of end user's usage. The hackers monitor and even change the integrity of transmitted data in current wireless technology. Lack of rigid security standards has caused companies to invest millions on securing their wireless networks.

Securing Next Generation Wireless Networks is an extremely challenging and interesting area of research. Unprotected wireless networks are vulnerable to several security attacks including eavesdropping and jamming that have no counterpart in wired networks. Moreover, many wireless devices are resource limited, which makes it challenging to implement security protocols and mechanisms.

The main objective of this chapter to study and analyze use of Cryptographic Algorithms for Next Generation wireless networks Security in terms confidentiality, Confidentiality, Integrity, Availability, Anti-virus, anti-spyware software, firewall, Authentication, Access control, and Cryptanalysis.

BACKGROUND

Information is an important asset and resource for business and needs to be protected like any other asset. The protection of information is usually known as information security. A basic and classical model of security objectives include the so called CIA triad which stands for confidentiality, integrity and availability. In this chapter we tried to notify security issues and services and use of cryptographic algorithms.

1. SECURITY ISSUES AND SERVICES

Signal fading, mobility, data rate enhancements, minimizing size and cost, user security and (Quality of service) QoS are the key challenges in wireless networks (Kumar & Jain, 2012).Handheld devices which are used in embedded application have not generally been viewed as posing security threats, their increased computing power and the ease with which they can access networks and exchange data with other handheld devices introduce new security risks to an agency's computing environment. This section describes how the security requirements for confidentiality, integrity, authenticity, and availability for handheld device computing environments can be threatened. In a sizable geographic area, Wireless Mesh Networks (WMNs) presents a good solution to provide wireless Internet connectivity (Siddiqui & Seon, 2007).

2. SECURITY ISSUES

2.1 Loss of Confidentiality

Confidentiality assures only the knowing recipients is accessible to the information transmitted across the network. On the handheld device, the storage module, or the PC or while being sent over one of the Bluetooth, 802.11, IR, USB, or serial communication ports; confidentiality of information can be compromised. Moreover, most handheld devices are shipped with connectivity that is enabled by default. These default configurations are typically not in the most secure setting and should be changed to match the agency's security policy before being used.

Confidentiality is the privacy of a useful thing. Specifically, confidentiality can be defined as which people, under what conditions are authorized to access a useful thing. The confidentiality of this information is extremely important because the subjection of this information could bring embarrassment and heavy penalties. Confidentiality can be achieved through strong asymmetric cryptographic solutions in wired networks (Anjum & Salil, 2009)

2.2 Loss of Integrity

Information secrecy, data integrity and resource availability of users are to be provided by security services. To prevent improper modification of data and resource availability preventing denial of services is data integrity (Liang & Wang, 2004).

The integrity of the information on the handheld device and of the handheld device hardware, applications, and underlying operating system are also security concerns. Information stored on, and software and hardware used by, the handheld device must be protected from unauthorized, unanticipated, or unintentional modification. Information integrity requires that a third party be able to verify that the content of a message has not been changed in transit and that the origin or the receipt of a specific message be verifiable by a third party

Integrity is evaluated by two primary properties. First, there is the notion that a useful thing should be trusted; that is, there is an expectation that a useful thing will only be modified in appropriate ways by appropriate people. For example, a hospital patient's allergy information is stored in a database. The doctor should be able to trust that the allergy information is correct and up-to-date.

If data is damaged or incorrectly altered by authorized or unauthorized personnel then you must consider how important it is that the data be restored to a trustworthy state with minimum loss; which is the second part of integrity. For example, suppose a nurse who is authorized to view and update a patient's allergy information is upset with his/her employer and wants to disrupt a patient's data to make the hospital look bad. How important is it that the hospital be able to catch this error and trace it back to the person(s) who caused it?

Data integrity is guaranteed because required keys may be generated during the authentication process for data encryption and message authentication.

2.3 Loss of Availability

In the modeling and design of fault tolerant wireless systems, availability for wireless mobile systems has presented great challenges. There are high expectations from customer for level of availability and performance from wireless communication system as with the rapid growth of wireless communication services (Kishor, Trivedi, Ma & Dharmaraja, 2003).

Authorized person, entity, or device can access a useful thing represented by availability. For example, an organization has a system which provides authentication services for critical systems, applications, and devices on campus. An interruption in this service could mean the inability for customers to access computing resources and staff to access the resources they need to perform critical tasks. Therefore a loss of the service could quickly translate into a large financial loss in lost employee time and potential customer loss due to inaccessibility of resources. Because of this, the availability of this authentication system would be considered 'High'.

The outage-and-recovery of its supporting functional units can affect the performance and availability of a wireless system.

2.4 Anti-Virus and Anti-Spyware Software, and a Firewall

Now a day's antivirus solution became a normal component of computer system. Just like other components or services of computer system antivirus software can be targeted. Anti-virus software is the most cumbersome implementation. It has to pass with hundreds of file types and formats like executables, documents, compressed archives, executable packers and media files. Such formats are quite complex. Hence to implement such software on these formats is extremely difficult (Feng, 2008). Install anti-virus and anti-spyware software, and keep them up-to-date.

For any program that tracks user's online activities and secretly transmits information to a third party is a spyware. Annoying interruptions like pop-ups ads to security breaches and loss of intellectual property are the effects of spyware (Webroot Software Inc., 2004)

Unauthorized access to or from a network is prevented by firewall which is hardware or software system. It can be implemented in both hardware and software, or a combination of both. Firewalls are used to prevent unauthorized Internet users from accessing private networks connected to the Internet. All data entering or leaving the Intranet pass through the firewall, which examines each packet and blocks those that do not meet the specified security criteria (Okumoku-Evroro & Oniovosa, 2005) If your firewall was shipped in the "off" mode, turn it on.

3. SECURITY SERVICES

3.1 Authentication

Entity authentication (or "peer entity authentication" as it is referred to in ISO 7498-2) provides corroboration to one entity that another entity is as claimed. This service provides confidence, at the time of use only, that an entity is not attempting to impersonate another entity or that the current communications session is an unauthorized replay of a previous connection.

To protect credit card transactions on the Internet the Secure Electronic Transaction is used. IBM, Microsoft, Netscape, RSA, Terisa and Verisign these companies are collaborated in the development of SET. This system is designed for

Wired networks and does not meet all the challenges of wireless network. Multifactor Authentication techniques can be used to provide secure web transactions using cell phones. This multifactor technique is based on TIC's and SMS confirmation (Tiwari & Sudip, 2007)

Authentication is a process which a user gains the right to identify him or her. Passwords, biometric techniques, smart cards, certificates, etc. are the key techniques to authenticate a user. These techniques namely come under Multifactor Authentication techniques. Usually within one institute, a user may have a single identity; however, if a user has rights to identify himself in several different organizations or systems, more than one identity from a person may cause problems.

There are four major scenarios based on different degrees of trust:

1. The right of an individual to self-determination as to the degree to which personal information will be shared among other individuals or organizations to control the collection, storage, and distribution of personal or organizational information.

2. The right of an individual to self-determination as to the degree to which the individual is willing to share with others information about himself that may be compromised by unauthorized exchange of such information among other individuals or organizations
3. The right of individuals and organizations to control the collection, storage, and distribution of their information or information about themselves.
4. The right of individuals to control or influence what information related to them may be collected and stored and by whom and to whom that information may be disclosed.

Authentication mechanisms differ in the assurances they provide:

1. Data was generated by the Principal at some point in the past.
2. The Principal was present when the data was sent.
3. The data received was freshly generated by the Principal.

Mechanisms also differ in the number of verifiers:

1. Support for single verifier per message.
2. Support for multiple verifiers.

Whether the mechanism supports the ability of the verifier to prove to a third party that the message originated with the Principal. We divide the authentication policy into three major categories:

1. Personal/system;
2. Internet;
3. Network authentication.

3.2 Access Control

To control the flow of information between subject and object where subject is always an active entity while object is a passive entity this mechanism is called access control. Access control is a three-step process which includes identification, authentication and authorization. There are three access control modes which have their own merits and demerits. These are

- Discretionary Access Control (DAC).
- Mandatory Access Control (MAC).
- Role Based Access Control (RBAC).

The first step in any access control solution is identification or authentication.

Authentication are often discussed in terms of "factors" of proof, such as a PIN, a smart card and a fingerprint. Access Control Techniques and Technologies are Rule Based Access Control, Menu Based Access Control (Vinay, 2007).

Over the years security practitioners have developed a number of abstractions in dealing with access control. Protection of objects is the crucial requirement, which in turn facilitates protection of other resources controlled via the computer system. Access control provides a secure solution for web services (WS). To find syntactic and semantic errors administrators of WS can specify access control policies and validate them (Yague & Javier, 2005).

- For the transmission of data the users belonging to the same multicast session form a Data Group (DG). One DG contains the users that can access to a particular resource. According to access privilege the users are also divided into non-overlapping Service Groups (SG).
- Access control on manipulation of resources via "Hyper Text Transfer Protocol" Adam's user agent attempts the reading, writing, or deletion of an information resource identified by a Universal Research Locator (URL);
- Adam's user agent attempts the use of a processing resource to execute programs.

3.3 Data Confidentiality

Robust Security Network Association (RSNA) provides two data confidentiality protocols, called the Temporal Key Integrity Protocol (TKIP) and the Counter-mode/ CBC-MAC Protocol (CCMP). Confidentiality protection schemes are:

- **Data Privacy:** The data produced by each sensor node should be only known to itself.
- **Data Confidentiality:** In addition to data privacy, partially or fully aggregated data should only be known by the sink.
- **Efficiency:** After the confidentiality protection schemes are introduced, the system overhead should be kept as small as possible (Taiming, 2001).

Guidelines have been issued in a number of specific areas to help to protect the confidentiality of personal data held in the department.

1. The first stage in establishing policies and procedures to ensure the protection of personal data is to know:

a. What data is held?
b. Where it is held, and
c. What are the consequences would be when data is lost or stolen.

With that in mind, as a first step identifying the types of personal data held within the department, identifying and listing all information repositories holding personal data and their location should conduct an audit.

The storage, handling and protection of this data should be included in the Department's risk register these are associated with risks. The security measures in place are appropriate and proportionate to the data being held can be established by department.

2. All data centers and server rooms used to host hardware and software on which personal data is stored should be restricted to access. This can be done using entail swipe card and/or PIN technology to the room(s) in question – such a system should record when, where and by whom the room was accessed. Such access records and procedures should be reviewed by management frequently.
3. Those computer systems which are no longer in active use and which contain personal data should be removed.
4. Passwords used to access PCs, applications, databases, etc. should be of sufficient strength. A password should include numbers, symbols, upper and lowercase letters. If possible, password length should be around 12 to 14 characters but at the very minimum 6 to 8 characters. Repetition, dictionary words, letter or number sequences, usernames, or biographical information like names or dates must be avoided as Passwords. They should be changed on a regular basis (CMOD, 2008).

It defines four types of data confidentiality service:

3.4 Connection Confidentiality

This service provides for the confidentiality of all user data transferred using a connection.

- **Connectionless Confidentiality:** This service provides for the confidentiality of all user data transferred in a single connectionless data unit (i.e. a packet).

- **Selective Field Confidentiality:** This service provides for the confidentiality of selected fields within user data transferred in either a connection or a single connectionless data unit.
- **Traffic Flow Confidentiality:** This service provides for the confidentiality of information which might be derived from observation of traffic flows.

4. CRYPTOLOGY AND ITS CLASSIFICATION

Cryptology has two main branches cryptography and cryptanalysis.

4.1 Cryptanalysis

The mathematical science that deals with analysis of a cryptographic system to gain knowledge needed to break or circumvent the protection that the system is designed to provide. Attacks, in the context of network security, can be classified in two main classes, active and passive. The many known attacks against WEP can be categorized into different groups according to their goals:

Cryptosystems come in 3 kinds:

1. Those that have been broken (most).
2. Those that have not yet been analyzed (because they are new and not yet widely used).
3. Those that have been analyzed but not broken. (RSA, Discretelog cryptosystem, AES).

Most common ways to turn cipher text into plaintext:

1. Steal/purchase/bribe to get key
2. Exploit sloppy implementation/protocol problems (hacking/cracking) examples someone used spouse's name as key, someone sent key along with message. The main goal of a cryptanalyst is to obtain maximum information about the plaintext (original data).
    a. **Message Decryption:** Allows the attacker to obtain the plaintext corresponding to the cipher-texts of messages intercepted in the network.
    b. **Message Injection:** Allows the attacker to actively generate new valid messages and send them to stations associated with the network.
    c. **Key Recovery:** Is the process of obtaining the pre-shared WEP key. This is the most interesting attack type, because successfully executed it allows the attacker full access to the network

5. LINEAR CRYPTANALYSIS

High probability occurrences of linear expressions involving plaintext bits, "ciphertext" bits, and sub-key bits will be the advantage of linear cryptanalysis. It is a known plaintext attack: that is, it is premised on the attacker having information on a set of plaintexts and the corresponding ciphertexts. However, the attacker has no way to select which plaintexts (and corresponding ciphertexts) are available. The attacker has knowledge of a random set of plaintexts and the corresponding ciphertexts in many applications and scenarios.

The basic idea is to approximate the operation of a portion of the cipher with an expression that is linear where the linearity refers to a mod-2 bit-wise operation (i.e., exclusive-OR denoted by "⊕"). Such an expression is of the form:

$$Xi_1 \oplus Xi_2 \oplus \ldots \oplus Xi_n \oplus Yj \oplus Yj_1 \oplus Yj_2 \oplus \ldots . Yj_n = u\ v \quad (1)$$

where Xi represents the i-th bit of the input X = [X1, X2, ...] and Yj represents the j-th bit of the output Y = [Y1, Y2, ...]. This equation is representing the exclusive-OR "sum" of u input bits and v output bits.

6. DIFFERENTIAL CRYPTANALYSIS

High probability of certain occurrences of plaintext differences and differences into the last round of the cipher exploits the Differential cryptanalysis. For example, consider a system with input X = [X1 X2 ... Xn] and output Y = [Y1 Y2 ... Yn]. Let two inputs to the system be X' and X'' with the corresponding outputs Y' and Y'', respectively. The input difference is given by $\Delta X = X' \oplus X''$ where "⊕" represents a bit-wise exclusive-OR of then-bit vectors and, hence,

$$[\Delta X] = [\Delta X_1\ \Delta X_2 \ldots \Delta X_n]$$

where $\Delta X_i = X'_i \oplus X''_I$ with Xi' and Xi'' representing the i-th bit of X' and X'', respectively.

Similarly, $\Delta Y = Y' \oplus Y''$ is the output difference and

$$[\Delta Y] = [\Delta Y_1\ \Delta Y_2 \ldots \Delta Y_n]$$

where $\Delta Yi = Y' \oplus Y''$.

In an ideally randomizing cipher, the probability that a particular output difference ΔY occurs given a particular input difference ΔX is $\frac{1}{2}^n$ where n is the number of bits of X.

A particular ΔY occurs given a particular input difference ΔX with a very high probability pD (i.e., much greater than $(1/2^n)$) seeks to exploit a scenario in Differential cryptanalysis

The pair $(\Delta X, \Delta Y)$ is referred to as a differential.

The Differential cryptanalysis which is a chosen plaintext attack, the attacker is able to select inputs and examine outputs in an attempt to derive the key. For differential cryptanalysis, the attacker will select pairs of inputs, X' and X'', to satisfy a particular ΔX, knowing that for that ΔX value, a particular ΔY value occurs with high probability (Heys, 2001)

7. CRYPTOGRAPHY

Cryptography is the study of how to design algorithms that provide confidentiality, authenticity, integrity and other security related services for data transmitted in insecure communication environments. Confidentiality protects data from leaking to unauthorized users. Authenticity provides assurance regarding the identity of a communicating party, which protects against impersonation. Integrity protects data against being modified (or at least enables modifications to be detected).

For the encryption and decryption of algorithm if we are using same key (i.e. for encryption and decryption) then it is called symmetric key cryptography or it is also called one key algorithm, where as in asymmetric or public key cryptography

Figure 1. Evolution of cryptography

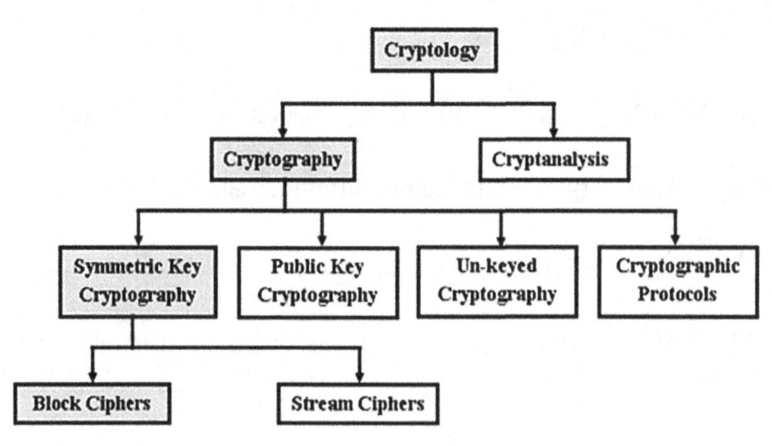

requires two keys one is used to encrypt the plaintext and other to decrypt the cipher text. One of these keys is published or public and the other is kept private.

Cryptography is classified into four categories.

7.1 Symmetric Key Cryptography

Symmetric-key encryption is that both the sender and the receiver of an encrypted message have a common secret key k (see Figure 2.). In order to encrypt the message m, also referred to as plaintext, the sender uses the function E together with the key

$c = E_k(m)$

D and secret key k (m = message, c = encrypted message).

One should assume that E and D are known to the public in that way an encryption scheme is designed, and obtaining the message m from ciphertext c merely depends on the secret key k (principle of Kerckhoff). In practice, the principle of Kerckhoff is not always used. That means that the encryption scheme is kept secret. There are two reasons for this: one can adopt an even higher security through this additional secrecy, to protect a system not only against cryptographic attacks but also against attacks on the hardware. Secondly, the use of a weak and inadequately examined algorithm is concealed through secrecy. Based on symmetric keys a new

Figure 2. Symmetric-key encryption system with encryption function E, decryption function

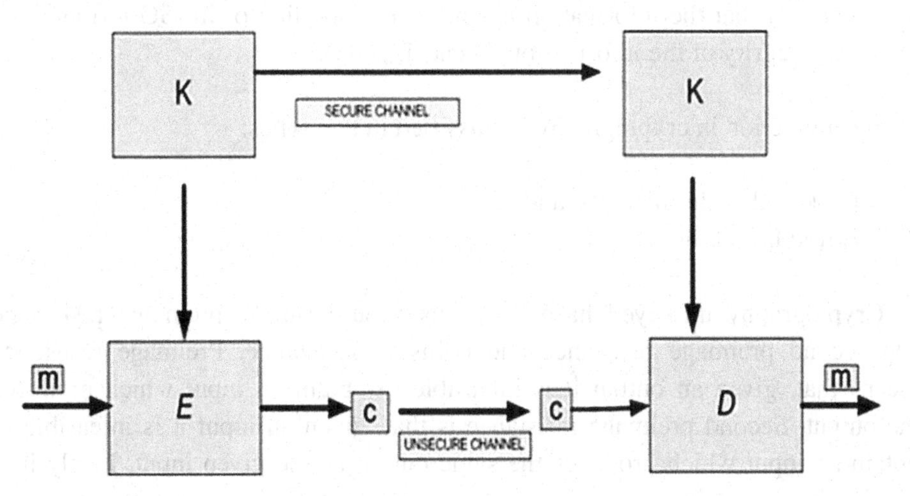

robust cryptography algorithm to increase security and prevent from unauthorized access to the contents of encrypted files is developed. It depends on structure of files; creation method of keys and resultantly the secret file cryptography using each of them are the key factors (Mohammad, 2013).

7.2 Public Key Cryptography

From the last 300-400 years Public-key cryptography (PKC) is the most significant new development in cryptography. Modern PKC was first described publicly by Stanford University professor Martin Hellman and graduate student Whitfield Diffie in 1976. Public-key introduces another concept involving key pairs: one for encrypting, the other for decrypting. This concept is very clever and attractive, and provides a great deal of advantages over symmetric-key:

- Simplified key distribution;
- Digital Signature;
- Long-term encryption.

7.3 Un-Keyed Cryptography

Un-keyed cryptography study gives details of hash function. Hash functions are very important primitives in cryptography. Hash functions can be used to protect the authenticity of information and to improve digital signature schemes.

The protection of the authenticity of information includes two aspects:

- The protection of the originator of the information, or in ISO terminology data origin authentication,
- The fact that the information has not been modified or in ISO terminology the integrity of the information (Preneel, 2003).

Hash function in cryptography is classified in two types:

- Un-keyed hash functions, and
- Keyed hash functions.

Cryptography un-keyed hash functions should satisfy preimage resistance and second preimage resistance and collision resistance. Preimage resistance means that, given an output it is infeasible to obtain an input which produces the output. Second preimage resistance is that, given an input it is infeasible to obtain an input which produces the same output as the given input. Lastly it is infeasible to obtain two different inputs which produce the same output as given input is called collision resistance.

8. CRYPTOGRAPHIC PROTOCOLS

A security protocol (cryptographic protocol or encryption protocol) is that performs a security-related function and applies cryptographic methods, often as sequences of cryptographic primitives. Security protocols are small programs that aim at securing communications over a public network, like Internet. A variety of such protocols has emerged and is seeing increasing use. SSL/TLS, SSH, and IPsec are used in internet traffic (Como, Cortier, & Zalinescu, 2009). How the algorithms should be used describes a protocol. A sufficiently detailed protocol includes details about data structures and representations, at which point it can be used to implement multiple, interoperable versions of a program.

Cryptographic protocols are widely used for secure application-level data transport, some of these aspects:

- Key agreement or establishment;
- Entity authentication;
- Symmetric encryption and message authentication material construction;
- Secured application-level data transport;
- Non-repudiation methods;
- Secret sharing methods;
- Secure multi-party computation.

9. STREAM CIPHERS AND BLOCK CIPHERS

Secret key cryptography schemes are generally categorized as stream and block cipher.

9.1 Stream Cipher

Stream ciphers, which belong to the symmetric encryption techniques. Design and analysis of stream cipher systems as well as the most well-known encryption systems are introduced. When a block cipher is used, a long message m is divided into blocks $m = m_0, m_1, \ldots, m_{N-1}$ of the same length. Here the blocks have usually a length of n = 64, 128 or 256 bits, depending on the processing length n of the block cipher.

When stream ciphers are used, the message to be encrypted m is also divided into blocks. Here, however, only short blocks of length n occur. In this case we do not speak of a division into blocks, but into symbols. Usually, $n = 1$ or $n = 8$ bit. The encryption of the single symbols mt is carried out through a state dependent unit. Stream ciphers are slower than block and the Transmission error in one cipher text block have effect on other block such that if a bit lost or a altered during transmission the error affect the nth character and cipher re-synchronous itself after n correct cipher text characters as well as not suitable in the software. In synchronous stream cipher if a cipher text character is lost during transmission the sender and receiver must re-synchronous their key generators before they can proceed further.

9.1.1 Classification of Stream Ciphers

The symmetric stream encryption systems are classified in

1. Synchronous stream ciphers, and
2. Self-synchronizing stream ciphers.

The sender can be found on the left and the receiver on the right side. When a synchronous stream cipher is used, the sender and the receiver of an encrypted message have to compute the keystream z_t synchronously at any time $t \geq 0$ for encryption and decryption

9.1.1.1 Synchronous stream ciphers

Figure 3 depicts a symmetric, synchronous stream encryption system. The sender can be found on the left and the receiver on the right side. When a synchronous stream cipher is used; the sender and the receiver of an encrypted message have to compute the keystream z ciphers for time t, synchronously at any time $t \geq 0$ for encryption and decryption.

The keystream zt is generated independently from the plaintext message and the ciphertext. The encryption of the message symbols m_t, $t \geq 0$, can be described by the following equations:

$\sigma_t+1 = f(\sigma_t, k)$,

$z_t = g(\sigma_t, k)$,

$c_t = h(z_t, m_t)$, Where $t \geq 0$ is valid.

Figure 3. Synchronous, symmetric stream cipher

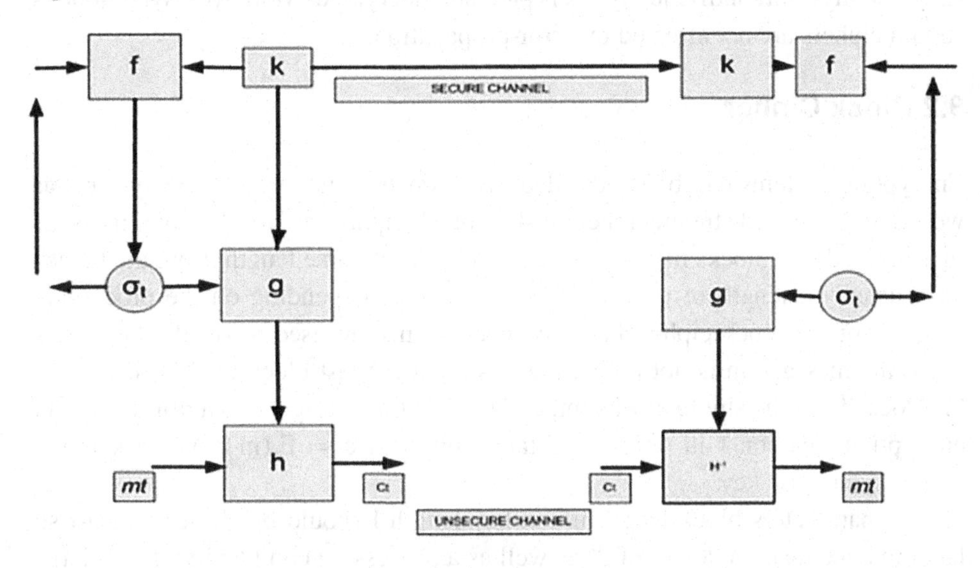

The system has a state variable σ_t whose initial state σ_0 can either be known publicly or be determined from the secret key k. In order to be able to carry out the encryption, the encryption function h must clearly be invertible. The function f is called the next state function and g is called the output function. The functions f, g and h are known publicly.

9.1.1.2 Self-synchronizing stream ciphers

Besides synchronous stream ciphers, there are also self-synchronizing stream ciphers, but they are hardly used in information and communication systems. In this case the keystream z_t depends on the key k and a fixed number l of previously generated ciphertext symbols.

The encryption of a sequence of plaintext symbols m_t, $t \geq 0$, can be described by the following equations:

$$\sigma t = (ct-l, ct-l+1, ..., ct-1),$$

$$zt = g(\sigma t, k),$$

$$ct = h(zt, mt).$$

Stream ciphers have several advantages which make them suitable for some applications. Most notably, they are usually faster and have a lower hardware complexity than block ciphers. They are also appropriate when buffering is limited, since the digits are individually encrypted and decrypted. Moreover, synchronous stream ciphers are not affected by error-propagation.

9.2 Block Cipher

Encryption systems can be subdivided in symmetric and asymmetric systems as well as in block and stream ciphers. When block ciphers are used, a long message m is divided into blocks $m = m_0, m_1, \ldots, m_{N-1}$ of the same length. Here the blocks have usually a length of n = 64, 128 or 256 bits, depending on the processing length n of the block cipher. Padding mechanisms are used to fill the last block when the message m is not long enough so that the last block m_{N-1} is also an n-bit block. Then the single blocks mt, $0 \leq t \leq N-1$ are assigned to a time-invariant encryption function f in order to obtain ciphertext $c_t = E_k(m_t)$, where k is the secret, symmetric key.

The parameters block length n and key length l should be chosen at least so large that a data complexity of 2^n as well as a processing complexity of 2^l is large enough not to allow an attacker to carry out an exhaustive key search in 10 or 20 years. Today, a block length of n = 64, 128 and 256 bits and an equally sized key length are used. The Feistel cipher is based on the idea of using the same function

$$G: GF(2)^l \times GF(2)^{n/2} \rightarrow GF(2)^{n/2}$$

for encryption as well as for decryption. The function G, for example, consists of a product cipher. Here we assume that n is even and l is the length of the key k or a sub-key derived from it. The plaintext block m of length n bit is split into two equally sized blocks L and R, each having a length of n/2 bit: m = (L,R).Thenthe ciphertext block c is, as shown in Fig. 4.2, put together from the block R andthe bitwise XOR operation of block L with the function value G(k,R):

$$c = (R, L + G(k,R)) = (R,X) .$$

Block ciphers are somewhat faster than stream cipher each time 'n' characters executed, transmission errors in one cipher text block have no effect on other blocks. Block ciphers can be easier to implement in software, because the often avoid time-consuming bit manipulations and they operate on data in computer-sized blocks. In the real world block ciphers seem to be more general (i.e. they can be used in any of the four modes, the modes are ECB, CBC, OFB, CFB).They have different structure like Feistel Network, for e.g. Kasumi and Clefia. In a Feistel cipher (see Figure 4), the plaintext is split into two halves. The round function is applied to one half, and the output of the round function is bitwise ex-or-ed with

Figure 4. Encryption principle of a Feistel cipher with one round

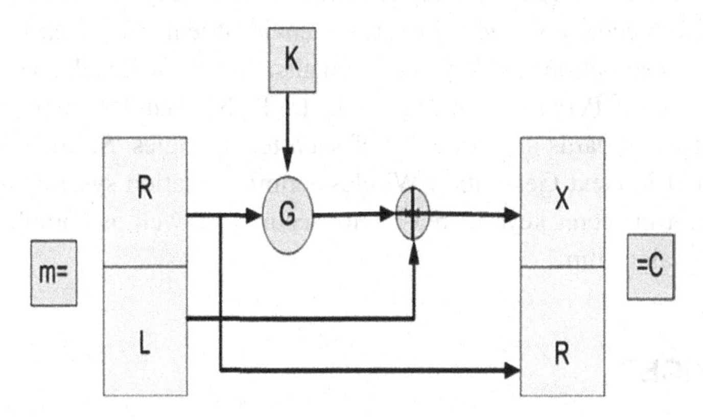

the other half finally, the two halves are swapped, and become the two halves of the next round. Another is Substitution-permutation network e.g. AES. In Substitution-Permutation (SPN) cipher, the round function is applied to the whole block, and its output becomes the input of the next round.

Although both stream ciphers and block ciphers belong to the family of symmetric encryption ciphers, there are some key differences. Block ciphers encrypt fixed length blocks of bits, while stream ciphers combine plain-text bits with a pseudorandom cipher bits stream using XOR operation. Even though block ciphers use the same transformation, stream ciphers use varying transformations based on the state of the engine. Stream ciphers usually execute faster than block ciphers. In terms of hardware complexity, stream ciphers are relatively less complex. Stream ciphers are the typical preference over block ciphers when the plain-text is available in varying quantities (for e.g. a secure wifi connection), because block ciphers cannot operate directly on blocks shorter than the block size. But sometimes, the difference between stream ciphers and block ciphers is not very clear. The reason is that, when using certain modes of operation, a block cipher can be used to act as a stream cipher by allowing it to encrypt the smallest unit of data available.

10. CONCLUSION AND FUTURE WORK

A cryptographic algorithm is an essential part in network security. Most of the sensitive information in the wireless communication has latent security problems. An End-to-end security has been an issue in next generation wireless networks and hence a solution has to be proposed for the same using Secure Socket Layer/ Transport Layer Security (SSL/TLS), Virtual Private Network (VPN), or a similar mechanism should be provided for security of data. Cryptographic algorithms are utilized for security services in various environments in which low cost and low power consumption are key requirements. Wireless Local Area Networks (WLAN), Wireless Personal Area Networks (WPAN), Wireless Sensor Networks (WSN), and smart cards are examples of such technologies. Security is the most important part in Next Generation Wireless communication system, where more randomization in secret keys increases the security as well as complexity of the cryptography algorithms.

REFERENCES

Al Tamimi. (2006). *Security in Wireless Data Networks: A Survey Paper*. Academic Press.

Anjum, N., & Salil, K. (2009). Article. *Authentication and Confidentiality in Wireless Ad Hoc Networks*, *21*, 28.

Ayu, T., & Sudip. (2007). A multi-factor security protocol for wireless payment-secure web authentication using mobile devices. *IADIS International Conference Applied Computing*.

Chandra, D.V., Shekar, V.V., Jayarama, & Babu. (2008). Wireless security: A comparative analysis for the next generation networks. *Journal of Theoretical and Applied Information Technology*.

Como, L. H., Cortier, V., & Zalinescu, E. (2009). Deciding security properties for cryptographic protocols. Application to key cycles. *ACM Transactions on Computational Logic*, *5*, 1–38.

Daemean, J., & Rijmen, R. (1999). *AES Proposal: Rijndeal version 2*. Available at http://www.esat.kuleuveb.ac.be/rijmen/rijndeal

Feng. (2008). *Attacking Antivirus*. Nevis Networks,Inc.

Heys, H. M. (2001). *A Tutorial on Linear and Differential Cryptanalysis*. Academic Press.

Kishor, S., Trivedi, Y. Z., Ma, & Dharmaraja, S. (2003). Performability modeling of wireless communication systems. *International Journal of Communication Systems, 16*, 561–577.

Kumar, Y., & Prashant. (2010). Hardware Implementation of Advanced Encryption Standard. In *Proceedings of International Conference on Computational Intelligence and Communication Networks.* IEEE. doi:10.1109/CICN.2010.89

Liang & Wang. (2004). *On performance analysis of challenge/responsebased authentication in wireless networks*. Elsevier.

Mohammad, S. (2013). A New Secure Cryptography Algorithm Based on Symmetric Key Encryption. *Journal of Basic and Applied Scientific Research*. Retrieved from www.textroad.com

Muhammad & Seon. (2007). Security Issues in Wireless Mesh Networks. In *Proceedings of International Conference on Multimedia and Ubiquitous Engineering* (MUE'07). IEEE.

Okumoku-Evroro &Oniovosa. (2005). *Internet Security: The Role Of Firewall System*. Department Of Computer Science Delta State Polytechnic Otefe-Oghara.

Preneel, B. (2003). *Analysis and Design of Cryptographic Hash Functions*. Academic Press.

Protecting the confidentiality of Personal Data. (2008). CMOD Department of Finance.

SANS Institute (2005) "SANS Institute Info Sec Reading Room"

Singh & Jain. (2012). Research Issues in Wireless Networks. *International Journal of Advanced Research in Computer Science and Software Engineering, 2*(4).

Sun & Liu. (2004). *Scalable Hierarchical Access Control in Secure Group Communications*. IEEE.

Taiming, F., Chuang, W., Wensheng, Z., & Lu, R. (2001). *Confidentiality Protection for Distributed Sensor Data Aggregation*. Retrieved from http://www.cs.iastate.edu/

Vinay. (2007). *Authentication and Access Control The Cornerstone of Information Security*. Trianz White Paper.

Webroot Software Inc. (2004). *Anti-spyware software: Securing the corporate network*. Academic Press.

Yagu¨e, Mana, & Lopez. (2005). *A metadata-based access control model for web services*. Retrieved from www.emeraldinsight.com/researchregister

Chapter 10
Security in Ad-Hoc Networks (MANETS)

Piyush Kumar Shukla
UIT RGPV, India

Kirti Raj Bhatele
UIT RGPV, India

ABSTRACT

Wireless Networks are vulnerable in nature, mainly due to the behavior of node communicating through it. As a result, attacks with malicious intent have been and will be devised to exploit these vulnerabilities and to cripple MANET operation. In this chapter, we analyze the security problems in MANET. On the prevention side, various key and trust management schemes have been developed to prevent external attacks from outsiders. Both prevention and detection method will work together to address the security concern in MANET.

BACKGROUND

Security is an essential requirement of mobile Ad-hoc network. In the past few years, we have seen a rapid expansion in the field of mobile computing due to the proliferation of inexpensive, widely available wireless devices. However, current devices, applications and protocols are solely focused on cellular or wireless local area networks (WLANs), not taking into account the great potential offered by mobile ad hoc networking. The mobile ad-hoc network consists of the nodes that are mobile in nature such as laptops, sensors, smart phones etc. The ad-hoc network does not have any pre-defined or fixed infrastructure. The nature of the ad-hoc network is wireless in nature which operates in distributed manner. The network can oper-

DOI: 10.4018/978-1-4666-8687-8.ch010

ates as standalone network or can have multiple points for connection. Application scenarios include, but are not limited to: emergency and rescue operations, conference or campus settings, car networks, personal networking, etc. Due to openness MANET is vulnerable to several attacks i.e. external or internal attacks. To Secure MANET several security solutions have been proposed by various scientists. In this chapter, we discuss the security issues, solution and research issues that need to be addressed in mobile ad-hoc network Overview.

Nature of MANET brings new security challenges to network design. Because node in mobile ad-hoc network generally communicates with each other via open and shared broadcast wireless channels, they are more vulnerable to security attacks. In addition, their distributed and infrastructure less nature means that centralized security control is hard to implement and the network has to rely to individual security solution from each mobile node. Furthermore, as ad-hoc network are often designed for specific environments and may have to operate with full availability even in adverse condition, security solution applied in more traditional networks may not be directly suitable.

DESCRIPTION

Mobile ad-hoc networks (MANETs) are spontaneously deployed over a geographically limited area without well-established infrastructure. The networks work well only if the mobile nodes behave cooperatively. MANETs is very vulnerable to various attacks from malicious nodes. In order to reduce the hazards from such nodes and enhance the security of network several security solutions have been introduced. Both prevention and detection method will work together to address the security concern in MANET.

INTRODUCTION

Today one of the major problems that the ad-hoc network is facing is the Security (Yi, Dai, Zhang, Zhong, 2010). Ad-hoc network requires the robust security schemes for transmitting information over the network. As the ad-hoc network are having dynamic topology, mobile nodes, wireless transmission media there the privacy and security of the node is to be maintained. There are two levels of security schemes that can be applied to the network: Low Level and High Level. The low level security scheme is applied to the network where the security has to be kept of the minimum level. The main focus is the low level scheme is to make the network secure applying security mechanisms. This implies key exchange, privacy, capture node attack, secured routing, trust establishment etc. The High level security is implied to the network where it is necessary to make network secure but also to identify the intruder or attacker. There are various mechanism that are applied in the high level security scheme such as intrusion detection, secure data aggregation, secure group policy etc.

As the security level are of two types the attacks is also classified in two categories active attack and passive attack. In the active attack the attacker listen, monitors and modifies the data packet according to his/her interest. There are various type of attack included in the active attack such as broadcasting fake information, routing attacks etc. The second type of attack is passive attack in which the attacker monitors and modifies the exchange packet. It includes the attack against the privacy of the node. This includes the attack such as eavesdropping, traffic analysis etc. The security of the network becomes more important because the nodes are mobile in nature and can affect the network more prominently then the fixed attacker node. Security mechanisms that are applies to the network must assure that when the malicious node are detected then the normal node are to be secured from them and the malicious link must be blocked.

MOBILE AD-HOC NETWORK

Mobile Ad Hoc Networks or MANETs (Bruschi, Rosti, 2002) are a collection of dynamic self-configuring networks of mobile nodes. These nodes can be groups of computers, laptops, and sensors. Every node in the network acts as host and router which provide the connectivity to the subsequent nodes in the network. The communication in the network must be able to adapt any changes in the network such change in the location of the node or changes due to the surrounding environment, as the mobility is the main feature of the network. It is important to find the max. route that exist between source and destination. As the route has been discovered each node will forward the packets to next node in the route till the packet reaches the target node. The importance of mobile ad hoc network is rapidly increasing in the field of military applications for e.g. in connecting autonomous robots etc.

A mobile ad-hoc network (MANET) (Rao, Moyer, Rohatgi, 1999) is composed of a group of mobile, wireless nodes which cooperate in forwarding packets in a multi-hop fashion without any centralized administration. Applications of MANETs occur in situations like battlefields, major disaster areas, and outdoor assemblies. The user always want to be connected to wireless network irrespective of their geographical position for example connectivity to internet always on their wireless devices such mobiles and laptops. A working group called "manet" has been formed by the Internet Engineering Task Force (IETF) (Kong, Hong, Gerla, 2010) to study the related issues and stimulate research in MANET.

Figure 1. Nodes in MANET

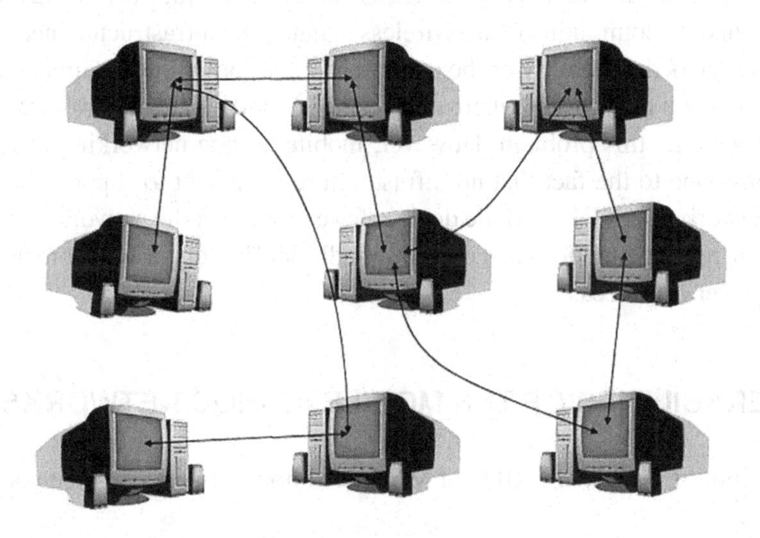

In MANET (Perrig, Hu, Johnson, 2003), each mobile node works as a router and also as an end node which is a source or a destination, as a result the failure of some nodes' action can greatly slow down the performance of the network and even influence the basic accessibility of the network. Energy depletion of nodes has been one of the main threats to the availability of Ad-Hoc network. As the mobile nodes have restricted battery power, it is very essential to use energy proficiently in MANET.

The threat if attack is increasing with the time and user on the mobile ad-hoc networks. MANETs (Nguyen, Nguyen, 2008) should have a protected way for transmission and communication; it is rather challenging and vital subject. To provide secure communication and transmission, researchers are working exclusively on the security issues in MANETs. There are many secure routing protocols and security measures for the mobile ad-hoc networks have been proposed by the researcher. Previously there are many works done on security issues in MANET which are specially based on reactive routing protocol like Ad-Hoc On Demand Distance Vector (AODV).

In recent years, a number of studies have been done in different layers, such as MAC layer and application layer, to achieve energy conservation. As wireless and mobile communications and services become more pervasive, appropriate networking technologies must be developed to support the user requirements to connect "anyplace, anytime, anywhere". While the IEEE 802.11 (Bruschi, Rosti, 2002) standard allows devices to communicate in a wireless manner, the infrastructure needed to be in place restricts the mobility of the devices and does not allow a wireless network to be present "on the fly" anywhere one desires. Mobile ad hoc networking will be able to overcome this problem. However, mobile ad hoc networking has its own limitations. Due to the fact that no infrastructure is present to support the mobile ad hoc network (MANET), mobile devices have to support the network themselves, performing routing functions that consume additional battery power from the node's limited power resources.

VULNERABILITIES OF THE MOBILE AD-HOC NETWORKS

Vulnerabilities (Visalakshi, 2011) of Mobile Ad-hoc Networks has been described as follows,

- **Wireless Links:** As, the mobile ad hoc network uses wireless link the node are connected to each other with the wireless link. The attackers do not need physical access to the network to carry out the attacks such as eavesdropping and active interference in the wired network.
- **Dynamic Topology:** The node in the MANET can travel independently, unlink and link the network. Therefore the network topology is dynamic in nature. In this dynamic environment it is very difficult to identify the normal behavior of the network form the malicious behavior.
- **Cooperativeness:** In mobile ad-hoc network the node are considered to be non-malicious and cooperative by the routing algorithms. Therefore a malicious attacker can easily become a significant routing representative and interrupt network operations by disobeying the protocol specifications.
- **Lack of a Clear Line of Defense:** The lines of defense are not clear in the mobile ad-hoc network. The threat of the attack on the node can come from all directions. The boundary that separates the inside network from the outside world are not specified on MANETs.
- **Limited Resources:** The devices connected on the mobile network ranges from the laptops, PADs and mobile phones. These devices generally have different computing and storage capacities that can be center of new attack. There are no of attributes that have be considered such as availability, integrity, authentication, confidentiality, non-repudiation, and authorization.

MALICIOUSNESS AND ATTACKS IN MOBILE AD-HOC NETWORKS

Malicious nodes (Johnson, Perrig, Hu, 2003) aim to deliberately disrupt the correct operation of the routing protocol, denying network services if possible. Hence, they may display any of the behaviors shown by the other types of failed nodes. The impact of a malicious node's actions is greatly increased if it is the only link between groups of neighboring nodes.

A node in MANET (Annadurai, Sivakumar, 2009) which undergo any attack exhibits an anomalous behavior called malicious behavior. A node is said to be a malicious node if and only if it undergoes with one or more of following characteristics:

1. Packet drop;
2. Battery drained;
3. Buffer over Flow;
4. Bandwidth consumption;
5. Stale packets;
6. Delay of packets;
7. Link break;
8. Message tampering;
9. Fake or wrong routing;
10. Stealing information;
11. Session capturing.

ATTACK TYPES IN MOBILE AD-HOC NETWORKS

As the nature of the MANET (Asokan, Zapata, 2002) medium is open wireless in nature, there is a number of possible attacks by which the MANET (Lee, Su, W, Gerla, 2002) is exposed. Hence, the chances for possible attacks are very high. The main purpose of the attacker is to create problems for legitimate users, and as a result services are not accessible.

1. **Denial of Service (DoS):** One of the serious problems for the ad-hoc networks is the Denial of Services Attack. As the name suggest this attack prevents the legitimate user from accessing the network services and resources. This type of attack is done by jamming the channel system so that no authenticate user can access it. The users are not able to communicate with each other, so no information is shared between them. This could be more devastating in case

of life critical application. There are three ways by attacker can affect the network:

a. In the first stage the attacker overcomes the node resources. It cannot performs the other necessary jobs this makes the node continuously busy and is not able do other tasks.
b. In the second stage the malicious node jams the channel. This is done by generating high frequency in the channel so that no other node is able to perform any other thing.
c. The third is drop the packet in the network.

2. **DDoS:** The Distributed Denial of Service Attack is the more devastating form of DoS attack. In this attack the malicious node uses different location to affect the network. It may also uses different time slot for transmitting the messages. This may differ for each other depending on the node to node. The main aim is to make the network down so that the node in the network are not able to use it. The DDoS attack has two possibilities:
   a. Node to node,
   b. Node to infrastructure.
3. **Impersonation:** Each node in the MANET has a unique id that helps other nodes in the network to identify each other. In impersonation attack the malicious node is able to change his identity. An attacker node can change his identity to same as the source node and can act as original generator of the message and can modify the message (Zhong, Zhang, Yi, 2008) according to his benefits, and transmit it into the network.
4. **Eavesdropping:** Eavesdropping attack (Shrivastava, 2013) is the process of gathering information by snooping on transmitted data on legitimate network. Eavesdrop secretly overhear the transmission. However, the information remains intact but privacy is compromised. This attack is much easier for malicious node to carry on as evaluate to wired network. Eavesdropping attack in MANET shared the wireless medium, as wireless medium make it more vulnerable for MANET malicious node can intercept the shared wireless medium by using promiscuous mode which allow a network device to intercept and read each network packet that arrives.

SECURITY SCHEMES IN THE MOBILE AD-HOC NETWORKS

Wormhole attack is a type of tunneling attack. In a tunneling attack there are two are more attacker is required. In wormhole attack there are two or more attacker are required. The special property of these attacker nodes is that they have better communication channel then other regular node in the network. The attacker node

Figure 2. Eavesdropping attack

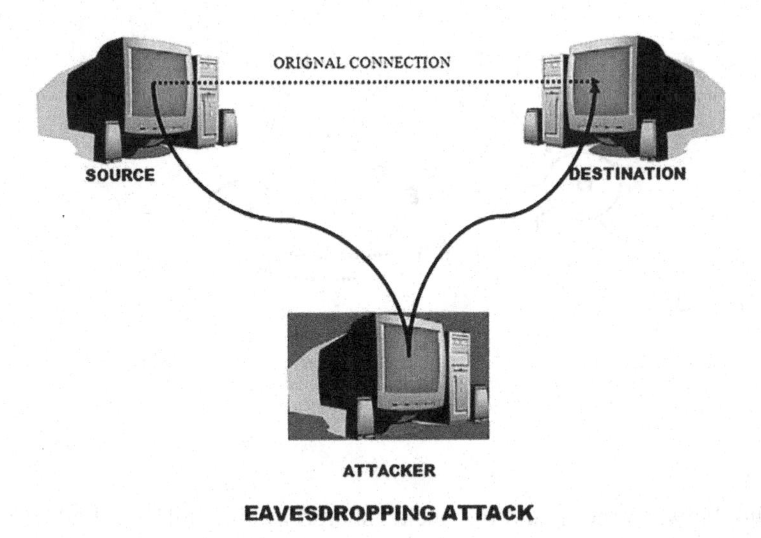

creates a special link between them with the property of high bandwidth and low latency. This link is called as "Tunnel". The malicious node promotes this "Tunnel" as a high quality route in the network. The other nodes in the network take on this link as in there communication paths. The data transmitted through this tunnel is being captured by the attacker as the link is being created by the attackers. The data packets are being collected by the attacker node at the one end and then it is forwarded using tunnel to the other end. It is very hard to detected the Wormhole are hard to detect because the link used for the communication does not belong actual to the network. The wormhole (Arya, Arya, 2012) does not seem to be harmful as it lowers the time taken for the data packet to reach its destination. But this is also harmful as the route which is used for the communication is fake as it promotes the route as shorter than the original route. The routing mechanism get confused due to the fake information of the route between the nodes is promoted in the network. The protocol or services offered in the network could not even detect the damage caused by the attack. It is very use to eavesdrop the communication in the wireless network. The packet is forwarded to other nodes in the network. This might be harmful if the information in the packet is being altered. There are two modes for the attack to be launched. The first mode is the hidden mode. In the hidden mode the attacker does not uses his identity thus they remain hidden in the network form the genuine users. The attacker appears to be simple node that receives the data and forward it to the other node in the row. But the attacker node captured the data packet at the one end of the wormhole and replicates at the other end. They established a

Figure 3. Wormhole attack

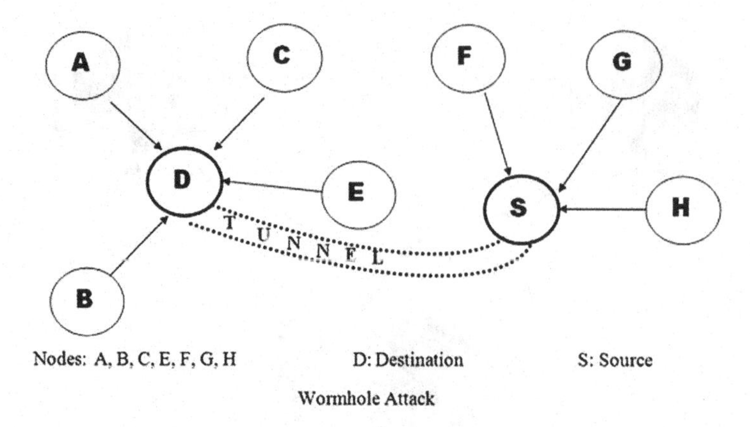

virtual link between them, and the packet shared by this link is called "Tunneling of the message". The main characteristic of the attacker node in the hidden mode is that it does not require any cryptographic keys to launch. Another mode is the participation mode in which the attackers uses the cryptographic keys and launch the attack more powerfully. The attacker does not establish the virtual link rather than participates in the communication as the genuine user. They use the wormhole attack either to deliver the packet faster or with smaller no. of hops. The attacker when included in the route drops the packet in the hidden mode between the source and destination. The systems that are most affected by the wormhole attack are location based security systems also clustering protocol, data aggregation and network routing are most affected.

PROPOSED DETECTION METHOD

There are different methods that are used to detect wormhole in MANET. These methods are as follows:

1. Distance and Location Based Solution;
2. Hop Count and Delay Based Solutions;
3. Key Based Solution;
4. Neighbor-Based solutions;
5. Special Hardware Based Approaches;
6. Synchronized Clock Based Solution;
7. Topology Based Solution.

Distance and Location Based Techniques:

Distance and Location Based Techniques (Nguyen, Nguyen, 2008) proposed Packet Leashes Method based on location and time. The key concept is to authenticate an exact timestamp or information of the location of the node with loose timestamp. The distance travelled by the data packet could be determined by the receiver. The packet leashing method is added on each packet at every link to restrict the distance of the transmission of the packet. If a packet travels any unrealistic distance in the network it is determined by the receiver. Generally there are two types of packet leashes:

1. Geographical; and
2. Temporal.

In the Geographical Leash the position and the sending time of the packet is inserted by the sender. The receiver estimates the maximum distance between the sender and receiver on the bases of the position of the receiver and receiving time. If the transmission range of the packet is more than estimated distance then it is discarded. The temporal leash mechanism assumes that the maximum transmission speed of radio signal is the speed of light. On the basis of the maximum range of the transmission and speed of the light the packets expiration time is estimated. Every packet contain this expiration time. On receiving the packet the receiver checks the expiration time and except or discard the packet on the basis of it. The leashes method has the drawback that it requires GPS and very strict time synchronization.

Hop Count and Delay Based Solutions

The Hop Count measures and reconstructs the mapping details in each node and it finally exploits the diameter feature to detect the distortion caused by the malicious node. One of the algorithms used for distributed detection of the Wormhole is Wormhole Geographic Distributed Detection (WGDD) (Shrivastava, 2014). The parameter used for the detection of Wormhole in the hop count is used in the algorithm. The algorithm detects the wormhole attack on the basis of the damage caused by the attack and this algorithm is very effective in finding the exact location of the Wormhole.

The wormhole is classified in two divisions Hidden and Exposed. DELPHI (Dalal, Vats, Loura, Rohila, 2012), is the effective method that implies the delay based method to detect the Wormhole. It provides the solution for the Exposed wormhole. It uses various path delays for detection. In every path the delay per hop is determined. For the genuine path the delay per Hop is minimum then the wormhole path. If the delay per hop is higher for a path then it is considered that it affected by the wormhole attack. End to End Detection of Wormhole Attack (EDWA) (Ghaffari, 2006) is another method for detection of the wormhole attack that is based on the hop count method. It is a method for end to end detection of wormhole attack. First the shortest hop count is estimated between the sender and receiver. If the hop count is less than the estimated hop count than it is said the path is affected by the wormhole attack.

Key Based Solutions

The Key based solution uses the key exchange mechanisms for the security of the network. The key based scheme has node to node (Zhang, Wang, Shin, 2002) authentication scheme. The nodes are able to localize the impact of compromised within their vicinity. It facilitates the establishment of pair wise keys between neighbour nodes. The main idea is to receive the packet only from the authenticated (Luo, Yang, Zhang, 2004) user. The node discard those messages tunneled from multi-hop-away locations preventing the wormhole attack.

Neighbor-Based Solutions

The Neighbour Based Solution technique is the method in which each node maintains two lists. The first list is 'Neighbour List' and the second list is 'Wormhole Node List' (Luna, Madruga, 2001). Whenever nodes transmits the RREQ (Krsul, Schuba, Kuhn, Spafford, Sundaram, Zamboni, 1997) message to the destination, every node monitors its neighborhood nodes behavior. In this phase the nodes generates the trust value for their neighboring nodes and maintain the information in the 'Neighboring List'. This trust value is generated on the basis of the packet drop pattern by the neighbour. The frequency of the link in the different routes is analyzed in the network when routing takes place. If any link is found suspicious, then the trust value of the node is used to check whether the node is wormhole or not. The algorithm implies neighbour node no. which helps them to find whether there is wormhole attack. Every node is capable of identifying reply packet that are being forwarded by the attacker. As the source node receives the RREP packet it can detect which route is affected by the wormhole attack. After the source node identifies the attacker node it record them in the another special list 'Wormhole Node List'. This information is stored with the source node so that every time these nodes are blocked from taking part in the routing.

Special Hardware Based Techniques

The main advantage of this technique is that it does not depend on the time synchronization. In a multi hop wireless network (Johnson, Hu, Perrig, 2003) each node encounters secure tracking in the Special Hardware Based Technique. To estimates the distance between the nodes in the network Mutual Authentication with Distance-bounding (MAD) protocol is used. There is a 'Transceiver' equipped with every node in the network, a special hardware device. This transceiver accepts single bit, apply bit XOR and broadcast it. The Directional Antenna detects the existence of the Wormhole nodes in the network. In this system the direction information is shared between the source and destination. The wormhole is detected in the network on the basis of the signal received from the attacker node and the direction information from the source. The wormhole is detected when there is difference between the signals from the source and intermediate node.

Synchronized Clock Based Solution

In the Synchronized Clock Based solution the True Link detection techniques are used which is based on the time mechanisms. It determines whether there is a direct link between the node and to its neighbour. The method has two phases. In the first phases the two nodes exchanges the nonce with the stiff timing factor. The nonce is a arbitrary number which is used in the cryptography communication. The second phase is the authentication phase in which each node authenticate the other that they are the original generator of the nonce. The only disadvantage of this technique is that is work on the device which has facility of firmware update with backward compatibility. To overcome this disadvantage, The Round Trip Time (RTT) (Luo, Yang, Zhang, 2004) is used which involves the use of additional hardware. The RTT is the total time taken from sending the RREQ by the source to the receiving of RREP form the destination node. Each node calculates the RTT between itself and its neighboring nodes. Malicious/Attacker nodes have higher Round Trip Time value then the other nodes. Therefore the source node could identify the legitimate and malicious node. This method is very effective in the hidden attacker case.

Topology Based Solution

The Topology Based Solution work in two phases (Shahrani, 2011) to analyze the wormhole attack. In the first phase the attacker try to a maximum extent to disrupt the communication of the network. The second phase is that the attacker target to a particular node. The communication in the network is under the full control of the attacker. The first phase has strict timing constraints, but in the second phase there are no time constraints. The nodes do not communicate with the adjacent nodes. The data packets are stamped with the transmission time instead of the signature.

DEFENSE MECHANISM AGAINST RUSHING ATTACKS IN MOBILE AD HOC NETWORKS

The Rushing attack (Palanisamy, Annadurai, 2009) is the DoS, Denial of Service attack (Luo, Yang, Zhang, 2004). In the DoS attack the main objective is that the legal user are prevented from accessing the network resources. In this type of attack the attacker jam the channel system so that no authenticate nodes can access the network. Rushing Attack (Su, Lee, Gerla, 2002) is the extended level attack, the attacker generates the high frequency in the channel for creating jam in the channel. The nodes are not able to communicate with each other. In this attack, the source transmits the Route Request (RR) packet in the network to another node. If an attacker is present in the network then it will accept the packet and then transmits it with the greater speed to the neighboring node in the network. Due to high transmission speed of the RR packet, the packet will reach the destination node first. The destination node will accept the RR packet forwarded by the attacker and drop the other packet which will reach late. The destination finds the route as the route and will use it for further communication. On this way the attacker gain the successful access over the communication between the source and destination.

In the Figure 4 The Rushing Attack (Hong, Kong, Gerla, 2006)The S is the Source Node and The D is the Destination Node. The F is The Attacker Node. The Source S transmits the RR packet in the network. The attacker node F after receiving the RR packet transmits it with the higher speed then the other entire node in the network.

Position of the Attacker

According to the position of the attacker (Johnson, Perrig, Hu, 2003) the scenario can be classified in the three different ways:

1. Attacker near to The Source node.
2. Attacker near to The Destination Node.
3. Attacker anywhere in the Network.

Attacker Near the Source Node

In this scenario the attacker node is present near to the Source node. As the Source node transmits the RR packet into the network the packet will reach the Attacker

Figure 4. Rushing attack

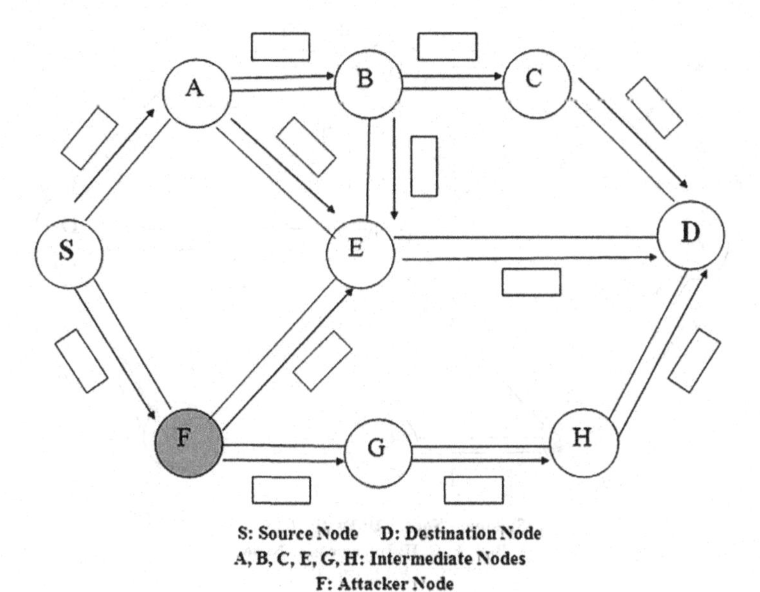

very early and then after the RR packet reaches the Attacker node (Johnson, Perrig, Hu, 2003) it will be transmitted with high transmission speed as compared to other node in the network. The packet transmitted by the attacker node will reach the destination first and the destination will discards the other entire packet that arrives late. It will consider the route as secure and thus the attacker has the access over the communication between the Source and the Destination. When the Rushing Attack happens near to The Source Node, the attack success rate is average; it only has to search the intermediate node. In the Figure 5 S is The Source Node, D is The Destination Node. F is The Attacker Node which is the nearer to the source node S. The RR packet transmitted by the S will reach the intermediate node A and attacker node F. On receiving the RR packet the node A and F will transmit it into the network. But as the node F is the attacker node it will transmit the RR packet with greater speed then the other entire node in the network. So the RR packet will reach to the destination faster as compared to other transmission in the network. The success of the attacker depends on the intermediate node in the transmission of the RR packet. The Attack Success Rate is Average as the transmission depends on the intermediate node. The attack success rate is average because the transmission of the RR packet to the destination depends on the intermediate nodes that are in-between the attacker and the source.

Figure 5. Rushing attack near the source node

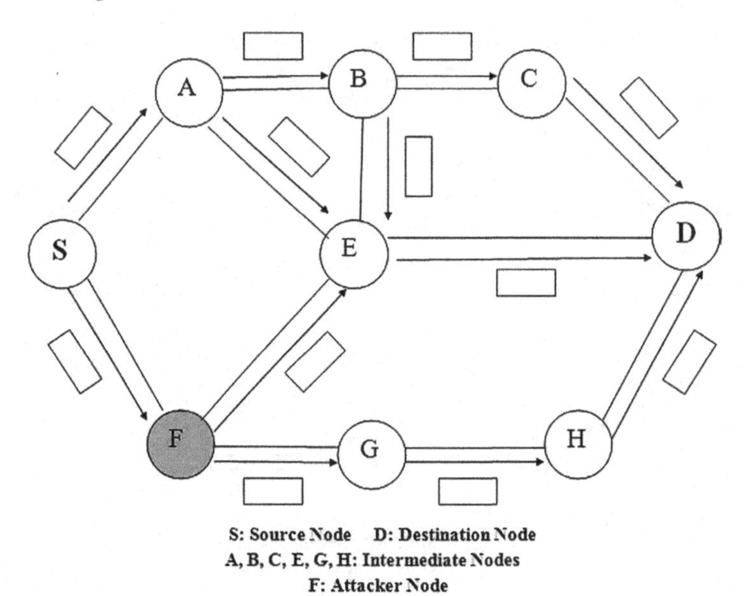

Attacker Near the Destination Node

In this scenario the attacker node is present near to the (Nguyen, Nguyen, 2006) Destination node. As the Source node transmits the RR packet into the network the packet will reach the Attacker, which is very nearer to the Destination Node and the Attacker has just to transmit the RR packet to Destination Node (Vijayalakshmi, Rabaral, 2014). This is very critical situation as the attacker is closer to destination node and the packet transmitted is revived by the destination node very quickly. The Attack success rate is very high in this situation. In the Figure 6 S is The Source Node, D is The Destination Node. C is The Attacker Node which is the nearer to the Destination node D. The RR packet transmitted by the S will reach the intermediate node A and node F. On receiving the RR packet the node A and F will transmit it into the network. Further on it will be transmitted in the network. As soon as it reaches the node C i.e. the attacker node it will transmit the RR packet with greater speed then the other entire node in the network. So the RR packet will reach to the destination D faster as compared to other transmission in the network. The success of the attacker does not depends on the any intermediate node in the transmission of the RR packet. The Attack Success Rate is Very High as the transmission does not depends on the intermediate node. The attack success rate is very high because

Figure 6. Rushing attack near the destination node

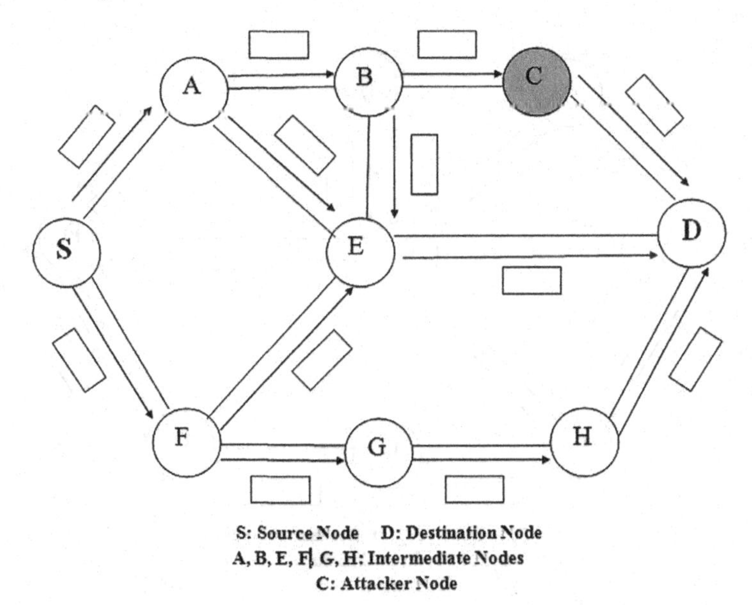

the transmission of the RR packet to the destination D is direct i.e. The attacker C transmits the RR packet to the destination D faster than the node E and H.

Attacker Anywhere in the Network

In this scenario the attacker node is present anywhere in the network not either nor closer to the Source Node or the Destination Node. As the Source node transmits the RR packet into the network the packet will reach the Attacker and after that RR packet reaches the Attacker node it will be transmitted with high transmission speed as compared to other node in the network to the next node. The possibilities of getting the packet form the attacker node depends on the intermediate node. The impact of the attack is least but greater to the Attack Success Rate of the Attacker near to the Source Node. In the Figure 7, S is The Source Node, D is The Destination Node. B is The Attacker Node which is at the intermediate location in the network. The RR packet transmitted by the S will reach the intermediate node A and attacker node F. On receiving the RR packet the node A and F will transmit it into the network. But as the node B is the attacker node it will transmit the RR packet with greater speed then the other entire node in the network. As compared to other transmission in the network the RR packet will reach to the destination faster. The success of the attack (Vyavahare, Rawat, Ramani, 2005) depends on the intermediate node in the transmission of the RR packet but the transmission is farter than the transmission

Figure 7. Rushing attack anywhere in the network

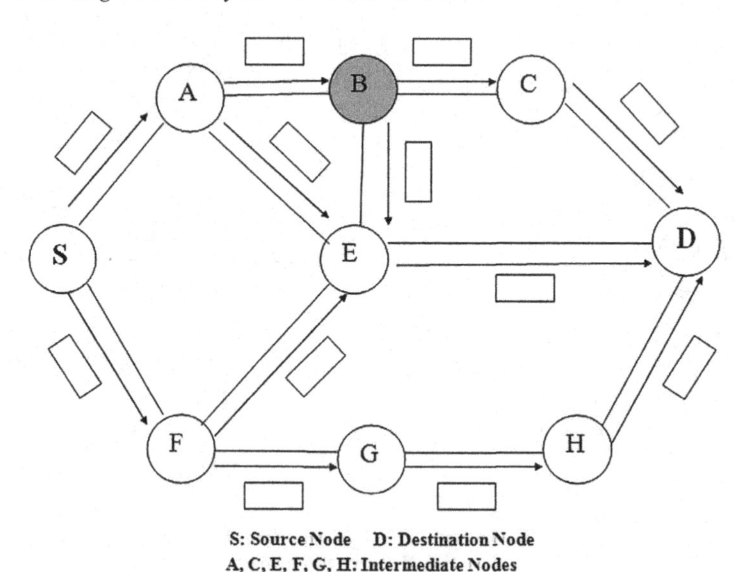

by all other node. The success rate of the attack is higher than the attack success rate of the attacker near to the source node. The Attack Success Rate is higher than the Average.

PREVENTION OF RUSHING ATTACK: SECURE PROCEDURE:

The Secure Procedure involves two phases:

1. Secure Neighbor Detection, and
2. Secure Route Discovery.

Secure Neighbor Detection

In secure neighbor detection (Shahrani, Saad, 2011) the nodes has the bidirectional links between themselves. A node broadcast a message for the neighbour to detect him. This message is called the Advertisement. Mostly On-Demand/Reactive routing protocols follows the secure neighbour detection mechanism. A node which receives the route request considers itself the neighbour of the previous node. A node when transmits the route request it declares a path between sender and receiver. The secure neighbour detection cannot prevent an attacker by receiving the request packet. But if the address of previous node (Jain, Shrivastava, 2013) is unauthorized, them the attacker can claim to be any other node in the network, and can propagate the request. The next nodes will belief the message. Therefore the concept of the Secure Route Discovery is applied.

Secure Route Discovery

In secure Route Discovery sender very rapidly broadcast the route request. For the reduction of the Rushing attack (Corson, Macker, 1999), a random path selection method is used. In traditional approach as soon the node revives the request, it immediately forward it, but in the in this method the node receives all the route request and from them randomly select a request and forward it. The two main parameter of this method are:

1. The total no of request packet received.
2. The algorithm chosen for timeout.

Threshold Values

To reduce the effect of the Rushing Attack in the network, the threshold value can be used. The threshold values are fixed values which are pre-defined on the basis of the average time required by the node to send the acknowledgement packet to source. The entire nodes in the network have the instruction that the packet must reach to the neighbour node at the fixed interval. Each node should check the RREQ of the neighboring node. If there is any Rushing Attacker then it will try to transmit the packet quickly and thus the neighboring node can identify the attacker node and inform about it. It can also blacklist the attacker node.

Watchdog and Pathrater: Watchdog Technique

In the network the method of Watchdog is used to find the malicious node. The Figure 8 shows the implementation of the simple network where S is the source node, D is the destination node, and A, B, C, are the intermediate nodes. The source S transmit the data to destination D. Node S forward the packet to node A. Node A will transmit the packet to the B, but cannot transmit it directly to C because there is no direct link between A and C is present. So it has to forward the packet via B. While transmitting data packet through B, it could be possible for A to verify that whether B has forwarded the packet to C or not. It could also be possible for A to identify that whether the packet header is modified by B or not. Encryption at every node could not be performed because it is very expensive. For this reason

Figure 8. Watchdog

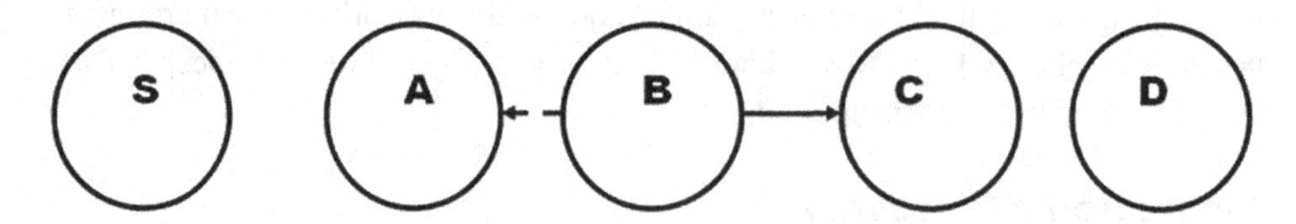

the method of Watchdog is implementation. The Watchdog maintains the buffer of recently sent data packet and compare each overhead packet with the buffer packet. If there is match between the buffer packet and overhead packet, it means that the packet has been delivered and then the packet stored in the buffer is removed and forgotten by the Watchdog. If match is not found then the packet remains in buffer itself. But if the packet remains for longer time in the buffer more than the timeout limit, the watchdog increment the malfunction, and finds the node that is responsible for packet forwarding. But if the computing in more than the certain threshold bandwidth, then the node is misbehaving and thus the watchdog can determines it and can notify to the source of the malicious node.

There are pro and cons of every method, therefore this technique also have the merits and demerits. The advantage of the watchdog with the DSR protocol is that it detect the misbehavior at the forwarding level With the DSR protocol has the advantages that it can detect the misbehavior at the forwarding level and not just at the link level. The main disadvantage of this method it that it is not able to detect the malicious node when there is Ambiguous Collisions, Collusion, False Misbehavior, Limited Transmission Power, Partial Dropping, Receiver Collisions.

In Figure 9 due to the ambiguous collision at A while it listen form B for packet forwarding. The node A could not identify that collision was caused due to packet forwarding to B or B has never forwarded the packet and collision was caused by the other neighboring node of A. This is an uncertainty situation. For this condition A could not accuse B for misbehaving immediately, but it should watch B for a period of time continuously. If A detect that B repeatedly failed to forward the packet, then it assume that B is misbehaving. The receiver collision problem is that where A can tell whether B send the packet to C, but cannot tell whether C receives

Figure 9.

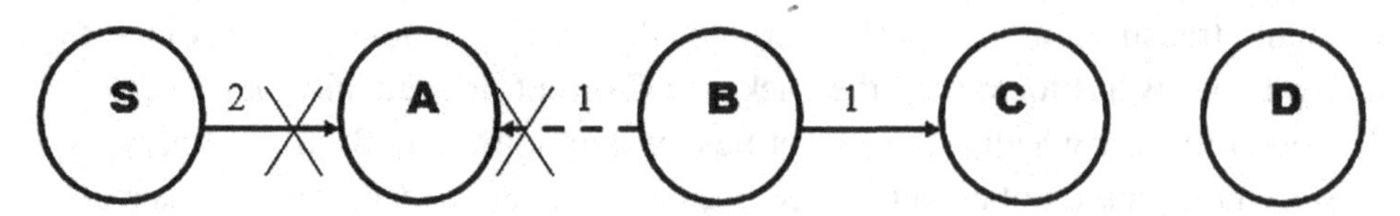

Figure 10.

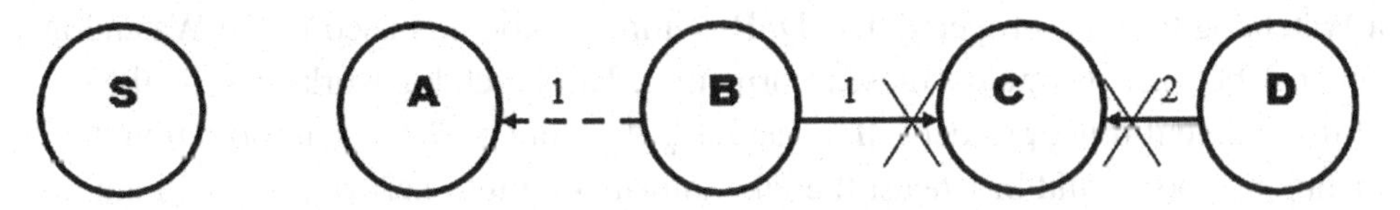

the packet or not. When collision occurs at C while B forwards the packet Figure 10, A could only see that B has forwarded the packet and assumes that C receives the packet successfully.

If C does not receives the packet then the B could skip the retransmission of the packet. Intensely also B could transmit the packet to collide at C by waiting until C is transmitting and forwarding on the packet. In the first case, the node could be selfish and doesn't want to waste power in packet retransmission. But B wastes the battery and CPU time therefore it could not be selfish. An overloaded node will also not engage in this situation. This is a rare occurrence condition. A misbehaving node could also report fake that there are few misbehaving node in the network. For e.g. B could report that C is not forwarding the packet but in real B is not forwarding the packet. Node A will mark C as attacker node whereas the B is the attacker node. But this could be detected later on. As B is passing message to C, any acknowledgement form D to S will go through B. When S will receive the acknowledgement form D which will go through B. If S receives the replies packet from the D then it will doubt because purportedly C drops the packet in the forward direction. If B drops the acknowledgements packets to hide them from A, this will be detected by C and will report the misbehavior to D.

The node that is misbehaving can control its transmission power can avoid the watchdog. This is another problem. A malicious node knows the transmission power required to reach to its neighbors. A node could limit it transmission power the signals are strong enough to be overheard by the previous node but they are too weak to be received by true recipient. If B forward the packet to C and C drops the packet, then B must report it to A. if it does not report it to A, and then it is necessary it ban these successive node from the network. The watchdog configured minimum misbehavior threshold and a node can get around the watchdog by dropping the packet at the lower rate than the minimum configured threshold. This node will not be detected by the watchdog as the malicious node but this node is forced to forward the packet at the configured threshold bandwidth. A great deal is required to maintain the state information at every node to ensure its neighbour that they will not transmits the packet that is being already forwarded. In the watchdog mechanism if a collision at the receiving node occurs, then it is necessary for a node to retransmit it. It is necessary to locate the packet between the two hop for a Watchdog to work properly. The DSR routing protocol is used in the Watchdog network because it contains these information. The watchdog works best on the top of the source routing protocol. If a watchdog does not have these information then malicious node could broadcast the fake information into network, it could not be possible for the watchdog to detect it.

Pathrater

The Pathrater is the combined knowledge of misbehaving nodes with the link reliability to pick the most reliable route. In a network each node runs the Pathrater (Vyavahare, Rawat, Ramani, 2005). Every node in the network maintains the rating of the other node it is connected. Path metric is calculated by the average of the nodes in the path. This path metric is used by the Pathrater as it gives a comparison of reliability of the different path and also allows Pathrater to emulate the shortest length path algorithm when no reliability information is there. But if there exist multiple paths for the same destination then the path with the highest metric is chosen. It is typically different form the standard DSR protocol where shortest path is chosen. The exact path that has been traveled by the packet must be known to the Pathrater. It must be implemented on the top of the source routing protocol. The Pathrater has the following steps for rating the node. In the route discovery phase the node become known to the pathrater in the network. A node rates itself with 1.0. To a newly discovered node the pathrater assigns a neutral rating 0.5. The path rate is calculated to insure that all the other nodes in the path are neutral node expect the suspected misbehaving node. The shortest length path is picked by the Pathrater. The rating is incremented by 0.01 at periodic interval of 200 ms on all actively used paths by the pathrater. An active path is that path that has been recently used by the node to sent packet with previous rate increment interval. The maximum value that a neutral node can attain is 0.8. During the packet forwarding if a link break is detected or if the node becomes unreachable it rating is decremented by 0.05. The lower bound rating of a neutral node is 0.0. The rating of the node that are inactive is not modified by the Pathrater. In the simulation the node suspected for misbehaving are assigned a very high negative value i.e. 100. When the path metric is calculated this negative value indicates the existence of misbehaving node in the path. If a node is marked as malicious node due incorrect information then it is better if that is not permanently excluded from the network. Thus the node that have negative value must increment there values to non-negative value. When the situation arises that the pathrater is not able to detect the path free of attacker then it sends a rout request to enable an extension it is called Send Route Request (SRR.)

SUMMARY

The main cause of the emergence of the mobile ad-hoc network is that the user now always wants to get connected to world. The mobile ad-hoc network has provides as with many convenience. The starting has been done with the introduction of the basic characteristics of the mobile ad-hoc network. The mobile ad-hoc network is getting now a day's more popular therefore there is an increase in the security threat for the network. Some of the threats are caused due to the characteristics of the mobile ad-hoc network such as dynamic topology, mobility, battery power limitations, open access. Some of the dangerous and typical vulnerabilities are also been discussed. These threats have made it necessary to find security solution to these threats also. So the network could be secure and the users are provided with genuine data in the network. The later discussion is on the few most typical threats and there solution to make the network free from these threats. In the end few protection mechanisms are mentioned which can be applied in the network to make it secure form all type of attacker such as internal attacker, external attacker.

CONCLUSION

The main feature of the mobile ad-hoc network is cooperativeness, dynamic topology, lack of defense line, lack of centralized control, limited resource, wireless link which make it insecure in nature. The nodes become selfish to save their power due to limited power supply. Due to absence of centralized control there are many security problems. The routing protocol must have high scalability and services due to dynamic topology. To ensure the security of the network robust security scheme is needed. There is several security solutions which provides solution to several threats in the network. But still there are much more to achieve to make the mobile ad-hoc secure. There are many security criteria requirements that are to be achieved to ensure complete security of the network. The security criteria must cover the location privacy and it should be specialized with application oriented. There is always need to develop the security criteria for the know threat which can detect them quickly and provide proper solution to these attacks with time, space and money bound. There are always possibilities for new attack that make the network misbehaving and we have the challenges to be prepared for all such type of new possible threats.

REFERENCES

Aad, I., Hubaux, H., & Knightly, E.W. (2000). *Impact of Denial of Service Attacks on Ad-Hoc Networks*. Academic Press.

Al-Shabi, M. A. (2012). Attack and Defense in Mobile Ad-Hoc Networks. *International Journal of Reviews in Computing, 12*.

Annadurai, P., & Sivakumar, K. (2009). Impact of rushing attack on multicast in Mobile Ad Hoc Network. *International Journal of Computer Science and Information Security*.

Arya, S., & Arya, C. (2012). Malicious Nodes Detection in Mobile Ad-Hoc Networks. *Journal of Information and Operations Management*, *3*, 210–212.

Asokan, N., & Zapata, M. (2002). *Securing Ad-Hoc routing Protocols. In ACM workshop, Wireless Security*. WiSe.

Birman, K., Ramasubramaniam, V., & Chandra, R. (2001). *Anonymous Gossip: Improving Multicast Reliability in Mobile Ad-Hoc Network*. ICDCS.

Bruschi, D., & Rosti, E. (2002). Secure Multicast in Wireless Networks of Mobile Hosts: Protocols and Issues||. *Mobile Networks and Applications*, *7*(6), 503–511. doi:10.1023/A:1020781305639

Corson, S., & Macker, J. (1999). *Mobile Ad-Hoc Networking (MANET)*. Routing Protocol Performance Issues and Evaluation Considerations.

Dalal, M., Vats, K., Loura, V., & Rohila, D. (2012). *OPNET based simulation and performance analysis of GRP routing protocol. International Journal of Advanced Research in Computer Science and Software Engineering*, 2.

Gagandeep, Aashima, Kumar, & Pawan. (2012). Analysis of Different Security Attacks in MANETs on Protocol Stack A-Review. *International Journal of Engineering and Advanced Technology, 1*.

Ghaffari, A. (2006). Vulnerability and Security of Mobile Ad-hoc Networks. *International Conference on Simulation, Modeling and Optimization WSEAS*, (pp. 22-24).

Hong, X., Kong, J., & Gerla, M. (2006). *Modeling Ad-hoc Rushing Attack in a Negligibility-based Security Framework*. Academic Press.

Jain, S., & Shrivastava, S. (2013). *A Brief Introduction of Different type of Security Attacks found in Mobile Ad-hoc Network. International Journal of Computer Science & Engineering Technology*, 4.

Johnson, D. B., Hu, Y. C., & Perrig, A. (2003). Efficient security mechanisms for routing protocols. *Network and Distributed System Security Symposium (NDSS).*

Johnson, D. B., Perrig, A., & Hu, Y. C. (2002). *A Secure On Demand Routing Protocol for Ad-Hoc Networks*. MobiCom.

Johnson, D. B., Perrig, A., & Hu, Y. C. (2003). *Rushing Attacks and Defense in Wireless Ad Hoc Network Routing Protocols*. WiSe.

Kamaljit, I. L., & Patel. (n.d.). Comparing Different Gateway Discovery Mechanism for Connectivity of Internet & MANET. *International Journal of Wireless Communication and Simulation, 2*(1), 51-63.

Kong, J., Hong, X., & Gerla, M. (2010). *A new set of passive routing attacks in mobile ad hoc network*. Academic Press.

Krsul, I., Schuba, C., Kuhn, M., Spafford, E., Sundaram, A., & Zamboni, D. (1997). Analysis of a Denial of Service Attack on TCP. *IEEE Symposium on Security and Privacy*.

Lakhtaria, K. I. (2010). *Study, analysis and modeling of IP multimedia systems on next generation networks providing mobile and fixed multimedia services*. Academic Press.

Lakhtaria, K. I. (2010). *Analyzing Zone Routing Protocol in MANET Applying Authentic Parameter*. arXiv preprint arXiv:1012.2510.

Lakhtaria, K. I. (2011). Protecting computer network with encryption technique: A Study. In *Ubiquitous Computing and Multimedia Applications* (pp. 381–390). Springer Berlin Heidelberg. doi:10.1007/978-3-642-20998-7_47

Lakhtaria, K. I. (2012). *Technological Advancements and Applications in Mobile Ad-hoc Networks: Research Trends*. Information Science Reference.

Lee, S.J., Su, W., & Gerla, M. (2002). On demand multicast routing protocol in multihop wireless mobile networks. *Mobile Networks and Application,* 441-453.

Luna, J. J. G., & Madruga, E. L. (2001). Scalable Multicasting: The Core-Assisted Mesh Protocol. *Mobile Networks and Applications*, *6*(2).

Luo, Yang, & Zhang. (2004). Security in Mobile Ad Hoc Networks: Challenges and Solution. IEEE Wireless Communication, 11, 38-47.

Lupu, T. G. (n.d.). *Main type of attack in wireless sensor network*. Academic Press.

Macker, J., & Corson, S. (1999). *Mobile Ad Hoc Networking: Routing Protocol Performance Issues and Evaluation Consideration*. RFC.

Nguyen, H. L., & Nguyen, U. T. (2008). A Study of different types of attack on multicast in mobile ad hoc networks. *Ad Hoc Networks*, *6*(1), 32–46. doi:10.1016/j.adhoc.2006.07.005

Nguyen, U. T., & Nguyen, H. L. (2006). Study of Different Types of Attack on Multicast in Mobile Ad Hoc Networks. *IEEE International Conferences on Networking*.

Palanisamy, V., & Annadurai, P. (2009). *Impact of rushing attack on multicast in mobile ad hoc network. International Journal of Computer Science and Informational Security*, 4.

Pateari, N. (2011). Performance Analysis of Dynamic Routing Protocol in Mobile Ad-Hoc Network. *Journal of Global Research in Computer Science, 2*.

Perrig, A., Hu, Y. C., & Johnson, D. B. (2003). *Efficient Security mechanisms for routing protocols.* Network and Distributed System Security Symposium.

Rao, M. J., Moyer, & Rohatgi, P. (1999). A Survey of Security Issues in Mlticast Communication. IEEE Network, 12-23.

Sankaranarayan, V., & Tamilselvan, L. (2006). *Solution to prevent rushing attack in wireless mobile ad hoc network*. IEEE.

Selvaraj, G., Sivakumar, K., (2013). Overview of various attacks in MANET and Counter Measures for Attacks. *International Journal of Computer Science and Management, 2*(1).

Shahrani, A., & Saad, A. (2011). Rushing Attack in mobile ad hoc Networks. *IEEE International Conference on Intelligent Network and Collaborative System*. doi:10.1109/INCoS.2011.145

Shahrani, A. S. A. (2011). *Rushing Attack in Mobile Ad Hoc Networks. In Intelligent Networking and Collaborative Systems*. INCS.

Sharma, K., Khandelwal, N., & Prabhakar, M. (2011). *An Overview of Security Problems in MANET*. Academic Press.

Shrivastava, S. (2013). *Rushing Attack and its Prevention Techniques. International Journal of Application or Innovation in Engineering Management*, 2.

Shrivastava, S. (2014). A New Technique to Prevent MANET against Rushing Attack. *International Journal of Computer Science and Information Technologies*, *5*, 3460–3464.

Singh, A., Goyal, P., & Batra, S. (2010). A Literature Review of Security Attack in Mobile Ad-hoc Networks. *International Journal of Computers and Applications, 9*.

Su, W., Lee, S. J., & Gerla, M. (2002). On-Demand Multicast Routing Protocol in Multihop Wireless Mobile Networks. *Mobile Networks and Applications, 7*.

Syrotiuk, V. R., Chalamtse, I., & Basagni, S. (1999). *Dynamic Source Routing for Ad Hoc Network Using The Global Positioning System*. IEEE.

Tripathi, R., Agarwal, R., & Tiwari, S. (2011). Performance Evolutation and Comparison of AODV and DSR under Adversarial Environment. *IEEE Conference on Computational Intelligence and Communication Systems*.

Vijayalakshmi, S., & Rabaral, S.A. (2014). Rushing Attack Mitigation in Multicast MANET (RAM). *International Journal of Research and Reviews in Computer Science, 1*(4).

Visalakshi, P. (2011). *Security issues and vulnerabilities in mobile ad hoc network- A survey. International Journal of Computational Engineering Research*.

Vyavahare, P. D., Rawat, A., & Ramani, A. K. (2005). *Evaluation of Rushing Attack on Secured Message Transmission Protocol (SMT/SRP) for Mobile Ad-hoc Network*. IEEE.

Yi, P., Dai, Z., Zhang, S., & Zhong, Y. (2010). A New Routing Attack in Mobile Ad Hoc Networks. *International Journal of Information Technology, 11*(2), 83–94.

Zhang, D., Wang, H., & Shin, K. G. (2002). *Detecting SYN Flooding Attacks. IEEE INFOCOM*. IEEE.

Zhong, Y., Zhang, S., & Yi, P. (2008). A New routing attack in mobile ad hoc network. *IJIT, 11*, 83–94.

KEY TERMS AND DEFINITIONS

Eavesdropping: Eavesdropping attack is the process of gathering information by snooping on transmitted data on legitimate network.

Hop Count: The Hop Count measures and reconstructs the mapping details in each node and it finally exploits the diameter feature to detect the distortion caused by the malicious node.

MANET: Stands for "Mobile Ad Hoc Network." A MANET is a type of ad hoc network that can change locations and configure itself on the fly. Because MANETS are mobile, they use wireless connections to connect to various networks.

Pathprater: The Pathrater is the combined knowledge of misbehaving nodes with the link reliability to pick the most reliable route.

Rushing Attack: Rushing attacks in MANETs cause system resources to become scarce and isolates legitimate users from the network. Therefore, this sort of attack significantly influences network connectivity and weakens networking functions and capabilities such as control and message delivery.

Watchdog: In the network the method of Watchdog is used to find the malicious node.

Chapter 11
Intrusion Detection and Tolerance in Next Generation Wireless Network

Deshraj Ahirwar
UIT RGPV, India

Kirti Raj Bhatele
UIT RGPV, India

P. K. Shukla
University Institute of Technology, India

Prashant Shukla
SIRT RGPV, India

Sachin Goyal
UIT RGPV, India

ABSTRACT

Organizations focuses IDPSes for respective purposes, e.g. identifying problems with security strategies, manually presented threats and deterring individuals from violating security policies. IDPSes have become a necessary technique to the security infrastructure of approximate each association. IDPSes typical record information interrelated to practical events, security administrators of essential observed events and construct write up. Many IDPSes can also respond to a detected threat by attempting to thwart it succeeding. These use several response techniques, which involve the IDPS restricting the attack, changing the security environment or the attack's content. Sensor node should diverge in size from a shoebox down to the small size, although functioning "motes" of genuine microscopic dimensions have to be formed. The cost of sensor nodes is variable, from a few to thousands of dollars, depend on the complexity of the sensor nodes. Size and cost constraints on sensor nodes represent in corresponding constraints on resources such as energy, memory, computational velocity and communications bandwidth. The arrangement of the WSNs alters itself from a star network to efficient multi-hop wireless mesh network. The proliferation technique between the hops of the network can be routing or flooding.

DOI: 10.4018/978-1-4666-8687-8.ch011

INTRODUCTION

An intrusion detection system is a software application that monitors system activities for malicious policy violations and generates reports to a management station. IDS come up to in a "flavors" and move toward the goal of detecting suspicious traffic in different types. There are network based and host based intrusion detection systems. System should try to stop an intrusion attempt but it is not expected of a monitoring system. Intrusion detection and prevention systems are primarily listening on identifying possible incidents, logging information, and reporting attempts (Scarfone & Mell, 2007). A wireless sensor network of spatially distributed autonomous sensors to test environmental conditions, for example temperature, sound, pressure, etc. and to cooperatively pass data through the network to prime location. Modern networks are bi-directional. Development of wireless sensor networks was aggravated by military applications such as battlefield surveillance; presently networks are second-hand in many industrial and consumer applications, e.g. industrial process monitoring and control, machine health monitoring. The various Topologies is illustrated in Figure 1.

The WSN is built of "nodes" – from a few to several hundreds or even thousands, Node is coupled to one sensors. Sensor network node has typically several parts: a radio transceiver with an internal antenna to an external antenna, a microcontroller, an electronic circuit for interfacing with the sensors and an energy source, the embedded form of energy harvesting. Sensor node may fluctuate in size from a shoebox down to the different size, although functioning "motes" of genuine microscopic dimensions to be displayed. The cost of sensor nodes is variable, A few to

Figure 1. Topologies of devices

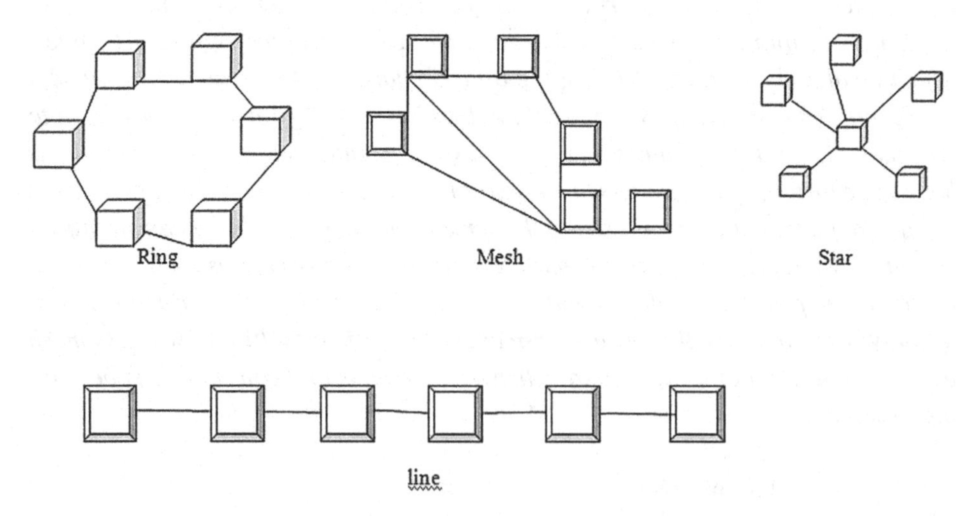

huge of dollars, depending on the difficult of the sensor nodes. Size and cost parameter (Nitin, Mattord, Verma, 2008) on sensor nodes generate in corresponding constraints on objects e.g. energy, memory, computational speed and communications bandwidth. Architecture of the WSNs be divert from a star network to powerful multi-hop wireless mesh network. The transmission strategies for hops of the network will be routing or flooding. Wireless sensor networks are rising research area.

ANOMALY-BASED INTRUSION DETECTION SYSTEM

Anomaly-Based Intrusion Detection System, is for detecting computer intrusions and false use by monitoring system activity and identify as either normal or anomalous. It is based on heuristics, rather than patterns, Try to define any type of cybercrime. It conflicts to signature based systems used to perceive attacks.

In order to find attack traffic, the system should train to be aware of system activity. It can be skilled in several ways, mostly artificial intelligence (Heberlein, 1990) type techniques. Systems using neural networks proposes great effect. Second process is to characterize normal usage of the system comprises mathematical model, and flag any deviation from this as an attack. It shows strict anomaly recognition.

Some faults are in anomaly-based Intrusion Detection, identified as high false positive speed and the ability to be fooled by a correctly delivered attack. Efforts prepared to address for issues through tactics used by PAYL and MCPAD. The Interconnection among computing devices in wireless network is illustrated with the help of Figure 2.

Figure 2. Interconnection of computing devices in wireless network

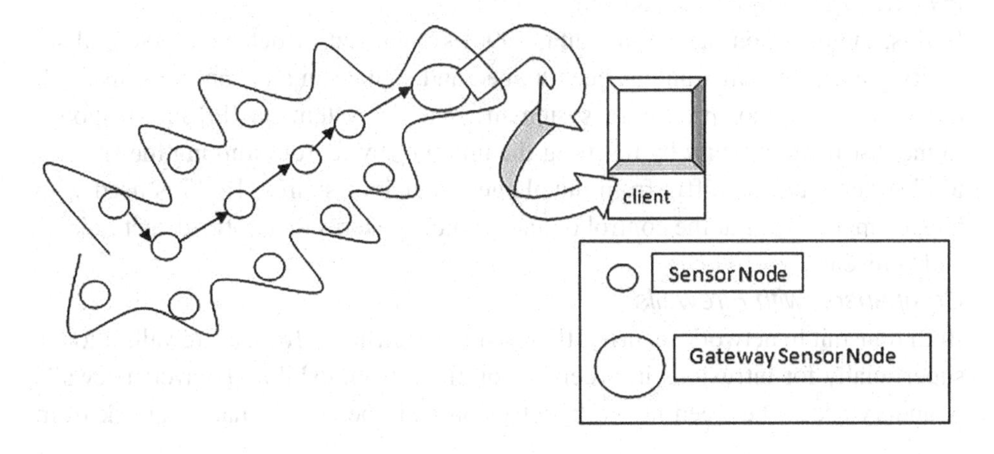

HIDS and NIDS

Intrusion detection systems types:

1. Network based;
2. Host based intrusion detection systems.

Network Intrusion Detection System

These are situated at a strategic point to examine traffic to and from connected devices on the network. It gives an analysis for a transient traffic on the complete subnet, acts in a promiscuous type, and links the traffic that is authorized on the subnets to the library of known attacks. Once the attack identified, or exceptional behavior can be observed, the security signals can be sent to the supervisor. For example of the NIDS would be installing it on the subnet (Denning, 1987) where firewalls are in order to see if someone is expecting to break into the firewall. Normally one would scan all inbound and outbound traffic, although it should construct a bottleneck that would destruct the overall irregularity of the network.

Host Intrusion Detection System

It focuses on individual network component. It invigilates the inbound and outbound packets from device only and gives signal the customer or administrator if doubtful activity is detected. It capture snapshot of system files and matches with inside snapshot. When serious system files were tailored, for inspection the vigilant is sent to the administrator .In HIDS usage are on mission significant machines, to alter their configurations that are not estimated.

System-specification with using custom tools and honeypots of IDS:

Passive and Reactive Systems

In this, the intrusion detection (Lunt, 1990) system sensor detects a potential security breach, Information has been logged and signals an alert on the console. It known as an intrusion prevention system in a reactive system, the IPS auto-responds to the distrustful activity by resetting the union or by reprogramming the firewall to obstruct network traffic from the alleged malicious source. IDPS is used with happen manually or at the control of an operator; systems that both "detect (alert)" and "prevent".

Comparison with Firewalls

Both transmit to network security, IDS is slightly different from a firewall. It looks superficially for intrusions in order to stop them from exhibit. To avert firewalls boundary access between networks intrusion and does not signal an attack from inside the network. IDS detect assumed intrusion and it has occurred and signals a beep. IDS vigilate for attacks that start from within. This is achieved by testing network communications, identifying heuristics and patterns of regular computer attacks, and taking action to observant administrators. It terminates relations is called an intrusion prevention system, and it points to an application layer firewall.

All Intrusion Detection Systems use one of two detection techniques:

1. Statistical anomaly;
2. Signature-based IDSs.

Statistical Anomaly-Based IDS

The IDS is anomaly based. It will monitor network traffic and compare it in opposition to an construct baseline. Baseline should determine type (Winkler, 1990) of network- bandwidth type is used, protocols definition, number of ports and devices adjacent to each other- and the administrator or device when traffic is occurred tends to anomalous, or different, with respect to baseline. A constraint may hoist a False Positive beep for a legitimate use of frequency peak value if the baselines are intelligently programmed.

Signature-Based IDS

An IDS with signature will examine packets on the network. It also evaluates them against a database of signatures. It provides security from known malicious threats. It shows analogous to the way most antivirus software detects malware. There will be a lag between a new threats being concealed. In IDS the signature for detecting that threat being applied. During that lag time your IDS would be unable to detect the new threat. The Figure 3 is illustrating the networking element of an intrusion detection system.

AAFID

Autonomous Agents for Intrusion Detection, a distributed intrusion detection system. Nodes of the IDS are approved in a hierarchical formation in a tree in this structure. In hierarchical building first invention of Distributed Intrusion Detection Systems.

The general types of distributed intrusion detection systems are as follows:

- Hierarchical;
- Network architecture;

Figure 3. Networking element of intrusion detection system

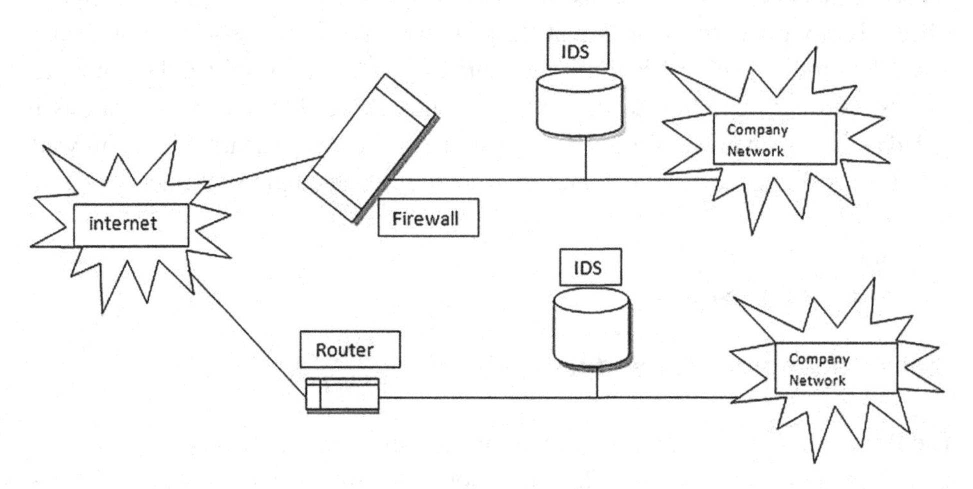

- Hybrid architecture;
- Mobile agent architecture;
- Disturbed.

Mobile Agents are not there in AAFID.

PROTOCOL-BASED INTRUSION DETECTION SYSTEM

On a web server a typically installed *protocol-based intrusion detection system* is an intrusion detection system, monitoring and analysis of the protocol are applications in use by the computing system. A PIDS might monitor the elegant behavior and protocol state and shall consist of a system that would located at server end, In shielding system monitoring and analyzing the communication between an attached device.

PIDS would be at the front end of a web server monitoring the HTTP stream. Since it understands the HTTP related to the web server. it is demanding to protect it can produce greater security than less in-depth strategies such as filtering by IP address, at the cost of improved computing on the web server huge security comes.

When system uses http then it would need to reside in the "shim" or boundary between. Where http is un-encrypted and instantly prior to it in coming the Web presentation layer.

MONITERING DYNAMIC BEHAVIOR

Initially PIDS seek for, and impose, the exact application of the protocol.

At an emerging level thc PIDS can seek or be qualified acceptable constraint of the protocol, and thus better determine anomalous manners.

Applications

Process Management

Field vigilance is a normal utility of WSNs. In this, the WSN is supported over a section where some occurrence is to be observed. In defense the use of sensors detects intrusion; a practical scenario is the geo-fencing of gas or oil pipelines. Field vigilance is advance part (Peiris, 2013).

Health Care Monitoring

Medical applications types:

1. Wearable, and
2. Entrenched.

Wearable devices are for the stiff plane of a human or just at shut propinquity of client. Transportable treatment equipment are that inserted inside human body. Remains location dimension and locality of anyone, for observing of ill patients in hospitals. Body-area networks can assemble information about an individual's health, strength, and power spending.

Environmental Sensing

Monitoring environmental parameters have many applications, for example they point out the additional challenges of ruthless environments and abridged power supply. The entire setup of interconnection among IDS devices to showcase environmental sensing is illustrated by the Figure 4.

- **Air Pollution Monitoring:** Wireless sensor networks have been deployed in several cities (Stockholm, London and Brisbane) to observe the attentiveness

Figure 4. Node interconnection among devices of IDS

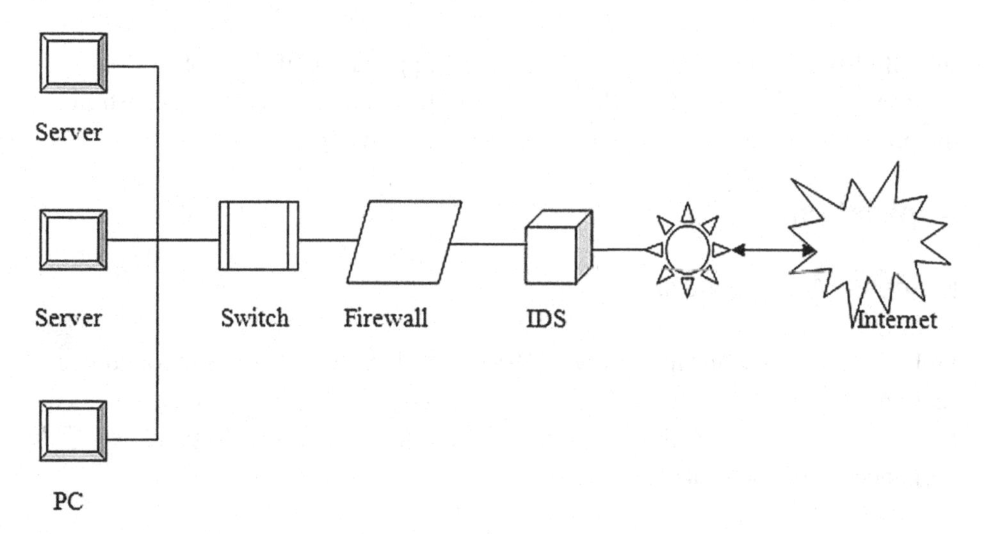

of hazardous gases for people. These can take gain of the ad hoc wireless links more willingly than wired installations, which also make them more portable for testing readings in dissimilar areas.

- **Forest Fire Detection:** Sensor Nodes can be constructed in a forest to identify when a fire has ongoing. To determine temperature the nodes can be connected with sensors, humidity and gases which are created by fire in the trees or vegetation. Initially detection is tuff for a victorious action of the fire fighters. Wireless Sensor Networks, the fire brigade will be able to know fire location.
- **Landslide Detection:** To determine the trivial activities of soil and changes in a range of parameters in landslide detection system makes use of a wireless sensor network .that may come about before or throughout a landslide. Through the data gathered it may be possible to know the happening of landslides long prior to it truly happens.
- **Water Quality Monitoring:** Monitoring of water is to analyze water properties in dams, rivers, lakes & oceans, as well as secretive water reserves. wireless distributed sensors facilitates the manufacture of a more precise map of the water status, and permits the consumption of observing stations in locations of complex access, devoid of the require of handbook data retrieval.

- **Natural Disaster Prevention:** Wireless sensor networks is use to avoid natural disasters, like floods. Wireless nodes have been applied in rivers. Monitoring of water level is in real time.

Industrial Monitoring

- **Machine Health Monitoring:** In machinery condition-based maintenance uses Wireless sensor networks since this preferred considerable cost hoard and gives new functionality in wired systems, the fixing of adequate sensors is regularly restricted by the cost of wiring. Unreachable locations, rotating machinery, perilous or constrained areas, and mobile resources can now be reached with wireless sensors.
- **Data Logging:** The gathering of data for monitoring of environmental information has wireless sensor networks, it is as the monitoring of the temperature in a fridge to the level of water in overflow tanks in nuclear power plants. The arithmetic information can then be used to explain how systems have been functioning.
- **Wastewater Monitoring:** Observing the performance and level of water includes behavior along with quality of underground water and ensuring a country's water infrastructure.

Structural Health Monitoring

Wireless sensor networks used to inspect the condition of construction infrastructure and related geo-physical processes which is real time, and for long periods through data logging, using properly interfaced sensors.

Characteristics

Buzzwords of a WSN Include:

- Power consumption constraints for nodes using batteries or energy harvesting.
- Ability to cope with node failures.
- Mobility of nodes.
- Heterogeneity of nodes.
- Scalability to large scale of deployment.
- Ability to withstand harsh environmental conditions.
- Ease of use.

POWER CONSUMPTION CONSTRAINTS FOR NODES USING BATTERIES OR ENERGY HARVESTING

The power which is being used by nodes and other clients is very limited and unstable with respect to time, now our aim is to identify power losses and applying stable energy sources so as not to break down the vigilance system.

- **Ability to Cope with Node Failures:** To identify the power failures nodes and causes and after tracing such weak point make them available again inside network for communication.
- **Mobility of Nodes:** The nodes and station are dynamic and changing their position deliberately so as to find out the desired location we will apply zone direction optimization.
- **Heterogeneity of Nodes:** The nodes and devices inside the network are not homogeneously distributed in spite they are clustered respect to requirement of packets from various location.
- **Scalability to Large Scale of Deployment:** The overall resources available for communication are generally time varying so we have to make them scalable along with unstable conditions.
- **Ability to Withstand Harsh Environmental Conditions:** When a particular station is trapped under intruder attack there should be rescue against the attack to protect that particular information resource from being steeled and withstand the network under such tuff conditions.
- **Ease of Use:** At any point of operation user should not feel any difficulty in establishing the communication under protocol defined for communication. User should easily track and identify the path and instruction essential for detection of attacker.

Cross-Layer Design

Cross-layer is becoming an implicative study area for wireless communications. With the traditional layered move toward brings three faults to us.

1. Conventional layered come close to cannot share unusual information among different layers, which leads to each layer not having entire information. The customary layered come near cannot warranty the optimization of the complete network.
2. The conventional layered come near to not have the capability to become accustomed to the environmental change.

3. The interference among the different users, access confliction, fading, and the change of environment in the wireless sensor networks, conventional layered advance for wired networks is not related to wireless networks. So we can use cross-layer to make the optimal modulation to recover the transmission performance, such as data rate, energy efficiency, QoS, etc. Sensor nodes expected as small computers which are exceeding fundamentally in terms of their interfaces. They usually consist of a processing unit through limited computational power and limited memory, sensors or MEMS, a communication device (alternatively optical), and a power source regularly in the form of a power source. Other possible inclusions are energy harvesting modules, secondary ASICs, and possibly secondary communication interface (e.g. RS-232 or USB).

Stations are one or more components of the WSN with much more computational, energy and communication resources. They act as a gateway involving sensor nodes and the end user as they uncharacteristically onward data from the WSN on to a server. Other special apparatus in routing based networks are routers, designed to compute, compute and distribute the routing tables.

Other Typical Properties of Distributed Systems

The system has to endure faults in personage computers. The organization of the system (network topology, network latency, number of computers) is not known in move ahead, the system may consist of dissimilar kinds of computers and network links, and the system may alter during the execution of a spread program (Silva, Ghanem, & Guo, 2012).

Restricted, incomplete view of the system is inside the system. Only one part of the input is known to a computer. The Figure 5 illustrated the various Attack trace engine model elements and its corresponding Datalog rule.

- **Sensor Node:** WSN is to fabricate low cost and tiny sensor nodes. Producing WSN hardware and the commercial position can be compared to home computing in growing number of little companies in the 1970s. Many of the nodes are at rest in the research and development phase, mainly their software. Additionally inherent to sensor network acceptance is the use of very low power methods for radio communication and data attainment.

Figure 5. Interconnection of subnets in sensor network

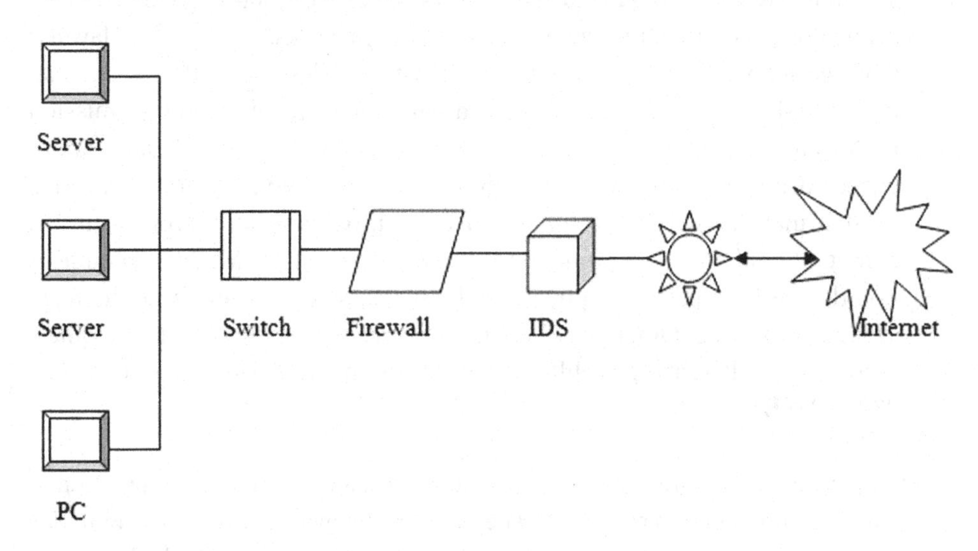

Table 1. Attack trace engine model element

Attack Trace Engine Model Element	Datalog Rule
Remote service exploration resulting in privilege escalation u_ig vulnerable services.	execCode(Attacker,Host,priv):- vulExists(Host,CVE_id,program) vulproperty(CVE_id,remoteExploit,privEscalation) networkService(Host,program,Protocol,Port,Priv)
Remote client exploration resulting in privilege escalation using vulnerable client programs.	execCode(Attacker,Host,priv):- vulExists(Host,CVE_id,program) vulproperty(CVE_id,remoteExploit,privEscalation)
Local client exploration resulting in privilege escalation using vulnerable client programs.	execCode(Attacker,Host,owner) vulExists(Host,CVE_id,program)
Local user exploration resulting in privilege escation using Trojan programs.	execCode(Attacker,Host,owner)
Local file access exploration	accessFile(Principal, Host,Access,Path)
Remote file access exploration using NFS	accessFile(Principal, Host,Access,Path) malicious(principal) execCode(Principal, Client root)
Multi-hop network access	netAccess (Principal, TargetHost,Protocol,Port)
Policy violation	Policy violation (principal access data)

Applications such as WSN communicate with a Local Area Network or Wide Area Network through a gateway. In the WSN and the other network the Gateway becomes a bridge. In a remotely positioned server this enables data to be stored and processed by devices with more resources (See Table 1).

Software

It determines the life span of WSNs, Energy is the scarcest resource of WSN nodes. Where ad hoc communications are a key constituent then WSNs may be deployed in huge numbers in a variety of environments, as well as remote and hostile regions. Due to this, algorithms and protocols necessitate to address the subsequent issues:

- Lifetime maximization;
- Robustness and fault tolerance;
- Self-configuration.

Lifetime Maximization

Sensor nodes ought to be energy efficient since their inadequate energy resource determines their lifetime. Power spending of the sensing device should be minimized. Usually turn off the radio transceiver when not in uses to preserve power the nodes.

Significant topics in WSN (Wireless Sensor Networks) software research are:

- Operating systems;
- Security;
- Mobility;
- Usability;
- Maintenance.

Sensor Node

In a wireless sensor network that is proficient of performing some processing, a sensor node, also identified as a remote, is a node gathering sensory information and communicating with other connected nodes in the network. A mote is a node but a node is not constantly a mote. The interconnection of subnets in sensor networks is illustrated with the help of figure 6.

Applications as diverse as earthquake measurements to warfare, wireless sensor nodes have existed for decades. Inside a cubic millimeter of space modern improvement of small sensor nodes dates back to the 1998 Smartdust project and the NASA Sensor Webs Project One of the objectives of the Smartdust project used to form autonomous sensing and communication. This project ended early on, it led to many more research projects. They converses most important research centres in Berkeley NEST and CENS. The researchers drawn in in these projects coined the term *mote* to submit to a sensor node. The corresponding term in the NASA Sensor Webs Project for a physical sensor node is *pod*, even if the sensor node in a Sensor Web can be an additional Sensor Web itself. Physical sensor nodes have been capable to add to their capability in conjunction with Moore's Law. The chip footprint contains more composite and lower powered microcontrollers. Thus, for the same node footprint, more silicon ability can be packed into it. These days, motes spotlight on provided that the longest wireless range, the lowly energy spending and the easiest advance process for the user. The Figure 5 is illustrating the Interconnection among subnets in a sensor network.

Controller

Processes data and controls the functionality of other mechanism in the sensor node in controller perform tasks. Controller is a microcontroller though the most common extra alternatives that can be used as a controller are: a common intention desktop microprocessor, digital signal processors, FPGAs and ASICs. A microcontroller is frequently used in many embedded systems such as sensor nodes since of its low cost, flexibility to connect to other devices, simplicity of programming, and low power consumption. A common purpose microprocessor normally has elevated power consumption than a microcontroller, consequently it is often not measured a fitting choice for a sensor node. but in Wireless Sensor Networks the wireless communication is regularly modest: i.e., simpler, easier to process modulation and the signal processing tasks of actual sensing of data is fewer difficult. Digital Signal Processors may be elected for broadband wireless communication applications, Therefore the recompense of DSPs are not frequently of much weight to wireless sensor nodes. FPGAs can be reprogrammed and reconfigured according to requirements, excluding this takes more time and energy than preferred.

Tranceiver

For free radio ISM band used in Sensor nodes regularly, spectrum allocation and global accessibility. The potential choices of wireless transmission media are radio frequency, optical communication (laser) and infrared. Lesser amount of energy is in Lasers, except need line-of-sight for communication and are sensitive to atmospheric conditions. Infrared, akin to lasers, needs no antenna but it is partial in its broadcasting capacity. The nearly all appropriate that fits most of the WSN applications similar to Radio frequency-based communication. WSNs use license-free communication frequencies: 173, 433, 868, and 915 MHz; and 2.4 GHz. The functionality of both transmitter and receiver are collective into a single device identified as a transceiver. Transceivers regularly lack inimitable identifiers. The operational states are transmitted, receive, idle, and sleep. Which is existing generation transceivers have built-in state machines to perform a few operations repeatedly.

Power consumption nearly equal to the power consumed in receive mode with Transceivers operating in idle mode. Hence, it is better to completely shut down the transceiver rather than abscond it in the idle mode where it is not transmitting or receiving. A sufficient power is obsessive when switching from sleep mode to transmit mode in order to transmit a packet.

External Memory

The most applicable memory are the on-chip memory of a microcontroller and Flash memory—off-chip RAM is rarely with respect to energy perspective, if ever, used. Due to their cost and storage capacity flash memories used. Memory requirements are very much application reliant. Two categories of memory based on the reason of storage are: user memory used for storing application related or personal data, and program memory used for programming the device. Program memory also contains detection data of the device.

Power Source

When node is a popular explanation a wireless sensor is complicated or not viable to run mains make available to the sensor node, while the wireless sensor node is regularly positioned in a hard-to-reach setting, changing the battery again and again can be problematic. A significant portion in the development of a wireless sensor node is making sure that there is for all time adequate energy existing to power the system. For sensing, communicating and data processing. More energy is compulsory for data communication than any other process the sensor node consumes power. The energy rate of transmitting 1 Kb a distance of 100 metres is around the same as that used for the execution of 3 million instructions by a 100 million commands per second/W processor. In batteries or capacitors power is stored whichever. Batteries, both rechargeable and non-rechargeable, are the main source of power supply for sensor nodes. These classifications according to electrochemical material used for the electrodes such as NiCd (nickel-cadmium), NiZn (nickel-zinc), NiMH (nickel-metal hydride), and lithium-ion. Sensors are able to renew their energy from solar sources, temperature differences, or vibration. Two power saving policies available are Dynamic Power Management (DPM) and Dynamic Voltage Scaling (DVS). DPM conserves power by shutting down parts of the sensor node which are not at present used or dynamic A DVS idea varies the power levels within the sensor node depending on the non-deterministic workload. The voltage next to with the frequency, to obtain quadratic diminution in power consumption it is doable.

Sensors

Sensors are hardware devices that construct a measurable reaction to a change in a physical form like temperature or pressure. Sensors measure physical data of the parameter to be monitored. The incessant analog signal created by the sensors is digitized by an analog-to-digital converter and sent to controllers for auxiliary processing. A sensor node should be petite in size, consume exceptionally low energy, activate in high volumetric densities, be sovereign and operate unattended, and be adaptive to the environment. When wireless sensor nodes are usually very small electronic devices, they can only be operational with some degree of power source of less than 0.5-2 ampere-hour and 1.2-3.7 volts.

Sensors are confidential into three categories: passive, omni-directional sensors; passive, narrow-beam sensors; and dynamic sensors. The data lacking truly manipulating the environment is sensed by Passive sensors by dynamic probing. That is self-powered; energy is required only to amplify their analog signal. on the go sensors actively probe the environment, for example, from a power source a sonar or radar sensor, and they have need of continuous energy. Narrow-beam sensors have a well-defined notion of direction of quantity, analogous to a camera. There is no notion of route implicated in their measurements in Omni-directional sensors.

WSNs works with reactive, omni-directional sensors. Each sensor node has a convinced area of exposure for which it can reliably and truthfully report the meticulous extent that it is observing. In sensors a number of sources of power expenditure are: signal sampling and conversion of physical signals to electrical ones, signal conditioning, and analog-to-digital conversion. Spatial density of sensor nodes in the field may be as high as 20 nodes per cubic meter.

PLATFORM

Simulation of WSNs

Simulation and Agent-based modeling is the barely paradigm which allows the simulation of compound behavior in the environments of wireless sensors (such as flocking). Comparatively new paradigm is in Agent-based simulation of wireless sensor and ad hoc networks. Agent-based modeling was initially based on social simulation (Niazi & Hussain, 2011).

OPNET, NetSim, WSNet and NS2 are Network simulators can be used to simulate a wireless sensor network.

Table 2. BI functionality by type of BI user

Information Consumer	Information Producer
View	Personalize
Navigate	Assemble
Modify	Craft
Explore	Source
Act	Develop

Development

IDS concept consisted of a set of tools intended to help administrator's analysis audit trails. User access logs, file access logs, and system event logs are examples of audit trails. The Table 2 is representing the BI functionally by Type of BI user in a tabular form below where as Table 3 is presenting BI Functionality: Smart Phones Vs Tablets.

Fred Cohen noted in 1984 that it is impractical to detect an intrusion in every case, with the total of usage that the resources needed to detect intrusions raise.

In 1986 Dorothy E. Denning, assisted by Peter G. Neumann, published a model of IDS that formed the basis for countless systems today. Model used information for anomaly detection, and resulted in early IDS at SRI International named the Intrusion Detection Expert System (IDES), which ran on Sun workstations and could reflect on both user and network level data. IDES had a dual draw near with a rule-based Expert System to perceive known types of intrusions plus a statistical anomaly detection component based on profiles of users, host systems, and target systems. Lunt proposed adding an artificial neural network as a third component. Three components could then report to a resolver. In 1993 with the Next-generation Intrusion Detection Expert System (NIDES) SRI followed IDES.

Table 3. BI functionality: smart phones vs. tablets

SmartPhone		Tablet	
View	Personalize*	View	Personalize
Nevigate	--	Nevigate	Assemble
Modify*	--	Modify	--
--	--	Explore	--
Act*	--	Act	--

* Not widely supported.

Table 4. Design and development of intrusion detection

Scalability	Flex 7500	WLC 5500/Wism-2/Wism-1
Total Access Points	2,000	500
Total Clients	20,000	7,000
Max HREAP Groups	500	100
Max APs per HREAP Groups	50	25
Max AP Groups	500	500

In 1988 based on the work of Denning and Neumann Haystack was also developed this year using statistics to diminish audit trails. The Multics intrusion detection and alerting system (MIDAS), an expert system using P-BEST and Lisp, was developed.

Wisdom & Sense (W&S) was a statistics-based anomaly detector. W&S created rules based on statistical analysis, and then used those rules for anomaly detection.

The Time-based Inductive Machine (TIM) did anomaly detection using inductive learning of sequential user patterns in Common Lisp on a VAX 3500 computer. For anomaly detection on a Sun-3/50 workstation the Network Security Monitor (NSM) performed masking on access matrices. The Information Security Officer's Assistant (ISOA) was a 1990 prototype that considered a variety of strategies together with statistics, a profile checker, and an expert system. at AT&T Bell Labs Computer Watch used statistics and rules for audit data diminution and intrusion detection.

In 1991, researchers at the University of California, Davis created a prototype Distributed Intrusion Detection System (DIDS), which was also an expert system. The Network Anomaly Detection and Intrusion Reporter (NADIR), also in 1991, was a prototype IDS developed at the Los Alamos National Laboratory's Integrated Computing Network (ICN), and was profoundly subjective by the work of Denning and Lunt. NADIR used a statistics-based anomaly detector and an expert system (see Table 4).

For packet analysis from libpcap data the Lawrence Berkeley National Laboratory announced Bro in 1998, which used its own rule language. Network Flight Recorder (NFR) in 1999 also used libpcap. APE was developed as a packet sniffer, also using libpcap, in November, 1998, and was renamed Snort one month later. APE has since become the world's largest used IDS/IPS system with over 300,000 active users. The Audit Data Analysis and Mining (ADAM) IDS in 2001 used tcpdump to build profiles of rules for classifications. In 2003, Dr. Yongguang Zhang and Dr. Wenke Lee argue for the importance of IDS in networks with mobile nodes. The Table 3 represents the Design and Development of Intrusion detection in terms of Scalability parameters such as Total access points, total Clients, Max HREAP Groups etc.

OPERATING SYSTEMS

Generally less complex Operating systems for wireless sensor network are use in nodes than general-purpose operating systems. For two reasons they more sturdily resemble embedded systems. earliest, with a particular application in mind wireless sensor networks are normally deployed, slightly than as a general platform. following, a need for low costs and low power leads most wireless sensor nodes to have low-power microcontrollers ensuring that mechanisms such as virtual memory are either unnecessary to implement.

However, such operating systems are repeatedly designed with real-time properties. Consequently promising to use embedded operating systems such as eCos or uC/OS for sensor networks.

For wireless sensor networks TinyOS is possibly the earliest operating system exclusively designed. TinyOS is based on an event-driven programming model instead of multithreading. TinyOS programs are tranquil of event handlers and tasks with run-to-completion semantics, for example an incoming data packet or a sensor reading, when an external event occurs, TinyOS signals the proper event handler to handle the event. Incident handlers can post tasks that are schedule by the TinyOS kernel time later.

For providing UNIX-like abstraction LiteOS is a lately developed OS for wireless sensor networks, and hold for the C programming language.

For advancements such as 6LoWPAN and Protothreads, Contiki is an OS which uses a simpler programming style in C while provided.

RIOT supports general IoT protocols such as 6LoWPAN, IPv6, RPL, TCP, and UDP. RIOT rigging a micro kernel architecture. This provide multithreading with standard API and allows for development in C/C++.

ERIKA Enterprise is an open-source and royalty-free OSEK/VDX Kernel present BCC1, BCC2, ECC1, ECC2, multi core, memory fortification and kernel fixed right of way adopting C programming language.

CONCLUSION

- **Fragmentation:** To detect the attack signature by sending fragmented packets, the attacker will be underneath the radar and can simply by pass the detection system's capacity.
- **Avoiding Defaults:** stipulation an attacker had reconfigured it to use a unlike port the IDS may not be capable to detect the presence of the Trojan. The TCP port utilized by a protocol does not forever afford an indication to the

protocol which is being transported. Such as, IDS may supposed to detect a Trojan on port 12345.

- **Low-Bandwidth Attacks:** coordinating a scan amongst numerous attackers (or agents) and to attach the captured packets and presume, allocating diverse ports or hosts to dissimilar attackers makes it complex for the IDS that a network scan is in progress.
- **Address Spoofing:** By means of poorly secured or inaccurately configured proxy servers attackers can amplify the obscurity of the ability of Security Administrators to resolve the source of the attack to rebound an attack. For IDS to detect the origin of the attack if the source is spoofed and bounced by a server then it makes it very complicated.
- **Pattern Change Evasion:** By varying the data used in the attack somewhat, it may be achievable to evade detection. IDS normally rely on 'pattern matching' to detect an attack. For instance, an IMAP server may be vulnerable to a buffer overflow, and An IDS is intelligent to detect the attack signature of 10 common attack tools. By means of modifying the payload sent by the tool, it may be possible to prevaricate detection. In order that it does not resemble the data that the IDS expects,

To permit sensor owners to register and attach their devices Online collaborative sensor data management platforms are on-line database services to feed data into an online database for storage and also agree to developers and build their own applications based on that data to fix to the database. Examples contain Xively and the Wikisensing platform. Such platforms shorten online collaboration among users over diverse data sets ranging from energy and environment data to that together from transport services. Services consist of allowing developers to set in real-time graphs & widgets in websites; from the data feeds analyze and process historical data pulled; send real-time alerts from whichever data stream to organize scripts, devices and environments.

In describes the key apparatus the architectural concept of the Wikisensing system is described of such systems to contain APIs and interfaces for online collaborators, for the sensor data management a middleware containing the business logic desired and dealing out and a storage model appropriate for the proficient storage and retrieval of huge volumes of data.

REFERENCES

Denning, D. E. (1987). An intrusion-detection model. *Software Engineering. IEEE Transactions on Volume, SE-13*, 222–232.

Heberlein, L. T., Dias, G. V., Levitt, K. N., Mukherjee, B., Wood, J., & Wolber, D. (1990). A network security monitor. In *Proceedings of the 1990 IEEE Symposium on Research in security and Privacy*. doi:10.1109/RISP.1990.63859

Lunt, T. F. (1990). IDES: An intelligent system for detecting intruders. In *Proceedings of the Symposium: Computer Security, Threat and Countermeasures*. Academic Press.

Mattord, N. V. (2008). Principles of Information Security. Course Technology, 290–301.

Niazi, M., & Hussain, A. (2011). A novel agent-based simulation framework for sensing in complex adaptive environments. *Sensors Journal, IEEE, 11*(2), 404–412. doi:10.1109/JSEN.2010.2068044

Niazi, M. A., & Hussain, A. (2011). A Novel Agent-Based Simulation Framework for Sensing in Complex Adaptive Environments. *IEEE Sensors Journal, 11*(2), 404–412. doi:10.1109/JSEN.2010.2068044

Peiris, V. (2013). Highly integrated wireless sensing for body area network applications. *SPIE Newsroom.*

Scarfone, K., & Mell, P. (2007). Guide to intrusion detection and prevention systems (IDPS). *NIST Special Publication, 94.*

Silva, D., Ghanem, M., & Guo, Y. (2012). WikiSensing: An online collaborative approach for sensor data management. *Sensors (Basel, Switzerland), 12*(10), 13295–13332. doi:10.3390/s121013295 PMID:23201997

Winkler, J. R. (1990). A unix prototype for intrusion and anomaly detection in secure networks. In *Proceedings of the 13th National Computer Security Conference*. Academic Press.

ADDITIONAL READING

Mafra, P. M., Fraga, J. S., & Santin, A. O. (2014). Algorithms for a distributed IDS in MANETs. *Journal of Computer and System Sciences, 80*(3), 554–570. doi:10.1016/j.jcss.2013.06.011

KEY TERMS AND DEFINITIONS

Alarm Filtering: The development of categorize attack alerts fashioned from an IDS in order to differentiate false positives from definite attacks.

Attacker or Intruder: An article which tries to locate a way to grow unauthorized access to information, impose harm or take on in other malicious actions.

Burglar Alert/Alarm: When a system signal suggesting that a system has been attacked.

Clandestine User: Individual who acts as a controller and tries to use his privileges so as to stay away from being captured.

Confidence Value: A value an organization places on an IDS based on past performance and examination to assist conclude its capability to efficiently classify an attack.

Data Integration and Sensor Web: In the shape of numerical data the data gathered from wireless sensor networks is frequently saved in an essential base station. In addition, the Open Geospatial Consortium (OGC) is specifying standards for interoperability interfaces and metadata encodings that facilitate real time addition of assorted sensor webs into the Internet, allowing any character Wireless Sensor Networks throughout a Web Browser to monitor it.

Detection Rate: The detection rate is distinct as the number of intrusion instances detected by the system (True Positive) separated by the total number of intrusion instances nearby in the test set.

Distributed Sensor Network: In a sensor network if a central construction is used and the central node fails, then the complete network will subside, nonetheless the consistency of the sensor network can be enlarged by using spread control construction. Distributed control is used in WSNs for the subsequent reasons: 1. Sensor nodes are prone to failure, 2. For better collection of data, 3. To provide nodes with backup in case of failure of the central node. There is also no centralized body to assign the resources and they have to be self-planned.

False Alarm Rate: In of 'normal' patterns definite as the number of 'normal' patterns classified as attacks (False Positive) separated by the total number.

False Negative: When no alarm is raised when an attack has taken place.

False Positive: An occurrence signaling an IDS to produce an alarm when no attack has taken place.

IDMEF: To classify data formats is the function of IDMEF (Intrusion Detection Message Exchange Format) is and replaces procedures for giving out information of attention to intrusion detection and response systems and to the management systems that may need to interrelate with them. For incidents reporting and exchanging this is used in computer security. It guarantees for simple automatic processing. Format details are described in the RFC 4765. An accomplishment of the data representation in the Extensible Markup Language (XML) is existing and XML Document Type explanation is developed.

In-Network Processing: Away from forward data to lessen communication costs a few algorithms eliminate or reduce nodes' redundant sensor information and stay that is of no use. Since nodes can look over the data they forward, they can determine averages or directionality for illustration of readings from other nodes. For illustration, in sensing and monitoring applications, most probably the case that nearest sensor node monitoring an environmental feature usually registers similar values. Without a job due to the spatial correlation this variety of data being among sensor observations inspires techniques for in-network data aggregation and mining.

Masquerader: Someone who attempts to grow unauthorized entrance to a system by pretending to be an authorized user. They are normally outside users.

Misfeasor: They are generally internal users and can be of two types: 1. An authorized user with partial permissions. 2. A customer with filled permissions and who misuses their powers.

Noise: Data or interference that can trigger a false positive or ambiguous a true positive.

Site Policy: Strategy within an association that control the rules and configurations of an IDS.

Site Policy Awareness: An IDS's capability to vigorously change its rules and configurations in reply to varying environmental action.

True Negative: An event when no attack has taken place and no detection is prepared.

True Positive: A legitimate attack which triggers an IDS to fabricate an alarm.

Compilation of References

3rd Generation Partnership Project. (1999). *3G Security; Security architecture* (Technical Specification TS 33.102 version 2.0.0 Release 99). Retrieved from www.3gpp.org/ftp/Specs/archive/33_series/33.102/33102-200.zip

3rd Generation Partnership Project. (2000). Counteracting envisaged 3G attacks. In *A Guide to 3rd Generation Security* (Technical Specification TR 33.900 version 1.2.0 Release 99). Retrieved from www.3gpp.org/ftp/Specs/archive/33_series/33.900/33900-120.zip

3rd Generation Partnership Project. (2000). Denial of Service. In *A Guide to 3rd Generation Security* (Technical Specification TR 33.900 version 1.2.0 Release 99). Retrieved from www.3gpp.org/ftp/Specs/archive/33_series/33.900/33900-120.zip

3rd Generation Partnership Project. (2014). Access link data integrity. In *3G Security; Security architecture* (Technical Specification TS 33.102 version 12.0.0 Release 12). Retrieved from www.3gpp.org/ftp/Specs/archive/33_series/33.102/33102-c00.zip

3rd Generation Partnership Project. (2014). Cipher key and integrity key lifetime. In *3G Security; Security architecture* (Technical Specification TS 33.102 version 12.0.0 Release 12). Retrieved from www.3gpp.org/ftp/Specs/archive/33_series/33.102/33102-c00.zip

3rd Generation Partnership Project. (2014). Cipher key and integrity key setting. In *3G Security; Security architecture* (Technical Specification TS 33.102 version 12.0.0 Release 12). Retrieved from www.3gpp.org/ftp/Specs/archive/33_series/33.102/33102-c00.zip

3rd Generation Partnership Project. (2014). Ciphering and integrity mode negotiation. In *3G Security; Security architecture* (Technical Specification TS 33.102 version 12.0.0 Release 12). Retrieved from www.3gpp.org/ftp/Specs/archive/33_series/33.102/33102-c00.zip

3rd Generation Partnership Project. (2014). Security mode set-up procedure. In *3G Security; Security architecture* (Technical Specification TS 33.102 version 12.0.0 Release 12). Retrieved from www.3gpp.org/ftp/Specs/archive/33_series/33.102/33102-c00.zip

3rd Generation Partnership Project. (2014). The Core Network (CN) entities. In *Network architecture*. (Technical Specification 23.002 version 12.4.0 Release 12). Retrieved from www.3gpp.org/ftp/Specs/archive/23_series/23.002/23002-c40.zip

3rd Generation Partnership Project. (2014). USIM. In *3G Security; Security architecture* (Technical Specification TS 33.102 version 12.0.0 Release 12). Retrieved from www.3gpp.org/ftp/Specs/archive/33_series/33.102/33102-c00.zip

Aad, I., Hubaux, H., & Knightly, E.W. (2000). *Impact of Denial of Service Attacks on Ad-Hoc Networks*. Academic Press.

Abowd, G.D., & Sterbenz, J.P.G. (2000). Final report on the interagency workshop on research issues for smart environments. *IEEE Personal Communications*, 36–40.

Adams, J. (2004). *Designing with 802.15.4 and ZigBee*. Paper presented at the Industrial Wireless Applications Summit, San Diego, CA.

Ahmed, K. (n.d.). *Techniques and challenges of data centric storage schemes in wireless sensor network*. Academic Press.

Akyildiz, I. F., Brunetti, F., & Blazquez, C. (2008). Nano networks: A new communication paradigm. *Computer Networks*, *52*(12), 2260–2279. doi:10.1016/j.comnet.2008.04.001

Akyildiz, I. F., & Jornet, J. M. (2010). Electromagnetic wireless nanosensor networks. *Nano Communication Networks*, *1*(1), 3–19. doi:10.1016/j.nancom.2010.04.001

Akyildiz, I. F., Melodia, T., & Chowdhury, K. R. (2006). A survey on wireless multimedia sensor networks. *Computer Networks*, *51*(4), 921–960. doi:10.1016/j.comnet.2006.10.002

Akyildiz, I. F., Melodia, T., & Chowdury, K. R. (2007). Wireless multimedia sensor networks: A survey. *IEEE Wireless Communications*, *14*(6), 32–39. doi:10.1109/MWC.2007.4407225

Akyildiz, I. F., Pompili, D., & Melodia, T. (2005). Underwater acoustic sensor networks: Research challenges. *Ad Hoc Networks*, *3*(3), 257–279. doi:10.1016/j.adhoc.2005.01.004

Akyildiz, I. F., & Stuntebeck, E. P. (2006). Wireless underground sensor networks: Research challenges. *Ad Hoc Networks Journal*, *4*(6), 669–686. doi:10.1016/j.adhoc.2006.04.003

Akyildiz, I. F., Sun, Z., & Vuran, M. C. (2009). Signal Propagation Techniques for Wireless nderground Communication Networks. *Physical Communication Journal*, *2*(3), 167–183. doi:10.1016/j.phycom.2009.03.004

Akyildiz, I. F., Su, W., Sankarasubramaniam, Y., & Cayirci, E. (2002). Wireless sensor networks: A survey. *Computer Networks*, *38*(4), 393–422. doi:10.1016/S1389-1286(01)00302-4

Akyildiz, I. F., Su, W., Sankarasubramaniam, Y., & Cayirci, E. (2002, August). (2002-2). A Survey on Sensor Networks. *IEEE Communications Magazine*, *40*(8), 102–114. doi:10.1109/MCOM.2002.1024422

Akykildiz, I. F., Su, W., Sankarasubramaniam, Y., & Cayirci, E. (2002). A Survey on Sensor Networks. *IEEE Communications Magazine*.

Al Tamimi. (2006). *Security in Wireless Data Networks: A Survey Paper*. Academic Press.

Alanri, A., Ansari, W. S., Hassan, M. M., Hossain, M. S., Alelaiwi, A. H., & Hossain, M. A. (2013). A Survey on Sensor-Cloud Architecture, Applications, and Approaches. *International Journal of Distributed Sensor Networks.*

Al-Fayoumi, M., & Al-Saraireh, J. (2011). An Enhancement of Authentication Protocol and Key Agreement (AKA) For 3G Mobile Networks. *International Journal of Security*, *5*(1), 35–51.

Alrajeh, N. A., Khan, S., & Shams, B. (2013). Intrusion Detection Systems in Wireless Sensor Networks: A Review. *International Journal of Distributed Sensor Networks*, 1–7.

Al-Shabi, M. A. (2012). Attack and Defense in Mobile Ad-Hoc Networks. *International Journal of Reviews in Computing, 12.*

Amundson, I. (n.d.). *A survey on localization for mobile wireless sensor network.* Academic Press.

Anjum, N., & Salil, K. (2009). Article. *Authentication and Confidentiality in Wireless Ad Hoc Networks*, *21*, 28.

Anliker, U., Ward, J. A., Lukowicz, P., Troster, G., Dolveck, F., Baer, M., & Vuskovic, M. et al. (2004). AMON: A werable multiparameter medical monitoring and alert system. *IEEE Transactions on Information Technology in Biomedicine*, *8*(4), 415–427. doi:10.1109/TITB.2004.837888 PMID:15615032

Annadurai, P., & Sivakumar, K. (2009). Impact of rushing attack on multicast in Mobile Ad Hoc Network. *International Journal of Computer Science and Information Security.*

Arya, S., & Arya, C. (2012). Malicious Nodes Detection in Mobile Ad-Hoc Networks. *Journal of Information and Operations Management*, *3*, 210–212.

Asokan, N., & Zapata, M. (2002). *Securing Ad-Hoc routing Protocols. In ACM workshop, Wireless Security*. WiSe.

Atakan, B., & Akan, O. (2010). Carbon nano tube-based nano scale ad hoc networks. *IEEE Communications Magazine*, *48*(6), 129–135. doi:10.1109/MCOM.2010.5473874

Atallah, L., Lo, B., Yang, G. Z., & Siegemund, F. (2008). Wirelessly accessible sensor populations (WASP) for elderly care monitoring. In *Proceedings of the Second International Conference on Pervasive Computing Technologies for Healthcare*. Tampere, Finland: Academic Press. doi:10.4108/ICST.PERVASIVEHEALTH2008.2777

Ayu, T., & Sudip. (2007). A multi-factor security protocol for wireless payment- secure web authentication using mobile devices. *IADIS International Conference Applied Computing.*

Bag, A., & Bassiouni, M. A. (2007). Hotspot preventing routing algorithm for delay-sensitive biomedical sensor networks. In *Proceedings of the IEEE International Conference on Portable Information Devices* (PORTABLE). Orlando, FL: IEEE. doi:10.1109/PORTABLE.2007.30

Bag, A., & Bassiouni, M. A. (2006). Energy efficient thermal aware routing algorithms for embedded biomedical sensor networks. In *Proceedings of the IEEE International Conference on Mobile Adhoc and Sensor Systems* (MASS). Vancouver, Canada: IEEE. doi:10.1109/MOBHOC.2006.278619

Bagga, S., & Adhikary, K. (2014). A Review on Various Protocols and Security Issues in MANET. *International Journal of Advanced Research in Computer and Communication Engineering*, *3*(7), 7478–7482.

Bao, S., Zhang, Y., & Shen, L. (2005). Physiological signal based entity authentication for body area sensor networks and mobile healthcare systems. In *Proceedings of the 27th Annual International Conference of the IEEE EMBS*. Shanghai, China: IEEE.

Bateni, G. H., & McGillem, C. D. (1992). Chaotic Sequences for Spread Spectrum: An Alternative to PN – Sequences. In *Proceedings ofIEEE International Conference on Wireless Communications*, (pp. 437 – 440). IEEE.

Bazaka & Mohan, V. (2013). Jacob: Implantable devices: Issues and challenges. *Electronics*, 2, 1–34.

Birman, K., Ramasubramaniam, V., & Chandra, R. (2001). *Anonymous Gossip: Improving Multicast Reliability in Mobile Ad-Hoc Network.* ICDCS.

Bischoff, R. (n.d.). *Wireless sensor network platforms*. Retrieved from http://www.ieee802.org/15/

Boubiche, D., & Bilami, A. (2011). HEEP (Hybrid Energy Efficiency Protocol) Based on Chain Clustering. *Int. J. Sensor Networks*, *10*(1/2), 25–35. doi:10.1504/IJSNET.2011.040901

Boukerche, A., Li, X., & Khatib, K. (2006). Trust based framework for secure data aggregation in wireless sensor networks. In *Proceedings of IEEE SECON* (pp. 718-725). IEEE.

Braem, B., Latre, B., Moerman, I., Blondia, C., & Demeester, P. (2006). The wireless autonomous spanning tree protocol for multihop wireless body area networks. In *Proceedings of the 3rd IEEE Annual Conference on Mobile and Ubiquitous Systems: Networking & Services*. San Jose, CA: IEEE.

Bruschi, D., & Rosti, E. (2002). Secure Multicast in Wireless Networks of Mobile Hosts: Protocols and Issues||. *Mobile Networks and Applications*, *7*(6), 503–511. doi:10.1023/A:1020781305639

Buchegger, S., & Boudec, Y. L. (2005). Self-policing mobile ad hoc networks by reputation systems. *IEEE Communications Magazine*, *43*(7), 101–107. doi:10.1109/MCOM.2005.1470831

Bulusu, N., Estrin, D., Girod, L., & Heidemann, J. (2001). *Scalable coordination for wireless sensor networks: self-configuring localization systems.* Paper presented at the International Symposium on Communication Theory and Applications (ISCTA 2001), Ambleside, UK.

Casas, R., Blasco, M. R., Robinet, A., Delgado, A. R., Yarza, A. R., & McGinn, J., ... Grout, V. (2008). User modelling in ambient intelligence for elderly and disabled people. In *Proceedings of the 11th International Conference on Computers Helping People with Special Needs.* Linz, Austria: Academic Press. doi:10.1145/1409635.1409663

Chandra, D.V., Shekar, V.V., Jayarama, & Babu. (2008). Wireless security: A comparative analysis for the next generation networks. *Journal of Theoretical and Applied Information Technology.*

Chengji, P., & Bo, W. (2012). New Optimal Design Method of Arbitrary Limited Period Spreading Sequences based on Logistic Mapping. *4th International Conference on Signal Processing Systems*, (vol. 58, pp. 246 – 250). doi:10.1109/ICCIS.2012.29

Cherukuri, S., Venkatasubramanian, K. K., & Gupta, S. K. S. (2003). Biosec: A biometric based approach for securing communication in wireless networks of biosensors implanted in the human body. In *Proceedings of 32nd Int Conf Parallel Processing* (pp. 432-439). Academic Press. doi:10.1109/ICPPW.2003.1240399

Chou, C. T. (2012). Molecular circuits for decoding frequency coded signals in nano-communication networks. *Nano Communication Networks*, *3*(1), 46–56. doi:10.1016/j.nancom.2011.11.001

Chou, C. T.Chun Tung Chou. (2013). Extended Master Equation Models for Molecular Communication Networks. *IEEE Transactions on Nanobioscience*, *12*(2), 79–92. doi:10.1109/TNB.2013.2237785 PMID:23392385

Como, L. H., Cortier, V., & Zalinescu, E. (2009). Deciding security properties for cryptographic protocols. Application to key cycles. *ACM Transactions on Computational Logic*, *5*, 1–38.

Corson, S., & Macker, J. (1999). *Mobile Ad-Hoc Networking (MANET).* Routing Protocol Performance Issues and Evaluation Considerations.

Cox, C. (2008). System architecture. In *Essentials of UMTS* (p. 37). New York, NY: Cambridge University Press. doi:10.1017/CBO9780511536731

Cui, J. (n.d.). *Data Aggregation in Wireless sensor networks: Compressing and forecasting.* Academic Press.

Cui, J.-H., Kong, J., Gerla, M., & Zhou, S. (2006). Challenges: Building Scalable Mobile Underwater Wireless Sensor Networks for Aquatic Applications. *IEEE Network. Special Issue on Wireless Sensor Networking*, *20*(3), 12–18.

Culpepper, J., Dung, L., & Moh, M. (2003). Hybrid indirect transmissions (HIT) for data gathering in wireless micro sensor networks with biomedical applications. In *Proceedings of the 18th IEEE Annual Workshop on Computer Communications* (CCW). Dana Point, CA: IEEE. doi:10.1109/CCW.2003.1240800

Daemean, J., & Rijmen, R. (1999). *AES Proposal: Rijndeal version 2.* Available at http://www.esat.kuleuveb.ac.be/rijmen/rijndeal

Dalal, M., Vats, K., Loura, V., & Rohila, D. (2012). *OPNET based simulation and performance analysis of GRP routing protocol. International Journal of Advanced Research in Computer Science and Software Engineering*, 2.

Dasgupta, S., Bhattacharya, I., & Bose, G. (2009). Energy-aware cluster based node scheduling algorithm in wireless sensor network for preserving maximum network life time. In *Proceeding of International Conference on Methods and Models in Computer Science (ICM2CS).*

Dash, S. K., Sahoo, J. P., Mohapatra, S., & Pai, S. P. (2012). Sensor-Cloud assimilation of Wireless sensor network and the cloud. *Advances in Computer Science and Information Technology Networks and Communications, Springer, 84*, 455–464. doi:10.1007/978-3-642-27299-8_48

Debar, H., Dacier, M., & Wespi, A. (1999). Towards a taxonomy of intrusion-detection systems. *Computer Networks, 31*(8), 805–822. doi:10.1016/S1389-1286(98)00017-6

Demirkol, I., Ersoy, C., & Alagoz, F. (2006). MAC protocols for wireless sensor networks: A survey. *IEEE Communications Magazine, 44*(4), 115–121. doi:10.1109/MCOM.2006.1632658

Denning, D. E. (1987). An intrusion-detection model. *Software Engineering. IEEE Transactions on Volume, SE-13*, 222–232.

Deying, L. (n.d.). *Sensor Coverage in Wireless sensor network.* Academic Press.

Dinan, E. H., & Jabbari, B. (1998, September). Spreading codes for direct sequence CDMA and wideband CDMA cellular networks. *IEEE Communications Magazine, 36*(9), 48–54. doi:10.1109/35.714616

Ding, Z., Guo, L., & Yang, Q. (2014). RDB-KV: A Cloud Database Framework for Managing Massive Heterogeneous Sensor Stream Data. In *Proceedings of IEEE Second International Conference on Intelligent System Design and Engineering Application* (ISDEA), (pp. 653-656). Retrieved from http://en.wikipedia.org/wiki/BigTable

Dinh, H. T., Loe, C., Niyato, D., & Wang, P. (2011). *A Survey of Mobile Cloud Computing: Architecture, Applications and Approaches.* Wireless Communications and Mobile Computing-Wiley Online Library.

Djenojuri, D., Khelladi, L., & Badache, N. (2005). A Survey of Security Issues in Mobile Ad hoc Networks. *IEEE Communications Surveys, 7*(4), 2–28. doi:10.1109/COMST.2005.1593277

Djenouri, D., & Balasingham, I. (2009). New QoS and geographical routing in wireless biomedical sensor networks. In *Proceedings of the 6th IEEE International Conference on Broadband Communications, Networks, and Systems* (BROADNETS). Madrid, Spain: IEEE. doi:10.4108/ICST.BROADNETS2009.7188

Durmus, Y., Ozgovde, A., & Ersoy, C. (2009). Event based queueing for fairness and on-time delivery in video surveillance sensor networks. In *Proceedings of IFIP Networking.* Aachen, Germany: IFIP.

Ellbruck, H. (n.d.). *Using and operating wireless sensor network testbeds with WISEBED*. Academic Press.

Elson, J. (n.d.). *Time synchronization for wireless sensor network*. Academic Press.

Essa, I. A. (2000). Ubiquitous sensing for smart and aware environments. *IEEE Personal Communications*.

ETSI 3rd Generation Partnership Project. (2001). Security threats. In *3G security; Security threats and requirements* (Technical Specification TS 21.133 version 4.1.0 Release 4). Retrieved from http://www.etsi.org/deliver/etsi_ts/121100_121199/121133/04.01.00_60/ts_121133v040100p.pdf

ETSI 3rd Generation Partnership Project. (2013). *Non-Access-Stratum (NAS) functions related to Mobile Station (MS) in idle mode* (Technical Specification TS 23.122 version 11.4.0 Release 11). Retrieved from http://www.etsi.org/deliver/etsi_ts/123100_123199/123122/11.04.00_60/

Falconi, C., Damico, A., & Wang, Z. (2007). Wireless Joule nano heaters. *Sensors and Actuators*, *127*(1), 54–62. doi:10.1016/j.snb.2007.07.002

Fall, K. (2003). *A delay-tolerant network architecture for challenged internets*. Paper presented at the SIGCOMM'2003, Karlsruhe, Germany. doi:10.1145/863955.863960

Farooq, M. O. (n.d.). *Operating system for wireless sensor networks: A survey*. Academic Press.

Farooq, M. O. (n.d.). *Wireless Sensor networks Testbeds and state of the art multimedia sensor nodes*. Academic Press.

Feng. (2008). *Attacking Antivirus*. Nevis Networks,Inc.

Fielding, R., & Reschke, J. (Eds.). (2014). *Request for Comments: 7235 Hypertext Transfer Protocol (HTTP/1.1):Authentication*. Internet Engineering Task Force (IETF). Retrieved from https://tools.ietf.org/html/rfc7235

Fok, C., Roman, G., & Lu, C. (2007). Towards A Flexible Global Sensing Infrastructure. *ACM SIGBED Review, 4*(3), 1-6.

Forouzan, B. (2011). *Data Communications and Networking*. New Delhi, India: Tata McGraw Hill Education Private Limited.

Francois, M., & Defour, D. (2013, February). A Pseudo-Random Bit Generator Using Three Chaotic Logistic Maps. *Hyper Articles en Ligne, 1*, 1–22.

Franks, J., Hallam-Baker, P., Hostetler, J., Leach, P., Lawrence, S., Luotonen, A., & Stewart, L. (1999). *Request for Comments: 2617 HTTP Authentication: Basic and Digest Access Authentication.* Available from http://tools.ietf.org/html/rfc2617

Frohlich, A. A. (n.d.). *Operating system support for wireless sensor network*. Academic Press.

Gagandeep, Aashima, Kumar, & Pawan. (2012). Analysis of Different Security Attacks in MANETs on Protocol Stack A-Review. *International Journal of Engineering and Advanced Technology, 1*.

Gao, T., Greenspan, D., Welsh, M., Juang, R., & Alm, A. (2005). *Vital SignsMonitoring and Patient Tracking over a Wireless Network.* Paper presented at IEEE 27th Annual International Conference of the Engineering in Medicine and Biology Society (EMBS), Shanghai, China.

Gao, T., Greenspan, D., Welsh, M., Juang, R. R., & Alm, A. (2005). Vital signs monitoring and patient tracking over a wireless network. In *Proceedings of IEEE-EMBS 27th Annual International Conference of the Engineering in Medicine and Biology*. Shanghai, China: IEEE.

Garg, V. K., & Wikes, J. E. (2005). *Principles and Applications of GSM.* Singapore: Pte. Ltd.

Ghaffari, A. (2006). Vulnerability and Security of Mobile Ad-hoc Networks.*International Conference on Simulation, Modeling and Optimization WSEAS*, (pp. 22-24).

Gopalsamy, C., Park, S., Rajamanickam, R., & Jayaraman, S. (2005). The wearable motherboard TM: The first generation of adaptive and responsive textile structures (ARTS) for medical applications. *Virtual Reality (Waltham Cross)*, *4*(3), 152–168. doi:10.1007/BF01418152

Gosalia, Weiland, Humayun, & Lazzi. (2004). Thermal elevation in the human eye and head due to the operation of a retinal prosthesis. *IEEE Tran. Biomedical Eng.*, *51*(8), 1469–1477.

Grandison, T., & Sloman, M. (2000). A survey of trust in internet applications. *IEEE Communications Surveys and Tutorials*, *3*(4), 2–16. doi:10.1109/COMST.2000.5340804

Grgen, L., Roncancio, C., Labb, C., & Olive, V. (2006). Transactional issues in sensor data management. In *Proceeding DMSN '06 Proceedings of the 3rd workshop on Data management for sensor networks: in conjunction.* doi:10.1145/1315903.1315910

Guo, S., Zhong, Z., & He, T. (2009). FIND: Faulty node detection for wireless sensor networks. In *Proceedings of the 7th ACM Confeence on Embedded Networked Sensor Systems (Sensys'09)*, (pp. 252-266). ACM.

Gupta & Schwiebert. (2001). Energy efficient protocols for wireless communication in biosensor networks. In *Proceedings of 12th IEEE Int'l Symp. Personal, Indoor and Mobile Radio Comm.* San Diego, CA: IEEE. doi:10.1109/PIMRC.2001.965503

Gürses, E., & Akan, Ö. B. (2005). Multimedia communication in wireless sensor networks. *Annales des Télécommunications*, *60*(7-8), 872–900.

Guyeux, C., Wang, Q., & Bahi, J. M. (2010). A Pseudo Random Numbers Generator Based on Chaotic Iterations. Application toWatermarking. *International Conference on Web Information Systems and Mining*, 6318, 202-211. doi:10.1007/978-3-642-16515-3_26

Halin, N., Junnila, M., Loula, P., & Aarnio, P. (2005). The life shirt system for wireless patient monitoring in the operating room. *Journal of Telemedicine and Telecare*, *11*(8), 41–43. doi:10.1258/135763305775124623 PMID:16375793

Han, Wang, Liu, & Li. (2013). Two Improved Pseudo-Random Number Generation Algorithms Based on the Logistic Map. *Research Journal of Applied Sciences, Engineering and Technology*, 2174 – 2179.

Havenith, G. (2001). Individualized model of human thremoregulation for the simulation of heat stress response. *Journal of Applied Physiology (Bethesda, Md.)*, *90*, 1943–1954. PMID:11299289

Heberlein, L. T., Dias, G. V., Levitt, K. N., Mukherjee, B., Wood, J., & Wolber, D. (1990). A network security monitor. In *Proceedings of the 1990 IEEE Symposium on Research in security and Privacy*. doi:10.1109/RISP.1990.63859

Hedetniemi, S., & Liestman, A. (1988). A survey of gossiping and broadcasting in communication networks. *Networks*, *18*(4), 319–349. doi:10.1002/net.3230180406

Heinzelman, W. R., Chandrakasan, A., & Balakrishnan, H. (2000). Energy-efficient communication protocol for wireless microsensor networks. *IEEE Proceedings of the Hawaii International Conference on System Sciences*, 1–10.

Heinzelman, W. B., Chandrakasan, A. P., & Balakrishnan, H. (2002). An application-specific protocol architecture for wireless microsensor networks. *IEEE Transactions on Wireless Communications*, *1*(4), 660–670. doi:10.1109/TWC.2002.804190

Heinzelman, W. R., Kulik, J., & Balakrishnan, H. (1999). Adaptive protocols for information dissemination in wireless sensor networks. In *Proceedings of the ACM MobiCom'99*. doi:10.1145/313451.313529

Herring, C., & Kaplan, S. (2000). Component-based software systems for smart environments. *IEEE Personal Communications*, 60–61.

Heys, H. M. (2001). *A Tutorial on Linear and Differential Cryptanalysis*. Academic Press.

Hirata, A., & Shiozawa, T. (2003). Correlation of maximum temperature increase and peak SAR in the human head due to handset antennas. *IEEE T. Microw. Theory*, *51*(7), 1834–1841. doi:10.1109/TMTT.2003.814314

Hoffman, L. J., Lawson-Jenkins, K., & Blum, J. (2006). Trust beyond security: An expanded trust model. *Communications of the ACM*, *49*(7), 95–101. doi:10.1145/1139922.1139924

Hong, X., Kong, J., & Gerla, M. (2006). *Modeling Ad-hoc Rushing Attack in a Negligibility-based Security Framework*. Academic Press.

Hori, T., Nishida, Y., Suehiro, T., & Hirai, S. (2000). SELF-network: Design and implementation of network for distributed embedded sensors. In *Proceedings of IEEE/RSJ International Conference on Intelligent Robots and Systems*. Takamatsu, Japan: IEEE. doi:10.1109/IROS.2000.893212

Hou, J. C. (n.d.). *Coverage in Wireless sensor networks*. Academic Press.

Hui, J. W., & Culler, D. E. (2008). *IP is dead, long live IP for wireless sensor networks*. Paper presented at the sixth ACM conference on embedded network Sensor System (SenSys '08), Raleigh, USA. doi:10.1145/1460412.1460415

Hu, Y., Johnson, D. B., & Perrig, A. (2003). SEAD: Secure Efficient Distance Vector Routing for Mobile Wireless Ad hoc Networks. *Ad Hoc Networks*, *1*(1), 175–192. doi:10.1016/S1570-8705(03)00019-2

Hu, Y., Johnson, D. B., & Perrig, A. (2005). Ariadne: A Secure On-demand Routing Protocol for Ad hoc Networks. *Wireless Communications and Mobile Computing*, *11*(1-2), 21–28.

IETF Working Group. (2013). *IPv6 over low power WPAN working group*. Available from http://tools.ietf.org/wg/6lowpan/

Intanagonwiwat, C., Govindan, R., & Estrin, D. (2000). Directed diffusion: a scalable and robust communication paradigm for sensor networks. In *Proceedings of the ACM Mobi-Com*. doi:10.1145/345910.345920

Intelligent Systems Centre Nanyang Technological University. (n.d.). Available: http://www.ntu.edu.sg/intellisys

Ishi, Y., Kawakami, T., Yoshihisa, T., Teranishi, Y., et al. (2012). Design and Implementation of Sensor Data Sharing Platform for Virtualized Wide Area Sensor Networks. In *Proceedings of Seventh International Conference on P2P, Parallel, Grid, Cloud and Internet Computing* (3PGCIC), (pp. 333-338). Academic Press.

Jain, S., & Shrivastava, S. (2013). *A Brief Introduction of Different type of Security Attacks found in Mobile Ad-hoc Network. International Journal of Computer Science & Engineering Technology*, 4.

Jasemian, Y. (2008). Elderly comfort and compliance to modern telemedicine system at home. In *Proceedings of the Second International Conference on Pervasive Computing Technologies for Healthcare*. Tampere, Finland: Academic Press. doi:10.4108/ICST.PERVASIVEHEALTH2008.2516

Jaydip, S. (2009, August). Security on Wireless Sensor Networks Security. *International Journal of Communication Networks and Information Security, 1.*

Johnson, D. B., Hu, Y. C., & Perrig, A. (2003). Efficient security mechanisms for routing protocols.*Network and Distributed System Security Symposium (NDSS).*

Johnson, D. B., Perrig, A., & Hu, Y. C. (2002). *A Secure On Demand Routing Protocol for Ad-Hoc Networks*. MobiCom.

Johnson, D. B., Perrig, A., & Hu, Y. C. (2003). *Rushing Attacks and Defense in Wireless Ad Hoc Network Routing Protocols*. WiSe.

Kaaranen, H., Ahtiainen, A., Laitinen, L., Naghian, S., & Niemi, V. (2005). Base Station (BS, Node B). In *UMTS Networks Architecture, Mobility and Services* (2nd ed.; pp. 101–103). Chichester, UK: John Wiley & Sons. doi:10.1002/047001105X

Kaaranen, H., Ahtiainen, A., Laitinen, L., Naghian, S., & Niemi, V. (2005). UMTS Core Network Architecture. In *UMTS Networks Architecture, Mobility and Services* (2nd ed.; pp. 146–152). Chichester, UK: John Wiley & Sons. doi:10.1002/047001105X.ch6

Kadir & Maarof. (2009). Randomness Analysis of Pseudorandom Bit Sequences. *International Conference on Computer Engineering and Applications, 2*, 390 – 394.

Kahn, J. M., Katz, R. H., & Pister, K. S. J. (1999). Next century challenges: mobile networking for smart dust.*Proceedings of the ACM MobiCom'99*. doi:10.1145/313451.313558

Kamaljit, I. L., & Patel. (n.d.). Comparing Different Gateway Discovery Mechanism for Connectivity of Internet & MANET. *International Journal of Wireless Communication and Simulation, 2*(1), 51-63.

Kanagachidambaresan & Chitra. (2014). Fail safe fault tolerant mechanism for wireless body sensor network. *Wireless Personal Communications, 78*(2), 247–260.

Kanagachidambaresan, SarmaDhulipala, & Udhaya. (2011). Markovian model based trustworthy architecture. *Procedia Engineering, 38*(4), 718-725.

Kanagachidambaresan, SarmaDhulipala, Vanusha, & Udhaya. (2011). Matlab based modeling of body sensor network using ZigBee protocol. In *Proceedings of CIIT* (pp. 773-776). CCIS.

Karlof, C., & Wagner, D. (2003). Secure routing in wireless sensor networks: Attacks and countermeasures. In *Proceedings of the 1st IEEE International Workshop on Sensor Network Protocols and Applications*. doi:10.1109/SNPA.2003.1203362

Karlof, C., & Wagner, D. (2003). Secure routing in wireless sensor networks: Attacks and countermeasures. *Ad Hoc Networks, 1*(2-3), 293–315. doi:10.1016/S1570-8705(03)00008-8

Katiyar, I., Chand, N., & Soni, S. (2010). Clustering Algorithms for Heterogeneous Wireless Sensor Network: A Survey. *International Journal of Applied Engineering Research, 1*(2), 273–287.

Kaushal, R., & Abhay, V. (2008). *HLR Interface to AUC*. Stockholm: Ericsson AB.

Khan, M., Ahmed, A., & Raza Cheema, A. (2008). Vulnerabilities of UMTS Access Domain Security. In *Proceedings of Architecture,Ninth ACIS International Conference on Software Engineering, Artificial Intelligence, Networking, and Parallel/Distributed Computing*, (pp. 350-355). doi:10.1109/SNPD.2008.78

Kifayat, K., Merabti, M., Shi, Q., & Llewellyn, D. (2010). Security in Wireless Sensor Networks. In Handbook of Information and Communication Security. Springer. doi:10.1007/978-3-642-04117-4_26

Kim, D. S., & Chung, Y. J. (2006). *Self-organization routing protocol supporting mobile nodes for wireless sensor network*. Paper presented at the First Int. Multi-Symp. on Computer and Computational Sciences, Hangzhou, China. doi:10.1109/IMSCCS.2006.265

Kim, K., Lee, I.-S., Yoon, M., Kim, J., Lee, H., & Han, K. (2009). An efficient routing protocol based on position information in mobile wireless body area sensor networks. In *Proceedings of the 1st International Conference on Networks and Communications (NETCOM)*. Chennai, India: Academic Press. doi:10.1109/NetCoM.2009.36

Kishor, S., Trivedi, Y. Z., Ma, & Dharmaraja, S. (2003). Performability modeling of wireless communication systems. *International Journal of Communication Systems, 16*, 561–577.

Kohad, H., Lngle, V. R., & Gaikwad, M. A. (2012, July – August). Security Level Enhancement In Speech Encryption Using Kasami Sequence. *International Journal of Engineering Research and Applications*, *2*, 1518–1523.

Kong, J., Hong, X., & Gerla, M. (2010). *A new set of passive routing attacks in mobile ad hoc network*. Academic Press.

Krishna, B. T. (2011, April7). Binary Phase Coded Sequence Generation Using Fractional Order Logistic Equation. *Circuits, Systems, and Signal Processing*, *31*(1), 401–411. doi:10.1007/s00034-011-9295-8

Krsul, I., Schuba, C., Kuhn, M., Spafford, E., Sundaram, A., & Zamboni, D. (1997). Analysis of a Denial of Service Attack on TCP.*IEEE Symposium on Security and Privacy.*

Kumar, G. S., Vinu, P. M. V., & Jacob, K. P. (2003). *Mobility Metric based LEACH-Mobile Protocol*. Paper presented at the 16th International Conference on Advanced Computing and Communications (ADCOM'08), Chennai, India.

Kumar, L. P. D., Grace, S. S., Krishnan, A., & Manikandan, V. M. (2012). Data Filtering in Wireless Sensor Networks using neural networks for storage in cloud. In *Proceedings of the IEEE International Conference on Recent Trends in Information Technology* (ICRTIT' 11). IEEE.

Kumar, R., Tsiatsis, V., & Srivastava, M. B. (2003). *Computation Hierarchy for In-Network Processing*. Paper presented at the 2nd Intl. Workshop on Wireless Networks and Applications, San Diego, CA.

Kumar, Y., & Prashant. (2010). Hardware Implementation of Advanced Encryption Standard. In *Proceedings ofInternational Conference on Computational Intelligence and Communication Networks*. IEEE. doi:10.1109/CICN.2010.89

Kumar, V., & Upadhyay, P. P. (2008). Next generation Network. *Telecommunications*, *58*(2), 18–27.

Kushida, M., & Yuriyama, T. (2012). Sensor-Cloud Infrastructure physical sensor management with virtualized sensors on cloud computing. Advance in Computer Science and Information Technology Networks and Communications, 84, 455-464.

Kweon, K., Ghim, H., Hong, J., & Yoon, H. (2009). *Grid-Based Energy-Efficient Routing from Multiple Sources to Multiple Mobile Sinks in Wireless Sensor Networks*. Paper presented at the 4th International Symposium on Wireless Pervasive Computing (ISWPC'09), Melbourne, Australia. doi:10.1109/ISWPC.2009.4800585

Lakhtaria, K. I. (2009). Enhancing QOS and QOE in IMS enabled next generation networks. In *Networks and Communications, 2009. NETCOM'09. First International Conference on* (pp. 184-189). IEEE. doi:10.1109/NetCoM.2009.29

Lakhtaria, K. I. (2010). *Analyzing Zone Routing Protocol in MANET Applying Authentic Parameter*. arXiv preprint arXiv:1012.2510.

Lakhtaria, K. I. (2010). *Study, analysis and modeling of IP multimedia systems on next generation networks providing mobile and fixed multimedia services*. Academic Press.

Lakhtaria, K. I. (2012). *Efficient detection of malicious nodes by implementing OpenDNS and statistical methods*. Paper presented at 4th International Conferences on IT and Businesses Intelligence.

Lakhtaria, K. I., & Jani, D. N. (2010). *Design and Modeling Billing solution to Next Generation Networks*. arXiv preprint arXiv:1008.1851.

Lakhtaria, K. I., & Nagamalai, D. (2010). Analyzing Web 2.0 Integration with Next Generation Networks for Services Rendering. In Recent Trends in Networks and Communications (pp. 581-591). Springer Berlin Heidelberg.

Lakhtaria, K. I. (2011). Protecting computer network with encryption technique: A Study. In *Ubiquitous Computing and Multimedia Applications* (pp. 381–390). Springer Berlin Heidelberg. doi:10.1007/978-3-642-20998-7_47

Lakhtaria, K. I. (2012). *Technological Advancements and Applications in Mobile Ad-hoc Networks: Research Trends*. Information Science Reference.

Lan, K. T. (2010). What's Next? Sensor+Cloud?. In *Proceeding of the 7th International Workshop on Data Management for Sensor Neworks*. ACM Digital Library.

Latre, B., Braem, B., Moerman, I., Blondia, C., Reusens, E., Joseph, W., & Demeester, P. A. (2007). Low-delay protocol for multihop wireless body area networks. In *Proceedings of the 4th IEEE Annual International Conference on Mobile and Ubiquitous Systems: Networking & Services* (MobiQuitous). Philadelphia, PA: IEEE. doi:10.1109/MOBIQ.2007.4451060

Lauterbach, C., Strasser, M., Jung, S., & Weber, W. (2002). Smart clothes self-powered by body heat. In *Proceedings of Avantex Symposium*. Frankfurt, Germany: Avantex.

Lawrance, A. J., & Wolff, R. C. (2003, June). Binary Time Series Generated by Chaotic Logistic Maps. *International Journal of Bifurcation and Chaos in Applied Sciences and Engineering*, *3*, 529–544.

Lee, S., & Gerla, M. (2001). Split Multipath Routing with Maximally Disjoint Paths in Ad hoc Networks. In *Proceedings of IEEE International Conference on Communications (ICC 2001)*. Helsinki, Finland: IEEE.

Lee, S.J., Su, W., & Gerla, M. (2002). On demand multicast routing protocol in multihop wireless mobile networks. *Mobile Networks and Application,* 441-453.

Lee, S., Noh, Y., & Kim, K. (2013). Key Schemes for Security Enhanced TEEN Routing Protocol in Wireless Sensor Networks. *International Journal of Distributed Sensor Networks*, *2013*, 1–8. doi:10.1155/2013/374796

Li, L., & Halpern, J. Y. (2001). *Minimum-energy mobile wireless networks revisited.* Paper presented at the IEEE International Conference on Communications ICC'01, Helsinki, Finland.

Liang & Wang. (2004). *On performance analysis of challenge/responsebased authentication in wireless networks*. Elsevier.

Liang, X., Balasingham, I., & Byun, S.-S. (2008). A reinforcement learning based routing protocol with QoS support for biomedical sensor networks. In *Proceedings of the 1st IEEE International Symposioum on Applied Sciences on Biomedical and Communication Technologies* (ISABEL). Aalborg, Denmark: IEEE.

Lindsey, S., & Raghavendra, C. (2002). *PEGASIS: Power-Efficient Gathering in Sensor Information Systems*. Paper presented at the IEEE Aerospace Conference, Montana, USA. doi:10.1109/AERO.2002.1035242

Liu, J., Chen, J., Peng, L., & Cao, X. (2012). An open, flexible and multilevel data storing and processing platform for very large scale sensor network. In *Proceedings of14th International Conference on Advanced Communication Technology* (ICACT), (pp. 926-930). Academic Press.

Liu, Z., Joy, A. W., & Thompson, R. A. (2004). A dynamic trust model for mobile ad hoc networks. In *Proceedings of 10th IEEE Int Workshop Future Trends of Distributed Computing Systems*. IEEE. doi:10.1109/RTCSA.2006.61

Li, Y., Panwar, S. S., Mao, S., Burugupalli, S., & Lee, J. (2005). A Mobile Ad Hoc Bio-Sensor Network. *Proceedings of the IEEE, ICC*, 2005.

Loo, C. E., Yong, M., Leckie, C., & Palaniswami, M. (2006). Intrusion Detection for Routing Attacks in Sensor Networks. *International Journal of Distributed Sensor Networks*, *2*(4), 313–332. doi:10.1080/15501320600692044

Lou, W., Liu, W., Zhang, Y., & Fang, Y. (2009). SPREAD: Improving Network Security by Multipath Routing in Mobile Ad hoc Networks. *Wireless Networks*, *15*(3), 279–294. doi:10.1007/s11276-007-0039-4

Lozi, R., & Cherrier, E. (2011). Noise-resisting Ciphering based on a Chaotic Multi-stream Pseudo-random Number Generator. *IEEE International Conference for Internet Technology and Secured Transactions (ICITST)*, (pp. 91 – 96). IEEE.

Luna, J. J. G., & Madruga, E. L. (2001). Scalable Multicasting: The Core-Assisted Mesh Protocol. *Mobile Networks and Applications*, *6*(2).

Lunt, T. F. (1990). IDES: An intelligent system for detecting intruders. In *Proceedings of the Symposium: Computer Security, Threat and Countermeasures*. Academic Press.

Luo, Yang, & Zhang. (2004). Security in Mobile Ad Hoc Networks: Challenges and Solution. IEEE Wireless Communication, 11, 38-47.

Luo, Z. (2008). Survey of Networking Techniques for Wireless Multimedia Sensor Networks. *International Journal of Recent Technology and Engineering*, *2*(2), 182–183.

Lupu, T. G. (n.d.). *Main type of attack in wireless sensor network*. Academic Press.

Macker, J., & Corson, S. (1999). *Mobile Ad Hoc Networking: Routing Protocol Performance Issues and Evaluation Consideration*. RFC.

Madoka, Y. T. K. (2010). Sensor-Cloud Infrastructure-Physical Sensor Management with Virtualized Sensors on Cloud Computing. *13th International Conference on Network-Based Information Systems*.

Mahdi, S. A., Othman, M., Ibrahim, H., & Desa, J. (2013). Protocols for Secure Routing and Transmission in Mobile Ad Hoc Network: A Review. *Journal of Computer Science*, *9*(5), 607–619. doi:10.3844/jcssp.2013.607.619

Malik, M.Y. (n.d.). An Outline of Security in Wireless Sensor Networks: Threats, Countermeasures and Implementations. *Wireless Sensor Networks and Energy Efficiency: Protocols, Routing and Management*. DOI: 10.4018/978-1-4666-0101-7.ch024

Mandi, Haribhat, & Murali. (2010). Generation of Large Set of Binary Sequences Derived from Chaotic Functions Defined Over Finite Field GF (2^8) with Good Linear Complexity and Pair wise Cross - correlation Properties. *International Journal of Distributed and Parallel Systems*, *1*, 93 – 112.

Manjeshwar, A., & Agrawal, D. P. (2002). *APTEEN: A Hybrid Protocol for Efficient Routing and Comprehensive Information Retrieval in Wireless Sensor Networks*. Paper presented at the 2nd International Workshop on Parallel and Distributed Computing Issues in Wireless Networks and Mobile Computing, Ft. Lauderdale, FL. doi:10.1109/IPDPS.2002.1016600

Mankar, V. H., Das, T. S., & Sarkar, S. K. (2012). Discrete Chaotic Sequence based on Logistic Map in Digital Communications. *National Conference on Emerging Trends in Electronics Engineering & Computing (E3C 2010)*, 1, 1016 – 1020.

Manna, G.C. (2010, August). *Quality of Service Monitoring*. Course material conducted at Information and Communication Authority (ICTA), Port Louis, Mauritius.

Maraiya, K., Kant, K., & Gupta, N. (2011). Application based Study on Wireless Sensor Network. *International Journal of Computers and Applications*, *21*(8), 9–15. doi:10.5120/2534-3459

Marcelloni, F., & Vecchio, M. (2009). An efficient lossless compression algorithm for tiny nodes of monitoring wireless sensor networks. *The Computer Journal*, *52*(8), 969–987. doi:10.1093/comjnl/bxp035

Marcelloni, F., & Vecchio, M. (2010). Enabling energy-efficient and lossy-aware data compression in wireless sensor networks by multi-objective evolutionary optimization. *Inf. Sci, 180*(10), 1924–1941. doi:10.1016/j.ins.2010.01.027

Marina, M. K., & Das, S. R. (2006). Ad hoc On-demand Multipath Distance Vector Routing. *Wireless Communications and Mobile Computing, 6*(7), 969–988. doi:10.1002/wcm.432

Marinkovic, S., & Popovici, E. (2009). Network coding for efficient error recovery in wireless sensor networks for medical applications. In *Proceedings of International Conference on Emerging Network Intelligence*. Sliema, Malta: Academic Press. doi:10.1109/EMERGING.2009.22

Marsh, S. (1994). *Formalizing trust as a computational concept*. (Unpublished PhD Thesis). University of Stirling, Stirling, UK.

Maskooki, A., Soh, C. B., Gunawan, E., & Low, K. S. (2011). Opportunistic routing for body area network. In *Proceedings of Consumer Communications and Networking Conference* (CCNC). Las Vegas, NV: Academic Press.

Mattord, N. V. (2008). Principles of Information Security. Course Technology, 290–301.

Mavropodi, R., & Douligeris, C. (2006). A Multipath Routing Protocols for Mobile Ad Hoc Networks: Security Issues and Performance Evaluation. In I. Stavrakakis & M. Smirnov (Eds.), *Autonomic Communication* (pp. 165–176). Springer Berlin Heidelberg. doi:10.1007/11687818_13

May, R. M., & Oster, G. F. (1980, July7). Chaos from Maps. *Journel of Physics Letters A, 78*, 1–124. doi:10.1016/0375-9601(80)90788-4

Meyer, U., & Wezel, S. (2004, October 1). *A Man-in-the-Middle Attack on UMTS*. ACM.

Misic, J. (2008). Enforcing patient privacy in healthcare WSNs using ECC implemented on 802.15.4 beacon enabled clusters. In *Proceedings of the Sixth Annual IEEE International Conference on Pervasive Computing and Communications*. Hong Kong: IEEE.

Misra, S., & Xue, G. (2006). Efficient anonymity schemes for clustered wireless sensor networks. *International Journal of Sensor Networks, 1*(1/2), 50–63. doi:10.1504/IJSNET.2006.010834

Mohammad, S. (2013). A New Secure Cryptography Algorithm Based on Symmetric Key Encryption. *Journal of Basic and Applied Scientific Research*. Retrieved from www.textroad.com

Mohapatra, P., & Krishnamurthy, S. (2005). *Ad Hoc Networks: Technologies and Protocols*. Springer Science.

Molisch, F. (2011). Wireless Communications (2nd ed.). Wiley and IEEE.

Momani, M. S. Challa, & Aboura, K. (2007). Modeling trust in wireless sensor networks from the sensor reliability prospective. In Innovative algorithms and techniques in automation, industrial electronics, and telecomm (pp. 317-321). Springer.

Momani, M., Challa, S., & Alhmouz, R. (2008). Can we trust trusted nodes in wireless sensor networks? In *Proceedings of International conference on Computer and Communication Engineering* (ICCCE 2008), (vol. 1, pp. 37-45). Kuala Lumpur, Malaysia: ICCCE.

Moneda, I., Ioannidou, M. P., & Chrissoulidis, D. P. (2003). Radio-wave exposure of the human head: Analytical study based on a versatile eccentric spheres model including a brain core and a pair of eyeballs. *IEEE Transactions on Bio-Medical Engineering*, *50*(6), 667–676. doi:10.1109/TBME.2003.812222 PMID:12814233

Mueller, S., Tsang, R., & Ghosal, D. (2004). Multipath Routing in Mobile Ad hoc Networks: Issues and Challenges. In Performance Tools and Applications to Networked Systems (pp. 209-234). Springer Berlin Heidelberg. doi:3_10 doi:10.1007/978-3-540-24663

Muhammad & Seon. (2007). Security Issues in Wireless Mesh Networks. In *Proceedings of International Conference on Multimedia and Ubiquitous Engineering* (MUE'07). IEEE.

Mundt, W., Montgomery, K. N., Udoh, U. E., Barker, V. N., Thonier, G. C., Tellier, A. M., & Kovacs, G. T. A. et al. (2005). A multiparameter wearable physiologic monitoring system for space and terrestrial applications. *IEEE Transactions on Information Technology in Biomedicine*, *9*(3), 382–391. doi:10.1109/TITB.2005.854509 PMID:16167692

Natarajan, A., de Silva, B., Kok-Kiong, Y., & Motani, M. (2009). To hop or not to hop: Network architecture for body sensor networks. In *Proceedings of the 6th Annual IEEE Communications Society Conference on Sensor, Mesh and Ad Hoc Communications and Networks* (SECON). Rome, Italy: IEEE. doi:10.1109/SAHCN.2009.5168978

Netto, F. S., & Eisencraft, M. (2008). Spread Spectrum Digital Communication System Using Chaotic Pattern Generator. *The 10th Experimental Chaos Conference.*

Ng, J. W. P., Lo, B. P. L., Wells, O., Sloman, M., Peters, N., Darzi, A., . . . Yang, G.-Z. (2004). *Ubiquitous Monitoring Environment for Wearable and Implantable Sensors*. Paper presented at the International Conference on Ubiquitous Computing (UbiComp), Tokyo, Japan.

Ng, H. S., Sim, M. L., & Tan, C. M. (2006). Security issues of wireless sensor networks in healthcare applications. *BT Technology Journal*, *24*(2), 138–144. doi:10.1007/s10550-006-0051-8

Nguyen, L. T., Defago, X., Beuran, R., & Shinoda, Y. (2008). *An Energy Efficient Routing Scheme for Mobile Wireless Sensor Networks*. Paper presented at the IEEE International Symposium on Wireless Communication Systems (ISWCS'08), Reykjavik, Iceland.

Nguyen, U. T., & Nguyen, H. L. (2006). Study of Different Types of Attack on Multicast in Mobile Ad Hoc Networks. *IEEE International Conferences on Networking.*

Nguyen, H. L., & Nguyen, U. T. (2008). A Study of different types of attack on multicast in mobile ad hoc networks. *Ad Hoc Networks*, *6*(1), 32–46. doi:10.1016/j.adhoc.2006.07.005

Niazi, M., & Hussain, A. (2011). A novel agent-based simulation framework for sensing in complex adaptive environments. *Sensors Journal, IEEE*, *11*(2), 404–412. doi:10.1109/JSEN.2010.2068044

Niemi, V., & Nyberg, K. (2003). *Set-up of UTRAN security mechanisms*. Chichester, UK: John Wiley & Sons.

Noury, N., Herve, T., Rialle, V., Virone, G., Mercier, E., Morey, G., & Porcheron, T. et al. (2000). Monitoring behavior in home using a smart fall sensor. *IEEE-EMBS Special Topic Conference on Microtechnologies in Medicine and Biology*. doi:10.1109/MMB.2000.893857

Okumoku-Evroro &Oniovosa. (2005). *Internet Security: The Role Of Firewall System*. Department Of Computer Science Delta State Polytechnic Otefe-Oghara.

Olariu, S., Xu, Q., Eltoweissy, M., Wadaa, A., & Zomaya, A. Y. (2005). Protecting the communication structure in sensor networks. *International Journal of Distributed Sensor Networks*, *1*(2), 187–203. doi:10.1080/15501320590966440

Osmocom. (2011). *OsmoSGSN*. Retrieved from http://openbsc.osmocom.org/trac/wiki/osmo-sgsn

Osmocom. (2014). *OpenBSC GPRS/EDGE Setup page*. Retrieved from http://openbsc.osmocom.org/trac/wiki/OpenBSC_GPRS

Osmocom. (2014). *Welcome to Osmocom OpenBSC*. Retrieved from http://openbsc.osmocom.org/trac/

Otal, B., Alonso, L., & Verikoukis, C. (2009). Highly reliable energy-saving MAC for wireless body sensor networks in healthcare systems. *IEEE Journal on Selected Areas in Communications*, *27*(4), 553–565. doi:10.1109/JSAC.2009.090516

Otto, C. A., Jovanov, E., & Milenkovic, E. A. (2006). WBAN-based system for health monitoring at home. In *Proceedings of IEEE/EMBS International Summer School, Medical Devices and Biosensors*. Boston, MA: IEEE. doi:10.1109/ISSMDBS.2006.360087

Padmavathi, G. & Shanmugapriya, D. (2009). A Survey of Attacks, Security Mechanisms and Challenges in Wireless Sensor Networks. *International Journal of Computer Science and Information Security, 4*.

Padmavathi, G., & Shanmugapriya, D. (2009). A Survey of Attacks, Security Mechanisms and Challenges in Wireless Sensor Networks. *International Journal of Computer Science and Information Security*, *4*(1-2), 1–9.

Palanisamy, V., & Annadurai, P. (2009). *Impact of rushing attack on multicast in mobile ad hoc network. International Journal of Computer Science and Informational Security*, 4.

Pandian, P. S., Mohanavelu, K., Safeer, K. P., Kotresh, T. M., Shakunthala, D. T., Gopal, P., & Padaki, V. C. (2007). Smart vest: Wearable multi-parameter remote physiological monitoring system. *Medical Engineering & Physics*, *30*(4), 466–477. doi:10.1016/j.medengphy.2007.05.014 PMID:17869159

Pandian, P. S., Safeer, K. P., Gupta, P., Shakunthala, D. T., Sundersheshu, B. S., & Padaki, V. C. (2008). Wireless sensor network for wearable physiological monitoring. *J. Netw.*, *3*, 21–28.

Pateari, N. (2011). Performance Analysis of Dynamic Routing Protocol in Mobile Ad-Hoc Network. *Journal of Global Research in Computer Science, 2.*

Pathan, A.-S. K., Dai, T. T., & Hong, C. S. (2006). A Key Management Scheme with Encoding and Improved Security for Wireless Sensor Networks. In S. Madria et al. (Ed.), *International Conference on Distributed Computing and Internet Technology (ICDCIT'06) (LNCS)* (Vol. 4317, pp. 102-115). Berlin, Germany: Springer. doi:10.1007/11951957_10

Patidar, V., Sud, K. K., & Pareek, N. K. (2009). A Pseudo Random Bit Generator Based on Chaotic Logistic Map and its Statistical Testing. *International Journal of Modern Physics, 251*, 441–452.

Peiris, V. (2013). Highly integrated wireless sensing for body area network applications. *SPIE Newsroom.*

Pereira, P., Grilo, A., Rochą, F., Nunes, M., Casaca, A., & Chaudet, C., … Johansson, M. (2007). End-to-end reliability in wireless sensor networks: Survey and research challenges. In *Proceedings of EuroFGI Workshop on IP QoS and Traffic Control.* Lisbon, Portugal: Academic Press.

Perrig, A., Hu, Y. C., & Johnson, D. B. (2003). *Efficient Security mechanisms for routing protocols.* Network and Distributed System Security Symposium.

Petriu, E. M., Georganas, N. D., Petriu, D. C., Makrakis, D., & Groza, V. Z. (2000).Sensor-based information appliances. *IEEE Instrumentation & Measurement Magazine*, ▪▪▪, 31–35.

Phatak, S. C., & Rao, S. S. (1994, July15). Logistic Map: A Possible Random Number Generator. *Physical Review. Statistical, Nonlinear and Soft matter Physics, 51*, 3670–3678.

Picó, J., & Pérez, D. (2011). *A practical attack against GPRS/EDGE/UMTS/HSPA mobile data.* Black Hat. Retrieved from https://media.blackhat.com/bh-dc-11/Perez-Pico/BlackHat_DC_2011_Perez-Pico_Mobile_Attacks-wp.pdf

Pirzada, A. A., & McDonald, C. (2004). Establishing trust in pure ad-hoc networks. In *Proceedings of 27th Australasian Computer Science Conf.* (ACSC '04), (*vol. 26*, pp. 47-54). ACSC.

Ponmagal & Raja. (2011). An extensible cloud architecture model for heterogenous sensor services. *International Journal of Computer Science and Information Security, 9.*

Poole, I. (n.d). UMTS Core Network. In *UMTS/WCDMA Network Architecture.* Retrieved from http://www.radio-electronics.com/info/cellulartelecomms/umts/umts-wcdma-network-architecture.php

Preneel, B. (2003). *Analysis and Design of Cryptographic Hash Functions.* Academic Press.

Protecting the confidentiality of Personal Data. (2008). CMOD Department of Finance.

Quwaider, M., & Biswas, S. (2009). On-body packet routing algorithms for body sensor networks. In *Proceedings of the 1st IEEE International Conference on Networks and Communications* (NETCOM). Chennai, India: IEEE. doi:10.1109/NetCoM.2009.54

Quyen, N. X., Yem, V. V., & Hoang, T. M. (2012, November21). A Chaos-Based Secure Direct-Sequence/Spread-Spectrum Communication System. *Journal of Abstract and Applied Analysis, 2013*, 1–11. doi:10.1155/2013/764341

Raghavendran, C. V., Satish, G. N., & Varma, P. S. (2013). A Study on Contributory Group Key Agreements for Mobile Ad Hoc Networks. *International Journal of Computer Network and Information Security, 5*(4), 48–56. doi:10.5815/ijcnis.2013.04.07

Rajesj, V., Gnanasekar, J. M., Ponmaga, R. S., & Anbalagan, P. (2010). Integration of Wireless Sensor Network with Cloud. *International Conference on Recent Trends in Information, Telecommunication and Computing*. doi:10.1109/ITC.2010.88

Rao, M. J., Moyer, & Rohatgi, P. (1999). A Survey of Security Issues in Mlticast Communication. IEEE Network, 12-23.

Rappaport, T. (1996). *Wireless Communications: Principles and Practice*. Englewood Cliffs, NJ: Prentice-Hall.

Rappaport, T. S. (1997). *Wireless Communications - Principles and Practice* (2nd ed.). Pearson Education.

Razzaque, M. A., Hong, C. S., & Lee, S. (2011). Data-centric multiobjective QoS-aware routing protocol for body sensor networks. *Sensors (Basel, Switzerland), 11*(12), 917–937. doi:10.3390/s110100917 PMID:22346611

Reddy, G. V. (2007). Performance Evaluation of Different DS-CDMA Receivers Using Chaotic Sequences.*International Conference on RF and Signal Processing Systems*, 32, 49-52.

Renaud, M., Karakaya, K., Sterken, T., Fiorini, P., Hoof, C. V., & Puers, R. (2008). Fabrication, modelling and characterization of MEMS piezoelectric vibration harvesters. *Sens. Actuat. A*, 380-386.

Rhee, S., Seetharam, D., & Liu, S. (2004). Techniques for Minimizing Power Consumption in Low Data-Rate Wireless Sensor Networks. In *Proc. of IEEE Wireless Communications and Networking Conference*. Atlanta, GA: IEEE.

Rodoplu, V., & Meng, T. H. (1999). Minimum energy mobile wireless networks. *IEEE Journal on Selected Areas in Communications, 17*(8), 1333–1344. doi:10.1109/49.779917

Rolim, C. O., Koch, F. L., Westphall, C. B., Werner, J., Fracalossi, A., & Salvador, G. F. (2010). A cloud computing solution for patient's data collection in health care institutions. In *Proceedings of the 2nd International Conference on Ehealth, Telemedicine, and Social Medicine*, (pp. 95-99).

Ruzzelli, A. G., Jurdak, R., O'Hare, G. M., & van Der Stok, P. (2003). Energy-efficient multi-hop medical sensor networking. In *Proceedings of the 1st ACM SIGMOBILE International Workshop on Systems and Networking Support for Healthcare and Assisted Living Environments*. ACM.

Sadler, C. M., & Martonosi, M. (2006). Data compression algorithms for energy-constrained devices in delay tolerant networks. In *Proceedings of the 4th ACM International Conference on Embedded Networked Sensor Systems*. Boulder, CO: ACM. doi:10.1145/1182807.1182834

Sanchez, J., & Thioune, M. (2007). PLMN selection. In *UMTS* (pp. 188–189). London, UK: ISTE Ltd.

Sangulagi, P., & Naveen, A. S. (2014). Efficient Security Approaches in Mobile Ad-Hoc Networks: A Survey. *International Journal of Research in Engineering and Technology*, *3*(3), 14–19.

Sankaranarayan, V., & Tamilselvan, L. (2006). *Solution to prevent rushing attack in wireless mobile ad hoc network*. IEEE.

SANS Institute (2005) "SANS Institute Info Sec Reading Room"

Sanzgiri, K., LaFlamme, D., Dahill, B., Levine, B. N., Shields, C., & Royer, E. M. (2005). Authenticated Routing for Ad hoc Networks. *IEEE Journal on Selected Areas in Communications*, *23*(3), 598–610. doi:10.1109/JSAC.2004.842547

Sara, G. S., Kalaiarasi, R., Pari, S. N., & Sridharan, D. (2010). Energy Efficient Mobile Wireless Sensor Network Routing Protocol. In N. Meghanathan et al. (Ed.), *Recent Trends in Networks and Communications: Proceedings of the International Conferences, NeCoM 2010, WiMoN 2010, WeST 2010*, (LNCS) (Vol. 90, pp 642-650). Berlin, Germany: Springer. doi:10.1007/978-3-642-14493-6_65

SarmaDhulipala, Kanagachidambaresan, & Chandrasekaran. (2012). Lack of power avoidance: A fault classification based fault tolerant framework solution for lifetime enhancement and reliable communication in wireless sensor network. *Information Technology Journal*, *11*(6), 247–260.

Sayrafian-Pour, K., Wen-Bin, Y., Hagedorn, J., Terrill, J., & Yazdandoost, K. Y. (2009). A statistical path loss model for medical implant communication channels. In *Proceedings of the 20th IEEE International Symposium on Personal, Indoor and Mobile Radio Communications*. Tokyo, Japan: IEEE. doi:10.1109/PIMRC.2009.5449869

Scarfone, K., & Mell, P. (2007). Guide to intrusion detection and prevention systems (IDPS). *NIST Special Publication, 94*.

Schoellhammer, T., Osterweil, E., Greenstein, B., Wimbrow, M., & Estrin, D. (2004). Lightweight temporal compression of microclimate datasets. In *Proceedings of the 29th Annual IEEE International Conference on Local Computer Networks*. Tampa, FL: IEEE. doi:10.1109/LCN.2004.72

Selvaraj, G., Sivakumar, K., (2013). Overview of various attacks in MANET and Counter Measures for Attacks. *International Journal of Computer Science and Management, 2*(1).

Sensor-Cloud. (n.d.). Available: hp://www.sensorcloud.com/system-overview

Seung, Y., Prasad, N., & Robin, K. (2001). Security-aware Ad hoc Routing for Wireless Networks. In *Proceedings of the 2nd ACM international symposium on Mobile ad hoc networking & computing*, (pp. 299-302). ACM.

Sevil, S., Clark, J. A., & Tapiador, J. E. (2010). Security Threats in Mobile Ad Hoc Networks. In Security of Self-Organizing Networks: MANET, WSN, WMN, VANET (pp. 127-146). CRC Press.

Shaerbaf, S., & Seyedin, S. A. (2011, September). Nonlinear Multiuser Receiver for Optimized Chaos-Based DSCDMA Systems. *Iranian Journal of Electrical & Electronic Engineering*, *7*, 149–160.

Shah, Khan, Ali, & Khan. (2013). A new Framework to Integrate Wireless Sensor Networks with Cloud Computing. In *Proceedings ofAeropace Conference*. IEEE.

Shahrani, A. S. A. (2011). *Rushing Attack in Mobile Ad Hoc Networks. In Intelligent Networking and Collaborative Systems*. INCS.

Shahrani, A., & Saad, A. (2011). Rushing Attack in mobile ad hoc Networks.*IEEE International Conference on Intelligent Network and Collaborative System*. doi:10.1109/INCoS.2011.145

Shaikh, J., Brian, A., & Heejo, L. (2009). Group-based trust management scheme for clustered wireless sensor network. *IEEE Transactions on Parallel and Distributed Systems*, *20*(11), 1698–1718. doi:10.1109/TPDS.2008.258

Shaikh, R. A., Jameel, H., Lee, S., Rajput, S., & Song, Y. J. (2006). Trust management problem in distributed wireless sensor networks. In *Proceedings of 12th IEEE Intl Conf Embedded Real-Time Computing Systems and Applications*. IEEE.

Sharma, K., Khandelwal, N., & Prabhakar, M. (2011). *An Overview of Security Problems in MANET*. Academic Press.

Sharma, N., & Sahni, P. (2014). Secure Routing & Data Transmission in Mobile Ad Hoc Networks. *International Journal of Innovative Research in Technology*, *1*(2), 407–412.

Shi, E., & Perrig, A. (2004). Designing secure sensor networks. *IEEE Wireless Comm*, *11*(6), 38–43. doi:10.1109/MWC.2004.1368895

Shrivastava, S. (2013). *Rushing Attack and its Prevention Techniques. International Journal of Application or Innovation in Engineering Management*, 2.

Shrivastava, S. (2014). A New Technique to Prevent MANET against Rushing Attack. *International Journal of Computer Science and Information Technologies*, *5*, 3460–3464.

Siegmund, N. (n.d.). *Towards Robust data storage in wireless sensor networks*. Academic Press.

Silva, A. P. R. D., Martins, M. H. T., Rocha, B. P. S., Loureiro, A. A. F., Ruiz, L. B., & Wong, H. C. (2005). *Decentralized intrusion detection in wireless sensor networks*. Paper presented at the 1st ACM international workshop on Quality of service & security in wireless and mobile networks, Montreal, Canada. doi:10.1145/1089761.1089765

Silva, A. R., & Vuran, M. C. (2010). *Communication with Aboveground Devices in Wireless Underground Sensor Networks: An Empirical Study*. Paper presented at the IEEE International Conference on Communications (ICC), Cape Town, South Africa. doi:10.1109/ICC.2010.5502315

Silva, D., Ghanem, M., & Guo, Y. (2012). WikiSensing: An online collaborative approach for sensor data management. *Sensors (Basel, Switzerland)*, *12*(10), 13295–13332. doi:10.3390/s121013295 PMID:23201997

Singh & Jain. (2012). Research Issues in Wireless Networks. *International Journal of Advanced Research in Computer Science and Software Engineering, 2*(4).

Singh, A., Goyal, P., & Batra, S. (2010). A Literature Review of Security Attack in Mobile Ad-hoc Networks. *International Journal of Computers and Applications, 9.*

Singh, P. K. (2007). *NGN Security. In Compendium on Next Generation Networks* (pp. 27–28). Delhi, India: The Gondals Press.

Singh, T. P., Kaur, S., & Das, V. (2012). Security Threats in Mobile Ad hoc Network: A Review. *International Journal of Computer Networks and Wireless Communications*, *32*(6), 27–34.

Sohrabi, K., Gao, J., Ailawadhi, V., & Pottie,G.J. (2000). Protocols for self-organization of a wireless sensor network. *IEEE Personal Communications*, 16–27.

Srinivasan, V., & Stankovic, J. (2008). Protecting your daily in home activity information from a wireless snooping attack. In *Proceedings of the 10th International Conference on Ubiquitous Computing*. Seoul, Korea: Academic Press.

Sun & Liu. (2004). *Scalable Hierarchical Access Control in Secure Group Communications*. IEEE.

Sundani, H. (n.d.). *Wireless Sensor network simulators A survey and comparison*. Academic Press.

Sun, Y. L., Yu, W., Han, Z., & Liu, K. J. R. (2006). Information theoretic framework of trust modeling and evaluation for ad hoc networks. *IEEE Journal on Selected Areas in Communications*, *24*(2), 305–317. doi:10.1109/JSAC.2005.861389

Su, W., Lee, S. J., & Gerla, M. (2002). On-Demand Multicast Routing Protocol in Multihop Wireless Mobile Networks. *Mobile Networks and Applications, 7.*

Syrotiuk, V. R., Chalamtse, I., & Basagni, S. (1999). *Dynamic Source Routing for Ad Hoc Network Using The Global Positioning System*. IEEE.

Taiming, F., Chuang, W., Wensheng, Z., & Lu, R. (2001). *Confidentiality Protection for Distributed Sensor Data Aggregation*. Retrieved from http://www.cs.iastate.edu/

Takahashi, D., Xiao, Y., & Hu, F. (2007). LTRT: Least total-route temperature routing for embedded biomedical sensor networks. In *Proceedings of Global Telecommunications Conference* (GLOBECOM). Washington, DC: IEEE. doi:10.1109/GLOCOM.2007.125

Talwar, S. K., Xu, S., Hawley, E. S., Weiss, S. A., Moxon, K. A., & Chapin, J. K. (2002). Behavioural Neuroscience: Rat Navigation Guided by Remote Control. *Nature*, *417*(6884), 37–38. doi:10.1038/417037a PMID:11986657

Tang, Q., & Tummala, N. (2005). TARA: Thermal-aware routing algorithm for implanted sensor networks. In *Distributed computing in sensor systems*. Springer. doi:10.1007/11502593_17

Tarique, M., Tepee, K. E., Adibi, S., & Erfani, S. (2009). Survey of Multipath Routing Protocols for Mobile Ad hoc Networks. *Journal of Network and Computer Applications*, *32*(6), 1125–1143. doi:10.1016/j.jnca.2009.07.002

Ting, C. K., & Liao, C.-C. (2010). A memetic algorithm for extending wireless sensor network lifetime. *Inf. Sci.*, *180*(24), 4818–4833. doi:10.1016/j.ins.2010.08.021

Tripathi, R., Agarwal, R., & Tiwari, S. (2011). Performance Evolutation and Comparison of AODV and DSR under Adversarial Environment. *IEEE Conference on Computational Intelligence and Communication Systems.*

Tse, D., & Viswanath, P. (2005). Fundamentals of Wireless Communications (2nd ed.). Cambridge. doi:10.1017/CBO9780511807213

Ullah, S., Higgins, H., Braem, B., Latre, B., Blondia, C., Moerman, I., & Kwak, K. S. et al. (2012). A comprehensive survey of wireless body area networks. *Journal of Medical Systems*, *36*(3), 1065–1094. doi:10.1007/s10916-010-9571-3 PMID:20721685

Undercoffer, J., Avancha, S., Joshi, A., & Pinkston, J. (2002). Security for Sensor Networks. *In Proceedings of CADIP Research Symposium.*

van Bosse, J. G. (1998). *Signaling in Telecommunication Networks.* New York, NY: John Wiley & Sons, Inc.

Vijayalakshmi, S., & Rabaral, S.A. (2014). Rushing Attack Mitigation in Multicast MANET (RAM). *International Journal of Research and Reviews in Computer Science, 1*(4).

Villatoro, J., & Monzon-Herńandez, D. (2005). Fast detection of hydrogen with nano fiber tapers coated with ultra thin palladium layers. *Optics Express*, *13*(13), 5087–5092. doi:10.1364/OPEX.13.005087 PMID:19498497

Vinay. (2007). *Authentication and Access Control The Cornerstone of Information Security.* Trianz White Paper.

Visalakshi, P. (2011). *Security issues and vulnerabilities in mobile ad hoc network- A survey. International Journal of Computational Engineering Research.*

Vyavahare, P. D., Rawat, A., & Ramani, A. K. (2005). *Evaluation of Rushing Attack on Secured Message Transmission Protocol (SMT/SRP) for Mobile Ad-hoc Network.* IEEE.

Wagner, M., Wagner, J., & Zucchini, W. (2010). *Authentication - AKA. In 3G Performance and Security, Evolution towards UMTS Network and Security Mechanism* (pp. 88–90). Saarbrücken, Germany: VDM Verlag Dr Müller Aktiengesellschaft & Co. KG.

Walters, L., & Shi, C. (2006). Wireless Sensor Network Security: A Survey. Security in Distributed, Grid and Pervasive Computing.

Walters, J. P., Liang, Z., Shi, W., & Chaudhary, V. (2006). Wireless sensor network security: A survey. *Security in Distributed Grid and Pervasive Computing*, *43*(5), 367–410.

Wang, H., Peng, D., Wang, W., Sharif, H., & Chen, H. (2008). *Energy-Aware Adaptive Watermarking for Real-Time Image Delivery in Wireless Sensor Networks*. Paper presented at theIEEE International Conference on Communications (ICC '08), Beijing, China. doi:10.1109/ICC.2008.286

Warneke, B., Liebowitz, B., Pister, K.S.J. (2001). Smart dust: communicating with a cubic-millimeter computer. *IEEE Computer.*

Webroot Software Inc. (2004). *Anti-spyware software: Securing the corporate network*. Academic Press.

Winkler, J. R. (1990). A unix prototype for intrusion and anomaly detection in secure networks. In *Proceedings of the 13th National Computer Security Conference*. Academic Press.

Wireless Sensor Network. (n.d.). In *Wikipedia, the free encyclopedia*. Available: http://en.widipedia.org/wiki/wireless_sensor_network

Wood, A., Virone, G., Doan, T., Cao, Q., Selavo, L., Wu, Y., . . . Stankovic, J. (2006). *ALARM-NET: Wireless Sensor Networks for Assisted-Living and Residential Monitoring*. Technical Report CS-2006-11. University of Virginia.

Xiang, L. (n.d.). *Compressed data Aggregation for energy efficient Wireless sensor network*. Academic Press.

Xueyi, Z., Lu, J., Kejun, W., & Dianpu, L. (2000). Logistic-Map Chaotic Spreading Spectrum Sequences Under Linear transformation. *The 3rd World Congress on Intelligent Control and Automation*, *4*, 2464 - 2467.

Yadav, S., Jain, R., & Faisal, M. (2012). Attacks in MANET. *International Journal of Latest Trends in Engineering and Technology*, *1*(3), 123–126.

Yagu¨e, Mana, & Lopez. (2005). *A metadata-based access control model for web services*. Retrieved from www.emeraldinsight.com/researchregister

Yang, H., Haiyun, L., Fan, Y., Songwu, L., & Zhang, L. (2004). Security in Mobile Ad hoc Networks: Challenges and Solutions. *IEEE Wireless Communications*, *11*(1), 38–47. doi:10.1109/MWC.2004.1269716

Yang, L., & Jun, T. X. (2012, May2). A new pseudorandom number generator based on complex number chaotic equation. *Chinese Physics B*, *21*.

Yazicioglu, R. F., Torfs, T., Merken, P., Penders, J., Leono, V., Puers, R., & VanHoof, C. et al. (2009). Ultra-low-power biopotential interfaces and their applications in wearable and implantable systems. *Microelectronics Journal*, *40*(9), 1313–1321. doi:10.1016/j.mejo.2008.08.015

Yibo, C., Hou, K.-M., Zhou, H., Shi, H.-L., Liu, X., Diao, X., . . . De Vaulx, C. (2011). *6LoWPAN stacks: a survey*. Paper presented at the 7th International Conference on Wireless Communications, Networking and Mobile Computing (WiCOM), Wuhan, China.

Yick, J., Mukherjee, B., & Ghosal, D. (2008). *Wireless Sensor Network Survey*. Elsevier.

Yi, P., Dai, Z., Zhang, S., & Zhong, Y. (2010). A New Routing Attack in Mobile Ad Hoc Networks. *International Journal of Information Technology*, *11*(2), 83–94.

Yonzon, C. R., Stuart, D. A., Zhang, X., McFarland, A. D., Haynes, C., & Vanduyne, R. (2005). Towards advanced chemical and biological nano sensors-An overview. *Talanta*, *67*(3), 438–448. doi:10.1016/j.talanta.2005.06.039 PMID:18970187

You, J., Lieckfeldt, D., Salzmann, J., & Timmermann, D. (2009). *Connectivity Aware Topology Management for Sensor Networks*. GAF & Co.

Younis, O., & Fahmy, S. (2004). HEED: A Hybrid Energy-Efficient Distributed Clustering Approach for Ad Hoc Sensor Networks. *IEEE Transactions on Mobile Computing*, *3*(4), 366–379. doi:10.1109/TMC.2004.41

Yu, L., Wang, N., Zhang, W., & Zheng, C. (2007). *Deploying a Heterogeneous Wireless Sensor Network*. Paper presented at the International Conference on Wireless Communications, Networking and Mobile Computing (WiCom'07), Shanghai, China.

Yuriyama, M., & Kushida, T. (2010). Sensor-Cloud infrastruture physical sensor management with virtualized sensors on cloud computing. In *Proceedings of the IEEE 13th International Conference on Network-Based Information Systems* (NbiS'10). IEEE.

Yuriyama, M., Kushida, T., & Itakura, M. (2011). A new model of accelerating service innovation with sensor-cloud infrastructure. In *Proceedings of the annual SRII Global Conference (SRII'11)*, (pp. 308-314). doi:10.1109/SRII.2011.42

Zeng, X. (n.d.). *GloMoSim A library for parallel simulator of large scale wireless networks*. Academic Press.

Zhang, Das, & Liu. (2006). A trust based framework for secure data aggregation in wireless sensor networks. In *Proceedings of IEEE SECON* (vol. 1, pp. 60-72). IEEE.

Zhang, P., Sadler, C., Lyon, S., & Martonosi, M. (2004). *Hardware design experiences in ZebraNet*. Paper presented at the ACM SenSys'04, Baltimore, MD. doi:10.1145/1031495.1031522

Zhang, D., Wang, H., & Shin, K. G. (2002). *Detecting SYN Flooding Attacks. IEEE INFOCOM*. IEEE.

Zhang, G. H., Poon, C. C. Y., Li, Y., & Zhang, Y. T. (2009). A biometric method to secure telemedicine systems. In *Proceedings of the 31st Annual International Conference of the IEEE Engineering in Medicine and Biology Society*. Minneapolis, MN: IEEE. doi:10.1109/IEMBS.2009.5332470

Zhang, P., Yan, Z., & Sun, H. (2013). A Novel Architecture Based on Cloud Compting for Wireless Sensor Network. In *Proceedings of the 2nd International Conference on Computer Science and Electronics Engineering*.

Zhao, F., & Guibas, L. (2004). *Wireless Sensor Networks: An Information Processing Approach*. San Francisco, CA: Morgan Kaufmann.

Zhong, Y., Zhang, S., & Yi, P. (2008). A New routing attack in mobile ad hoc network. *IJIT*, *11*, 83–94.

Zhou, Q., & Zhang, R. (2013). *A Survey on All-IP Wireless Sensor Network*. Paper presented at the 2nd International Conference on Logistics, Informatics and Service Science, Beijing, China. doi:10.1007/978-3-642-32054-5_105

About the Contributors

Kamaljit I. Lakhtaria is working as Associate Professor in Department of Computer Science, Gujarat University, India. He obtained a PhD in Computer Science, area of Research is "Next Generation Networking Service Prototyping & Modeling". He holds an edge in Next Generation Network, Web Services, Mobile Ad Hoc Networks, Network Security and Cryptography. He is author of 8 Reference Books in the area of Computer Science and has Published 2 chapters in International Editorial Volumes. He presented 23 Research Papers in International Conferences. His Papers are published in the proceedings of IEEE, Springer and Elsevier. He presented 32 Research Papers in National Conferences, published 23 Research Papers in Referred International Journals, and 6 Research Papers in National Journals. He is a Life time member of ISTE, IAENG, and many Research Groups. He hold the post of Editor and Associate Editor in many International Research Journals. He is Program Committee member of many International Conferences. He is reviewer in IEEE Wireless Sensor Network, Inderscience and Elsevier Journals. He is Ph.D. and M.Phil. Supervisor in Recognized Universities.

* * *

Ana Vázquez Alejos received the M.S. Electrical Engineering (Telecommunication Engineering) degree from the University of Vigo, Spain, in 2000, and in 2006 the Ph.D. Electrical Engineering in the same University. Granted in 2009 with Marie Curie International Outgoing fellowship she became a post-doctoral researcher in the New Mexico State University (USA). At present she works as Associate Professor in the University of Vigo being responsible for several radio communications courses.

Diego Fernández Alonso received the Telecommunication Engineer degree from the University of Vigo in 2013, Spain. He is working as Research Engineer from 2013 to present in the Radio System Group, Department of Theory of Signal and Communications, University of Vigo. He develops GSM and wireless security projects.

S. P. Anandaraj, Senior Assistant Professor, Department of CSE, joined SR Engineering College, Warangal in 2007. He teaches graduate and undergraduate courses in Web Technology, Compiler Design, Computer Networks, and Network Security in the areas of Computer Science and Engineering. In addition to teaching, he developed new curriculum and programs that takes a broad focus on the discipline of computer science education. He is a member of Board of Studies in SR Engineering College, Warangal. As part of Administrative duties, he is serving as Coordinator of SREC Spot Evaluation Centre. He received his Ph.D in Computer Science and Engineering from St. Peter's University, Chennai in 2014. His research, in the form of conceptual articles, case studies, surveys, and large sample empirical analysis, focuses on wireless sensor networks. It encompasses three streams – fault tolerance in sensor networks, Networking design issues, and Security Analysis. His research output was employed in the design of networks in real-world implementations, which resulted in efficient working performance. He has authored over 33 papers in various international, national journals and conferences, includes publishing in Elsevier, CIIT Journal, and Springer. He delivered several guest lectures in international seminars and served as keynote speaker in international conferences in the areas of Networking and Fault tolerance networks. He is a Life Member in ISTE, CSI, IRED, IACSIT.

Romeli Barbosa received Ph.D. and M.S. degrees in Energy Engineering option from National Autonomous University of Mexico, in 2012 and 2007, respectively. He received the B.Sc. degree in Engineering in Energy Systems from "Quintana Roo University", Mexico in 2004. He is currently an Assistant Professor at Sciences and Engineering department at the Quintana Roo University, Mexico. He has been awarded a national recognition (SNI member) as a researcher by CONACYT. His research interest includes numerical simulation and design of Catalyst layer and PEM Fuel Cells. Dr. Barbosa, at present is author of nine international papers and is involved in five innovative pilot projects, mainly in the fields of development of new devices for power generation with high power density.

Djallel Eddine Boubiche received his Ph.D in Computer Science from Batna University, Algeria, in 2013 and Magister degree in Industrial Informatics from El Hadj Lakhdar University, Algeria, in 2008. He received the B.Eng degree in Computer Engineering from UHL University, Algeria, in 2004. He is currently an Assistant Professor at the Computer Sciences Department of Batna University, Algeria. He is also the Vice Director of the Computer Sciences Department and a member of LaSTIC Laboratory. Prior to holding this position, he served as an Assistant Research at Computer Sciences Department of Tebessa University, Algeria.

His research interests include Wireless Communication, Emended System, Intelligent Multi-Agent System, Distributed Computing and Sensor Networks.

Ashraf Darwish is Associate Professor, Undergraduate and Postgraduate Programs Coordinator of the Department of Computer Science at the Faculty of Science, Helwan *University*, Egypt. He received the PhD degree in computer science from Saint Petersburg State University, Russian Federation in 2006. He received a BS.C and MS.C. in Mathematics from Faculty of Science, Tanta University, Egypt. He keeps in touch with his mathematical background through his research and his research interests include information security, data and web mining, intelligent computing, image processing (in particular image retrieval, medical imaging), modeling and simulation, intelligent environment, sensor networks, and theoretical foundations of computer science. Dr. Darwish is an editor for some prestigious international journals and member of many computing associations such as IEEE, Egyptian Mathematical Society (EMS). He is a board member of the Russian-Egyptian Association for graduates, Quality Assurance and Accreditation Project (QAAP) and Machine Intelligence Research Lab (MIR Lab) USA. Dr. Darwish was invited as reviewer, invited speaker, organizer, and session chair to many international conferences and journals. He has worked in a wide variety of academic organizations and supervises masters and PhD theses in different areas of computer science. He speaks four languages on a daily basis. Dr. Darwish has a wealth of academic experience and has authored multiple publications which include papers in international journals and conference proceedings. His publications have appeared in international journals, conferences and publishing houses such as IEEE Systems, Man & Cybernetics Society (SMC) on Soft Computing, Springer and Advances in computer science and engineering.

Faouzi Hidoussi has received his master degree in Computer Science from Batna University, Algeria, in 2013. He is currently a collaborating researcher at the laboratory Networking and Distributed Computing Laboratory of International Islamic University Malaysia (IIUM). His research interests include wireless communication, security and sensor networks. He has served as TCP member in many international conferences and as guest editor of some international journals. He is the founder of two international conferences; International Conference on Advanced Wireless, Information, and Communication Technologies (AWICT) and international conference on Intelligent Information Processing, Security and Advanced Communication (IPAC).

G. R. Kanagachidambaresan received his B.E degree in Electrical and Electronics Engineering from Anna University in 2010 and M.E Pervasive Computing Technologies in Anna University in 2012. He is a part time research scholar in PSG

College of Technology, pursuing a Ph.D in Anna University Chennai. He is currently an Assistant Professor in Department of Computer Science and Engineering in Dhanalakshmi Srinivasan College of Engineering, Coimbatore. His main research interest includes Body Sensor Network and Fault tolerant Wireless Sensor Network.

Katyayani Kashyap received her B.E. degree in Electronics and Telecommunications from Girijananda Chowdhury Institute of Management and Technology, Gauhati University, Guwahati, Assam, India in 2012. She received her M.Tech degree in Signal Processing and Communication from Gauhati University, Guwahati, Assam, India in 2014. Her areas of interests are error correcting codes, chaotic codes in wireless channels, implementation of chaotic code based structure in hardware level, Wireless Communication, VLSI design, Digital Signal Processing, and Mobile Communication.

Gopal Chandra Manna is working as Chief General Manager, BRBRA Institute of Telecom Training, BSNL under Department of Telecommunications (DoT), Govt. of India. He has carried out extensive research on coverage issues of GSM, CDMA, WCDMA and WiMAX radio access. Study of Wireless Traffic and QoS estimation of Cognitive Radio are his current areas of research. He is also a member of Wi-MAX forum and ITU-NWG-SG 9 and -SG12. Dr. Manna has developed and conducted one week course on Quality of Service Monitoring at Information and Communication Technologies Authority, Mauritius as International Expert through Commonwealth Telecom Organisation London during August 2010. From 1997 to 2002 Dr. Manna has worked as Deputy General Manager in a Telecommunication Training Centre of DoT. He had installed live training node for Internet Service Provider (ISP), conducted several training programs deputed by Asia Pacific Telecommunity (APT) and Sri Lankan Telecom. He had also conducted several seminars with international experts through UNDP/ITU projects. During 1995 and 1996, Dr. Manna was posted in Telecommunication Engineering Centre (TEC) and developed Artificial Intelligence (AI) based software for E10B telephone exchanges named E10B Maintenance Advisor (E10BMAD). He had worked as Development Officer in WEBEL (erstwhile PHILLIPS) Telecommunication Industries during 1983-1984 after which he joined DoT and worked in different executive capacities upto 1994. He was awarded National Scholarship in 1973 based on school level examination and silver medal for performance in college. He had both graduated and post graduated in RadioPhysics and Electronics Engineering from University of Calcutta and Ph.D. from Electronics and Telecommunications Engineering Department. He had undergone trainings at Beijing University of Post and Telecom China in 1990 and DARTEC, Montreal, Canada in 1999.

Prasanta K. Manohari is a M.Tech student in Silicon Institute of Technology. He did his Master of Science degree from Ravenshaw University, Cuttack and his research interest includes mobile ad hoc networks.

Chinmaya Kumar Nayak is an Assistant Professor in the Department of Computer Science & Engineering, Gandhi Institute for Technological Advancemen (GITA), Bhubaneswar, Odisha, India. He is an author of the book "Data Structure Using C". He published many papers in national seminars and international journals. His research area includes Computer network, WSN, adhoc networks, etc.

Manoranjan Pradhan holds a Ph. D Degree in Computer Science. He is presently working as a professor and Head of the Department of Computer Science & Engineering, Gandhi Institute for Technological Advancement (GITA), Bhubaneswar, Odisha, India. He has 15 years of teaching experience and has published many papers in national and international journals. His research interests include Computer Security, Intrusion Detection and Soft Computing.

Niranjan K. Ray received his PhD in Computer Science and Engineering (CSE) from NIT Rourkela in 2014. He did his Masters in Computer Science and Engineering from Utkal University Bhubaneswar in 2007. He is an Asst. Professor in the Department of Computer Science and Engineering, Silicon Institute of Technology, Bhubaneswar. His research interest includes wireless ad hoc and sensor networks. He was the publication chair of International Conference on Information Technology (ICIT-14). He is a member of IEEE.

Manuel García Sánchez received the Telecommunication Engineer degree from the University of Santiago de Compostela, Spain, in 1990 and Doctor in Telecommunication Engineer (Ph.D.) degree from the Universidad of Vigo, Spain, in 1996. In 1990, he joined the Department of Communication Technologies, University of Vigo. He was the Head of Department of Theory of Signal and Communications in the same University from 2004 to 2010. Currently, he teaches courses in Spectrum Management and Wideband Radio communication Systems, as Professor.

Kandarpa Kumar Sarma, presently with the Department of Electronics and Communication Technology, Gauhati University, Assam, India, completed M.Sc in Electronics from Gauhati University in 1997 and M.Tech in Digital Signal Processing from IIT Guwahati, Guwahati, India in 2005 where he further continued his research work and received his Ph.D. His areas of interest include Applications of ANNs and Neuro Computing, Document Image Analysis, 3-G Mobile Communication, Smart Antenna, Mobile and Wireless Communication, Soft-Computation, Speech

Processing, Antenna Design, Document Image Analysis, and Digital Signal Processing. His Academic Organizational Positions Held include Lecturer in Electronics in LCB College, Guwahati (past), Associate Professor, Department of Electronics and Communication Technology (former Department of Electronics Science), Gauhati University (present).

Manash Pratim Sarma received his M.Sc degree in Electronics from Gauhati University, Guwahati Assam, India in 2008. He also completed M.Tech in Electronics Design and Technology from Tezpur University, Assam, India. Presently he is an Assistant Professor in the Department of Electronics and Communication Engineering, Gauhati University, Assam, India. His field of interest includes Wireless Communication, VLSI design and ANN applications.

Lokesh Sharma is an Indian author who has done work over Wireless sensor Network. He was born in Mathura (Kosi Kalan) and studied in the Vidya Devi Jindal School in his hometown. He studied Computer Science engineering from GLA University, Mathura and then his Master degree from J. P. University, Solan. He worked as an Asst. Professor in Rajeev Gandhi proudhogiki University Madhya Pradesh and as a Consultant in Laurus Infosystem, Bangalore. Presently he is a Research Assistant in Chang Gung University, Taiwan. Sharma's keen interest to develop their own protocols in wireless sensor network, for this he has done his masters in Computer Science, which led to gain interest and develop their own backbone construction schemes. He has published journals and papers over Energy Efficient Schemes in Backbone Construction of Wireless Sensor Network. His writing style tends to be simple, and more elaborative. He is spiritually and intellectually believe in himself. His philosophy is to live life simply and always make other's to think in his own way. He believes in "A reasonable man adapts himself to the world and an unreasonable man adapts the world to himself".

Vishnu Suryawanshi received his M.tech degree in Vlsi and System Design in 2009 from J.N.T. University Kakinada with distinction. He is presently pursuing a Ph.D. from Rashtra Sant Tukdoji Maharaj University Nagpur under the guidance of Dr. G.C. Manna. His research interests include VLSI design, hardware implementations of cryptographic algorithms, and security protocols for wireless communication systems. He has more than twelve years of experience in teaching. He has published papers on Advanced Encryption Standard block cipher. Presently he is an Assistant Professor in the department of Electronics and Telecommunication at G.H.Raisoni Institute of Engineering and Technology Pune.

Homero Toral-Cruz received Ph.D. and M.S. degrees in Electrical Engineering, Telecommunication option from CINVESTAV, Jalisco, Mexico, in 2006 and 2010, respectively. He received the B.Sc. degree in Electronic Engineering from "Instituto Tecnológico de la Laguna", Coahuila, Mexico in 2002. He is currently an Assistant Professor at Sciences and Engineering department in University of Quintana Roo, Mexico. His research interest includes VoIP technologies, traffic modeling and WSN. He has served as Guest Editor of some international journals and TCP member of several international conferences. He has been awarded a national recognition (SNI member) as a researcher by CONACYT and has been elected as member of the Mexican Academy of Sciences (AMC).

Miroslav Voznak is an Associate professor and department chair with Department of Telecommunications, VSB-Technical University of Ostrava, Czech Republic. He received his Ph.D. degree in telecommunications, dissertation thesis "Voice traffic optimization with regard to speech quality in network with VoIP technology" in 2002. Topics of his research interests are Next Generation Networks, IP telephony, speech quality and network security.

Index

F

G

H

I

L

M

N

P

R

S

T

U

W

Printed in the United States
By Bookmasters